ADVANCES IN
WAVE TURBULENCE

WORLD SCIENTIFIC SERIES ON NONLINEAR SCIENCE

Editor: Leon O. Chua
University of California, Berkeley

*To view the complete list of the published volumes in the series, please visit:
http://www.worldscientific.com/series/wssnsa

WORLD SCIENTIFIC SERIES ON NONLINEAR SCIENCE

Series A   Vol. 83

Series Editor: Leon O. Chua

ADVANCES IN WAVE TURBULENCE

edited by

Victor Shrira
Keele University, UK

Sergey Nazarenko
University of Warwick, UK

World Scientific

NEW JERSEY · LONDON · SINGAPORE · BEIJING · SHANGHAI · HONG KONG · TAIPEI · CHENNAI

Published by

World Scientific Publishing Co. Pte. Ltd.

5 Toh Tuck Link, Singapore 596224

USA office: 27 Warren Street, Suite 401-402, Hackensack, NJ 07601

UK office: 57 Shelton Street, Covent Garden, London WC2H 9HE

British Library Cataloguing-in-Publication Data
A catalogue record for this book is available from the British Library.

World Scientific Series on Nonlinear Science, Series A — Vol. 83
ADVANCES IN WAVE TURBULENCE

ISBN 978-981-4366-93-9

Typeset by Stallion Press
Email: enquiries@stallionpress.com

Printed in Singapore by World Scientific Printers.

Preface

Wave turbulence (WT) is a branch of science concerned with the evolution of random wave fields of all kinds and scales, from waves in galaxies to capillary waves on water surface, from waves in nonlinear optics to quantum fluids. In spite of the enormous diversity of wave fields in nature, there is a common conceptual and mathematical core which allows us to describe the processes of random wave interactions within the same conceptual paradigm and in the same language. The development of this core and its links with the applications is the essence of the WT science which is an established integral part of nonlinear science.

WT is aimed at predicting statistical characteristics of random wave fields, which includes special and temporal spectra, fluxes of various quantities, probability density functions of various physical fields. For five decades, the WT was primarily just a mathematical theory able to predict certain power law spectra. The theoretical foundations of WT were explained in the classical monograph by V.E. Zakharov, V.S. L'vov, G. Falkovich *Kolmogorov Spectra of Turbulence*, Springer, 1992. In reality, throughout most of the WT history it was often difficult to measure the observed spectra with the accuracy sufficient to verify the theoretical predictions and, therefore, the beautiful edifice of mathematical theory was somewhat detached from the ground. There was also little feedback from the experimental side. Nowadays, on the experimental side the techniques have become much more advanced and sophisticated, while the theory has moved beyond the spectra and predicts and describes more subtle and sometimes easier to measure characteristics of random fields, which makes the interaction between the theoretical and experimental sides much more fruitful.

The present book aims at discussing new challenges arising in WT and perspectives of its development with special emphasis upon the links between the theory and experiment. The book does not aim at presenting a comprehensive picture of new advances in the WT; an overview, starting with the basics, can be found in the very recent book by S.V. Nazarenko, Wave Turbulence, Springer, 2011. The choice of selected topics in the present book is unashamedly subjective. We present what we believe, along with the authors we invited, to be most hot and exciting. Several most prominent players in the field contributed reviews on advances in the areas of their research. Most of the reviews have substantial experimental content, including water waves, where we have the strongest reality check of the WT theory, wave turbulence experiments in optical fibers, WT experiments on a metal plate, and observations of astrophysical WT. We wanted to focus more on the new challenges these results pose for the theory, not the other way around; to what extent we have succeeded is for the readers to judge. The seven reviews constituting the book each in its own way narrows the existing gap between the theory and experiment.

Each of the reviews in the book is devoted to a particular field of application (there is no overlap), or a novel approach or idea. As a whole, the book also covers a wide range of applications where WT is successfully used, with a strong emphasis on the areas where experiments were made or could be expected in the near future. This includes both laboratory experiments mentioned, as well as working with observational data, including the astrophysical applications. In the latter case, as well in some other areas of WT applications, one of the main tasks for theoreticians is to formulate definitive criteria for identifying WT in observational/experimental data. In particular, we should be able to know how to spot the manifestations of nonlinear wave resonances in the data, how to determine if nonlinearity is indeed weak, which processes or features could be considered as clear evidence of WT, how to recognize WT masked by extra physical factors, like medium inhomogeneity and anisotropy, external forces, presence of non-wave motions such as hydrodynamic shear or vortices. In other words, our book aims to address WT as a way for engaging with real world phenomena, not a purely academic exercise. Below, both the experimentalists and the theoreticians are presenting their results and vision of WT.

The book starts with a stage-setting review **Wave Turbulence; A Story Far From Over** by *A.C. Newell* and *B. Rumpf.* The focus is upon discussing the premises on which the theory is based and from which a natural asymptotic closure is obtained. This review contains some introductory material such as a tutorial on wave turbulence closure. At the same time, the review touches upon subtle points and issues that still remain challenging to the theoreticians. Particular attention is paid to the consequences of the fact that many of the most relevant solutions, those connected with finite flux spectra, are not uniformly valid over all scales; remedies are suggested and discussed. The authors discuss the current state of the wave turbulence science and compare it to that of the pattern formation theory at the end of 1960s, when much important theoretical work had already been done but predictions remained in the "looks like" category until the key experiments by Ahlers *et al.* in the mid to late 1970s.

Traditionally, measuring the wave-number dependence of the wave energy spectra has been the main testing method of the wave turbulence predictions. However, similar or even the same spectral exponents may arise for totally different reasons, e.g. due to the presence of coherent structures. On the other hand, another key quantity in wave turbulence, the energy input rate, has been much less studied. Note that there are clear and testable predictions about the dependence of the energy spectrum on the energy injection rate. It is only recently that measurements of the injected power into wave turbulence systems have been performed directly at the wave maker. The review **Fluctuation of the energy flux in wave turbulence** by *S. Aumaître, E. Falcon* and *S. Fauve* surveys recent efforts in this direction. It attracts attention to a puzzling discrepancy with theory for the observed scaling law of the power spectrum with the injected power. It also shows that observed fluctuations of the injected power are much larger than the mean value, frequent negative input events — effects remaining to be explained theoretically.

Besides carefully controlled laboratory settings, rich experimental data can be extracted from observation of natural phenomena. Astrophysics provides us with such a natural "wave turbulence laboratory", in particular in the interplanetary solar wind environment, solar surface and Earth's magnetosphere. The review **Wave turbulence in astrophysics** by *S. Galtier* outlines the recent progress on wave turbulence in astrophysical plasmas. The review covers studies based upon several relevant models of magnetohydrodynamics: incompressible and compressible MHD, Hall MHD and electron MHD. The interplanetary medium and the solar atmosphere are presented as two examples of media where anisotropic wave turbulence is relevant. The wave turbulence predictions and interpretations are presented and discussed in the context of space and simulated magnetized plasmas.

Fuelled by explosive growth of modern optical communications the problems of electro-magnetic wave interactions in optical lines have moved to the forefront of wave turbulence wave studies. The review **Optical wave turbulence in fibre lasers** *by S. K. Turitsyn, S. A. Babin, E. G. Turitsyna, G. E. Falkovich, E. V. Podivilov, D. V. Churkin* surveys recent progress in optical wave turbulence with a specific focus on the processes in fibre lasers. These wave-guiding fibre systems provide unique opportunities for experimental studies of wave turbulence, since, in contrast to all fluid dynamics situations, the physical properties of the systems can be tailored in the desired way and the high precision optical experiments involving large numbers of realisations are much easier and faster to carry out. The recent experimental, theoretical and numerical results on optical wave turbulence in fibre lasers ranging from weak to strong developed turbulence presented in the review show an immensely rich and very fast growing field. The reported achievements have important links with other areas of physics with one-dimensional wave turbulence.

We meet yet another recent creative experimental setup in the review **Wave turbulence in a thin elastic plate: the sound of the Kolmogorov spectrum?** by *G. During* and *N. Mordant*. This experiment is done by exciting waves in a human-size metal sheath, which can be heard by observers nearby because they emit sound. Similar plates were used in theatres for imitating the sound of thunderstorm. Yet again some puzzling disagreements with the wave turbulence predictions have been observed in this system, which calls for further thoughts.

"*Wave turbulence*" is a broader term that the often used as synonymous with "*weak turbulence*", the latter implies that all interacting waves are weakly nonlinear, while the former does not. In Nature, very often the reality is mixed: weakly nonlinear waves co-exist with strongly nonlinear coherent patterns (e.g. with vortices or breakers). In encountering wave turbulence in reality the fundamental question one has to deal with is to find what kind of turbulence it is: that is what features can be captured using the weak turbulence well developed machinery and what features require novel approaches. In their review "**Surface Gravity Wave Turbulence in the Laboratory**" *R. Bedard, S. Lukaschuk* and *S. Nazarenko* address this question by performing a series of experiments in a big water tank, for both steady and un-steady states (i.e. initial wave turbulence rise and final decay stages), and analysing the resulting data. The starting point is what to expect if the weak turbulence regime is dominant. To this end the authors provide a brief tailored overview of recent theoretical and numerical results for gravity wave turbulence in finite size tanks. On testing the weak turbulence theory predictions against the experimental data it becomes apparent that such a theory does not capture all the features of the observed wave field behaviour. To capture the features due to strongly nonlinear coherent patterns, such as wave breaking and sharp crests, the structure functions approach to data analysis proved to be extremely fruitful. It enables one to spot the presence of such coherent patterns, such as wave breaking, and grasp their significance in the overall picture. The developed approach is quite general; it can be applied to analysis of wide classes of wave turbulence situations independently of the physical nature of the wave field.

The prevalent tools for studying wave turbulence theoretically are the direct numerical simulations (DNS) and kinetic equations which describe evolution of wave spectra. The kinetic equations are traditionally derived by making implicit use of long-time asymptotics. Without resorting to the long-time asymptotics assumption the statistical description becomes qualitatively different: the kinetic equations become non-local in time, but able to describe much faster evolution of random wave fields. The review **Towards a new picture of wave turbulence** by *V. Shrira & S. Annenkov* discusses a novel DNS approach enabling one to tackle hitherto inaccessible problems and the profound implications for wave turbulence of the new kinetic description allowing description of "fast" evolution of wave statistical characteristics.

Having briefly outlined what is in the book, it is appropriate to say who might benefit most from reading it. The field of WT applications is so rapidly expanding, that even the scientists very active in the field of WT cannot follow all the recent developments, since many WT branches have grown quite far apart. Such scientists represent the prime audience for our book, which is designed to be helpful for researchers of all levels: from the PhD students upward, to aid their research on advancing the WT understanding. There is also a much wider circle of scientists working in the areas where WT forms a part, often a significant part, but not the whole of the physical picture. The book will help such researchers to learn more about WT and its role in the systems they study. In particular, this circle includes scientists studying MHD turbulence, quantum turbulence, geophysical fluid dynamics, nonlinear optics and Bose-Einstein condensates. We hope that all the readers, independently of their specific field, will enjoy reading the book.

Victor Shrira and Sergey Nazarenko

Contents

Chapter 1

Wave Turbulence: A Story Far from Over

Alan C. Newell

*Mathematics Department, University of Arizona,
Tucson, Arizona 85721, USA*

Benno Rumpf

*Mathematics Department, Southern Methodist University,
Dallas, Texas 75275-0156, USA*

The goals of this chapter are: To state and review the premises on which a successful asymptotic closure of the BBGKY hierarchy of moment/cumulant equations is based; to describe how and why this closure is attained; to examine the nature of solutions of the kinetic equation; to discuss obstacles which limit the theory's validity and suggest how the theory might then be modified; to compare the experimental evidence in a range of applications with the theory's predictions; and finally, and most importantly, to suggest open challenges and to encourage the reader to apply and explore wave turbulence with confidence.

Contents

1.1. Introduction

Turbulence is the irregular motion of fluids and plasmas in nonequilibrium, or more generally, the motions associated with energy-conserving nonlinear field equations (e.g. the forced high Reynolds number Navier–Stokes equations) with additional external forcing and dissipation. It makes little sense to try to follow individual trajectories of the representative point in phase space for such complex systems. They cannot be reproduced in experiments. Instead, one is interested in average behavior. Turbulence theory is, therefore, concerned with coming to grips with and understanding long time statistical behavior. The principal aims are to understand and to be able to calculate transport, such as the average flux of mass down a pipe as function of the pressure head, the spectral distributions which carry the energy or other conserved densities from the scales at which they are injected to the scales where they are dissipated, and structure functions from which physical quantities such as momentum and energy flux may be determined. Alas, despite some successes such as the four-fifth's law and predictions based on scaling arguments, quantitative results are hard to achieve. The main obstacle is the lack of a consistent statistical closure of the infinite hierarchy of moment equations (Frisch, 1996): The equations of motion of statistical quantities, such as moments, depend on the unknown behavior of higher moments.

In contrast, wave turbulence, the turbulence of a sea of weakly interacting dispersive wavetrains (the "eddies"), has a natural asymptotic closure (Benney *et al.*, 1966, 1967, 1969; Newell *et al.*, 2001). All of the long time statistical quantities, the energy density, the nonlinear frequency renormalization, the long time behaviors of the cumulants and the structure functions can be calculated from a set of core particle densities $\{n^{(r)}(\mathbf{k}, t)\}$ which are proportional to the Fourier transforms of two point averages. For the purpose of simplicity, in this chapter, we will use examples where there is only one such density, $n_{\mathbf{k}} \equiv n(\mathbf{k}, t)$, such as the waveaction density in the case of ocean gravity waves. This quantity is called the number or particle density in Bose gas condensation and the quantum Boltzmann equation (Nordheim, 1928; Peierls, 1929). It is the power in optical contexts (Dyachenko *et al.*, 1992). It is closely related to the energy density which to leading order is equal to $e_{\mathbf{k}} = e(\mathbf{k}, t) = \omega_k n_{\mathbf{k}}$, where ω_k is the linear dispersion relation of wavetrains. Moreover, it is the quantity which in many cases (gravity waves, Bose gases, optics, semiconductor lasers) undergoes an inverse cascade through which long waves are built from the resonant interactions of shorter waves. It is central for the success of wave turbulence theory that this density satisfies a closed (Boltzmann like, kinetic) equation (Hasselmann, 1962, 1963a, 1963b;

Zakharov *et al.*, 1992), i.e. an equation that expresses time derivatives of the density only as a function of this density. The form of this equation reveals that, to leading order, all transport is carried by N-wave resonances $(N = 3, 4, \ldots)$. Further, it is possible to compute solutions of the kinetic equation analytically. The first types of solution are stationary states that describe the statistical energy equipartition of isolated systems. Such states simply represent an entropy maximum under the condition of an energy constraint or an entropy maximum under the joint constraints of all the conserved densities if there is more than one. The second type of stationary solution is achieved in nonequilibrium, namely the finite flux Kolmogorov behavior of nonisolated systems that are driven by an external force and damped by viscosity on different length scales. The dynamics formally conserves quantities such as energy and particle number in the inertial range where the kinetic equation is valid. The spectral densities flow from sources in k space to sinks (Zakharov and Filonenko, 1967a, 1967b). These Kolmogorov–Zakharov (KZ) solutions are the analogues of the familiar Kolmogorov energy spectrum prediction $E(k) = cP^{2/3}k^{-5/3}$ of high Reynolds number hydrodynamics. In addition, the kinetic equation has time-dependent solutions of self-similar type which describe how the stationary solutions are accessed.

Furthermore, nature and laboratories abound with situations where wave turbulence theory can be applied. The most familiar example is that of ocean gravity waves on a wind stirred sea but, in principle, its signatures should also be found in magnetohydrodynamic waves in astrophysical contexts, Rossby-like waves in the atmospheres of rotating planets, in the formation of condensates, in capillary waves, in optical waves of diffraction, in acoustic waves, and in the music of vibrations on large, thin, elastic sheets. But does the hand of wave turbulence really guide the behavior of ocean waves, of capillary waves, of all the just quoted examples where one might expect the theory to apply? What we shall learn is that while there have been notable successes, the theory also has limitations. In short, both the good and bad news is that the wave turbulence story is far from over. One might compare its current standing, particularly with respect to laboratory experiments, to the situation regarding pattern formation in the late sixties. By that time, there had been many theoretical advances but the experimental confirmation of the predictions fell very much in the "looks like" category. It took the pioneering experimental works of Ahlers, Croquette, Fauve, Gollub, Libchaber and Swinney in the mid to late seventies (which overcame some extraordinary challenges of managing long time control of external parameters) to put some of the advances on a firm footing. For wave turbulence, we are only at the beginning of the experimental stage.

It is also a subject that has been presented from many points of view and sometimes in a language which seems more to confuse than enlighten. Terms such as random phase approximation or joint Gaussian approximation have entered the lexicon and their necessity as premises for wave turbulence closure are debated as fiercely as philosophical debates over the existence of deities. One of our goals in this chapter, therefore, is to present our point of view in as clear a language as possible. We will show that the hypotheses required for deriving the closure are mild.

We start by describing the model and listing the premises on which the wave turbulence closure is based. We take an infinite domain and assume that our physical fields are bounded everywhere and are sufficiently smooth so that all the terms in the partial differential equations governing their behavior are well defined. Later in the chapter, we discuss what size finite domains must be in order that the principal transfer mechanisms of wave turbulence, resonant wave interactions, are not missed because the Fourier space is too quantized. We then rewrite the governing equations in terms of a diagonalized set of Fourier amplitudes of the physical fields which, because we are dealing with bounded rather than with decaying fields, must be considered as generalized rather than ordinary functions. Only statistical averages of the Fourier amplitudes have any meaning.

There are at least three premises on which the wave turbulence closure rests.

Premise 1 (P1): First, we assume the fields are spatially homogeneous and that ensemble averages of fields evaluated at the set of points $\mathbf{x}, \mathbf{x} + \mathbf{r}_1, \mathbf{x} + \mathbf{r}_2, \ldots$ depend only on the separations $\mathbf{r}_1, \mathbf{r}_2, \ldots$.

Comment 1. This assumption suggests that the symmetry of translation invariance, while broken by ensemble members, is restored on the average. This leads to the Fourier amplitudes being Dirac delta correlated in the wavevector space. This property is vital to the success of the wave turbulence closure. The reason is that in the decomposition of moments of Fourier amplitudes into cumulants, the terms which are multiple products have more Dirac delta functions which, when executed in the multiple integrals in wavenumber space, directly lead to stronger secular time behaviors. The most potent of these combinations (in the case where the field means are zero) are products of two point functions, the waveaction densities. Indeed, we find that the long time behavior of all higher order cumulants are basically given as a product of the waveaction densities. This is one reason for the closure.

Comment 2. The spatial homogeneity assumption is widely used and is rarely debated in turbulence theories. It also means that ensemble averages can be replaced by spatial averages with respect to the base point $\mathbf{x}$.

But there is a rub! We introduce a new and stunningly surprising result. The premise may be broken by an instability. In certain rare cases, the spontaneous breaking of the spatial homogeneity symmetry can lead to the nucleation of coherent structures and a breakdown of the wave turbulence closure.

Premise 2 (P2): Second, we assume that at some initial point in time, the moment at which the external driving, e.g. the storm over the sea surface, is initiated, the fields at distant points are uncorrelated. This means that the physical space cumulants initially have the property that, as the separations $|\mathbf{r}_j|$ become large, the cumulants decay sufficiently rapidly with the result that their Fourier transforms are ordinary functions. This is a mild assumption but it is necessary because in the evaluation of the long time behavior of integrals such as $\int f(x)\frac{\sin xt}{x}dx$ we need to know that $f(x)$ is sufficiently smooth in wavenumber x so that this integral behaves in long time as $\pi\mathrm{sgn}(t)f(0)$.

Comment 1. It is important to stress that long distance correlations induced by the dynamics, namely resonant wave interactions, act through large distances and eventually introduce a weak singular behavior into higher order cumulants in Fourier space. We call these terms asymptotic survivors because they lead to nonvanishing long time contributions to the corresponding cumulants in physical space. We show how these behaviors are both critical for the energy transfer mechanisms and at the same time can be handled without any violation of the wave turbulence closure.

Comment 2. We discuss later in the chapter, how this mild assumption, namely premise P2, relates to the so-called random phase approximation on the Fourier amplitudes and we show in what sense the phases and the absolute values of the Fourier amplitudes can be considered random. Because of the dynamically induced singular behavior mentioned above, it is not true that products of the amplitudes can be expanded exactly in combinations of two point functions. We show, however, that the leading order behavior of such products can be so expanded. However, we stress that the so-called random phase approximation is, in fact, unnecessary to effect the natural closure. And, of course, if the theory is to describe reality and to be robust, one should not have to rely on any sorts of approximations such as random phase or initially joint Gaussian fields in order to effect closure. One has to doubt the logical consistency of those who employ such approximations. Either nature brings about such conditions as a result of its dynamics in which case one does not need such severe *a priori* restrictions, or it does not in which case the theory has extremely limited validity. What happens is the former. The linear dynamics naturally relaxes the system very quickly towards a state of joint Gaussianity. Physically, the relaxation is induced additively by interacting waves carrying uncorrelated information from distant points. Mathematically, it occurs because the waves are dispersive and because of the Riemann–Lebesgue (RL) lemma. But then, weak nonlinear interactions rebuild a weak nonjoint Gaussian behavior over long times. It is the fact that the rebuilding of non-Gaussianity only depends on the waveaction density $n_\mathbf{k}$ which gives rise to the natural asymptotic closure.

Premise 3 (P3): Third, we must ensure that various asymptotic expansions for the slow evolution of such two point functions as the waveaction density $n_\mathbf{k}$ remain uniformly valid in wavenumber. In its simplest form, this means that the ratio of linear (t_L) to nonlinear (t_{NL}) time scales is small at all wavenumbers. It also means that all asymptotic expansions for the slow evolution of the waveaction density $dn_\mathbf{k}/dt$, the frequency renormalization which accounts for the slow time behavior of the leading order, higher order cumulants, and the structure functions remain uniformly valid in wavenumber on almost all relevant solutions.

Comments. The reason that we require $\frac{t_L}{t_{NL}}(k) \ll 1$ (with $k = |\mathbf{k}|$) is that when we look for the long time behavior of integrals such as $\int f(x)\frac{\sin xt}{x}dx$, we want to know that the multiplying function $f(x)$, which will contain products of the waveaction densities $n_\mathbf{k}$, is not only smooth in x (wavenumber) but also that it varies slowly in time. The ratio $\frac{t_L}{t_{NL}}$ is given by $\frac{1}{\omega_k}\frac{1}{n_\mathbf{k}}\frac{dn_\mathbf{k}}{dt}$ which, on relevant solutions of the kinetic equation and in particular the KZ spectrum, will be k dependent. The fact that $\frac{t_L}{t_{NL}}(k)$ will fail to be small for very high or very low wavenumbers does not mean that the wave turbulence closure does not apply but rather that the KZ spectrum has to be amended in these regions. In Sec. 1.5, we discuss two examples of remedies for this partial breakdown.

The question remains: Are premises one, two and three sufficient to guarantee that we will obtain a wave turbulence closure in which, over long times, the behavior of all statistical quantities (cumulants, structure functions) can be inferred from solutions of a classical kinetic equation for $n_\mathbf{k}$ and from the frequency renormalization? What we find is that the three premises are enough to guarantee that all the steps in the calculations are valid but, alas, we do not know what are *a priori* conditions to guarantee that the ensemble of fields consists of uncorrelated wavetrains and not nonlinear coherent objects. We have evidence that in some cases, admittedly rare, usually in one dimension, the wave turbulence closure fails. Families of solitary waves dominate the dynamics. The kinetic equation's solutions are invalid at all wavenumbers. The search for the sufficient conditions on the governing equations, which would guarantee that spatial homogeneity is not temporarily broken by the emergence of coherent structures, remains one of the open challenges of wave turbulence. In Sec. 1.5, we suggest some candidates.

The outline of this chapter is as follows. In Sec. 1.2, we give a tutorial on the derivation of the wave turbulence closure, the kinetic equation and the frequency renormalization. We begin this section by emphasizing that the set up does not require *a priori* knowledge of the Hamiltonian structure. The appropriate Fourier amplitudes are chosen by the diagonalization of the linear part of the governing equations. We carry out the derivation for the governing equation for the case of gravity capillary waves in Appendix 1. We then carry out an asymptotic analysis of the BBGKY hierarchy of cumulant equations and show how, using the premises P1, P2, P3, and some analysis of the long time behavior of certain integrals and multiple time scale ideas, the wave turbulence closure is obtained. In Sec. 1.3, we summarize some of the principal properties of the equation and finite flux solutions of the kinetic equation, introduce such notions as capacity and breakdown, how KZ solutions are realized, the anomalies in finite capacity situations, and talk about structure functions and finite box size considerations. In Sec. 1.4, we discuss the existing experimental evidence in the contexts of capillary waves, ocean gravity waves, waves on vibrating elastic sheets and photorefractive crystals. The emphasis and strong message of this chapter and of our recent review in the Annual Reviews of Fluid Mechanics (Newell and Rumpf, 2011) is that more experiments are needed. In Sec. 1.5, we discuss two open questions: What must be done to amend the KZ solution in wavenumber ranges where the KZ spectra violate premise P3 and, how to identify *a priori* those situations for which the wave turbulence closure is invalid at all scales? We introduce two new premises as possible candidates which, along with P1, P2, P3, may be sufficient to guarantee self-consistent wave turbulence closure. We end in Sec. 1.6, with a list of open challenges.

1.2. A Tutorial on the Wave Turbulence Closure

The starting point for deriving the wave turbulence closure is the governing equation for the Fourier transforms $A_{\mathbf{k}}^s \equiv A^s(\mathbf{k}, t)$ of suitable combinations of the field variables $u^s(\mathbf{x}, t)$, combinations chosen so as to diagonalize the linearized equations of the system under study. The $A_{\mathbf{k}}^s$'s satisfy what we call the governing equations,

$$\frac{dA_{\mathbf{k}}^s}{dt} - i\omega_{\mathbf{k}}^s A_{\mathbf{k}}^s = \sum_{r=2} \epsilon^{r-1} \sum_{s_1 \ldots s_r} \int L_{\mathbf{k}\mathbf{k}_1 \ldots \mathbf{k}_r}^{s s_1 \ldots s_r} A_{\mathbf{k}_1}^{s_1} \ldots A_{\mathbf{k}_r}^{s_r}$$
$$\times \delta(\mathbf{k}_1 + \cdots + \mathbf{k}_r - \mathbf{k}) d\mathbf{k}_1 \ldots d\mathbf{k}_r. \qquad (1.1)$$

In (1.1), $0 < \epsilon \ll 1$ is a small parameter (e.g. the wave slope), $\delta(x)$ is the Dirac delta function, $\omega_{\mathbf{k}}^s$ is the linear dispersion relation where s enumerates the set of cardinality $\{s\}$ of frequencies associated with the wavevector $\mathbf{k}$. In this formulation, the coefficients $L_{\mathbf{k}\mathbf{k}_1 \ldots \mathbf{k}_r}^{s s_1 \ldots s_r}$ are pure imaginary, reflecting the fact that the system is conservative. For the purposes of this chapter, $\{s\} = 2$, $s = \pm 1$, $\omega_k^s = s\omega_k$; our working example for which the derivation of (1.1) is given in Appendix 1 is capillary waves where $\omega_k = \sqrt{gk + Sk^3/\rho} \simeq \sqrt{S/\rho}k^{3/2}$ for $k \gg (\rho g/S)^{1/2}$, S is the surface tension, g is gravity and ρ is water density. We make three comments in relation to (1.1) and the cumulant hierarchy formed from it.

Hamiltonian form: Although it has been customary in many presentations to prepare the system in the Hamiltonian form, this is a luxury, and sometimes even an obstacle to get going. Indeed in some cases, e.g. magnetohydrodynamic turbulence, it is better not to use canonical Hamiltonian variables at all. What is important is that the combinations $u^s(\mathbf{x}, t)$ are chosen so that the corresponding Fourier amplitudes $A_\mathbf{k}^s$ diagonalize the linear part of the dynamical system. We refer the reader to the water wave example in Appendix 1.

The nature of the Fourier amplitudes $A_\mathbf{k}^s$: For physical fields $u^s(\mathbf{x}, t)$ (e.g. the surface elevation $\eta(\mathbf{x}, t)$ and velocity potential $\phi(\mathbf{x}, t)$ at the free surface) which are bounded as $|\mathbf{x}| \rightarrow \pm\infty$, the functions $A_\mathbf{k}^s$ are generalized functions. In and of themselves, they have no meaning. However, suitable combinations of ensemble averages of products of the $u^s(\mathbf{x}, t)$ fields, e.g. $\langle u^s(\mathbf{x}, t)u^{-s}(\mathbf{x} + \mathbf{r}, t)\rangle - \langle u^s(\mathbf{x}, t)\rangle\langle u^{-s}(\mathbf{x} + \mathbf{r}, t)\rangle$, called cumulants, do have meaning in that they are ordinary functions of the separation $\mathbf{r}$ which, initially, from premise P2, decay sufficiently fast for their Fourier transforms, the Fourier cumulants, to be ordinary functions of $\mathbf{k}$. These Fourier cumulants are related to ensemble averages of suitable combinations of the products of Fourier amplitudes. Premise P2 is a mild one which simply states that at some initial time the fields at distant points are uncorrelated. We will see later that, in the long time limit $t \rightarrow \infty$, the Fourier cumulants develop Dirac delta contributions at higher orders in ϵ, reflecting the fact that resonances eventually induce long distance correlations. These contributions are easily handled within the framework of wave turbulence theory as long as they are properly taken into account.

Joint Gaussianity: An exact joint Gaussian field will have all cumulants of order three and higher identically zero. We will see that the linear dynamics of the wave field brings the system close to a state of joint Gaussianity in that, via the RL lemma, the initial non-Gaussian physical space cumulants decay in time. However, we shall also see that nonlinear interactions rebuild the weakly non-Gaussian field and it is this rebuilt non-Gaussianity and its particular nature which is responsible for the closure and all the nontrivial transfer processes. Thus, the dynamics naturally supplies all the ingredients to effect closure. Other than P2, no *a priori* assumptions on the nature of the statistics is required.

The cumulant hierarchy. For the purposes of this tutorial, we follow, in detail, the time evolution of cumulants of order two and three, and indicate what happens to the hierarchy. In particular, we define

$$\langle A^s_{\mathbf{k}} A^{s'}_{\mathbf{k}'} \rangle = \delta(\mathbf{k}+\mathbf{k}') Q^{ss'}(\mathbf{k}'),$$

$$\langle A^s_{\mathbf{k}} A^{s'}_{\mathbf{k}'} A^{s''}_{\mathbf{k}''} \rangle = \delta(\mathbf{k}+\mathbf{k}'+\mathbf{k}'') Q^{ss's''}(\mathbf{k},\mathbf{k}',\mathbf{k}''),$$

$$\langle A^s_{\mathbf{k}} A^{s'}_{\mathbf{k}'} A^{s''}_{\mathbf{k}''} A^{s'''}_{\mathbf{k}'''} \rangle = \delta(\mathbf{k}+\mathbf{k}'+\mathbf{k}''+\mathbf{k}''') Q^{ss's''s'''}(\mathbf{k},\mathbf{k}',\mathbf{k}'',\mathbf{k}''') \quad (1.2)$$
$$+ P^{00'0''} \delta(\mathbf{k}+\mathbf{k}') \delta(\mathbf{k}''+\mathbf{k}''')$$
$$\times Q^{ss'}(\mathbf{k},\mathbf{k}') Q^{s''s'''}(\mathbf{k}'',\mathbf{k}'''),$$

where $P^{00'0''}$ stands for the cyclic permutation over $(s,\mathbf{k})$, $(s',\mathbf{k}')$, $(s'',\mathbf{k}'')$.

In writing down (1.2), we have assumed that (i) the mean $\langle u^s \rangle$ is identically zero for all times (the condition $L^{ss_1\cdots s_r}_{0\mathbf{k}_1\ldots\mathbf{k}_r} = 0$ in (1.1) guarantees that if the mean is initially zero, it stays zero) and (ii) the property of spatial homogeneity, namely that the ensemble average of products such as $R^{ss'}(\mathbf{r}) = \langle u^s(\mathbf{x})u^{s'}(\mathbf{x}+\mathbf{r}) \rangle$ only depends on the relative geometry coordinate $\mathbf{r}$ and not on the base coordinate $\mathbf{x}$. This second property, which we have called premise P1, gives us that

$$\langle A^s(\mathbf{k})A^{s'}(\mathbf{k}') \rangle = \frac{1}{(2\pi)^{2d}} \int \langle u^s(\mathbf{x}_1)u^{s'}(\mathbf{x}_2) \rangle e^{-i\mathbf{k}\mathbf{x}_1 - i\mathbf{k}'\mathbf{x}_2} d\mathbf{x}_1 d\mathbf{x}_2$$

$$= \frac{1}{(2\pi)^{2d}} \int e^{-i(\mathbf{k}+\mathbf{k}')\mathbf{x}_1} d\mathbf{x}_1 R^{ss'}(\mathbf{r}) e^{-i\mathbf{k}'\mathbf{r}} d\mathbf{r}$$

$$= \frac{\delta(\mathbf{k}+\mathbf{k}')}{(2\pi)^d} \int R^{ss'}(\mathbf{r}) e^{-i\mathbf{k}'\mathbf{r}} d\mathbf{r}$$

$$= \delta(\mathbf{k}+\mathbf{k}') Q^{ss'}(\mathbf{k},\mathbf{k}'),$$

where $Q^{ss'}(\mathbf{k},\mathbf{k}')$ is the Fourier transform of $R^{ss'}(\mathbf{r})$ and $\mathbf{x}_2 = \mathbf{x}_1 + \mathbf{r}$. As we shall see, the delta function correlation of Fourier amplitudes is directly responsible for the nature of the long time secular behavior of iterates of the Fourier cumulants and thereby for the wave turbulence closure.

By multiplying (1.2) by $A^{s'}_{\mathbf{k}'}$ and the equivalent equation for $A^{s'}_{\mathbf{k}'}$ by $A^s_{\mathbf{k}}$ and adding and averaging, we obtain the first equation in the cumulant hierarchy,

$$\frac{dQ^{ss'}(\mathbf{k}')}{dt} - i(s\omega + s'\omega') Q^{ss'}(\mathbf{k}')$$

$$= \epsilon P^{00'} \sum_{s_1 s_2} \int L^{ss_1 s_2}_{\mathbf{k}\mathbf{k}_1 \mathbf{k}_2} Q^{s's_1 s_2}(\mathbf{k}',\mathbf{k}_1,\mathbf{k}_2) \delta(\mathbf{k}_1+\mathbf{k}_2-\mathbf{k}) d\mathbf{k}_1 d\mathbf{k}_2, \quad (1.3)$$

where $\mathbf{k} + \mathbf{k}' = 0$ and $\omega' = \omega(|\mathbf{k}'|)$; for $\mathbf{k} + \mathbf{k}' + \mathbf{k}'' = \mathbf{0}$, by a similar calculation, we obtain the second equation in the cumulant hierarchy to be

$$\frac{dQ^{ss's''}(\mathbf{k}, \mathbf{k}'\mathbf{k}'')}{dt} - i(s\omega + s'\omega' + s''\omega'')Q^{ss's''}(\mathbf{k}, \mathbf{k}', \mathbf{k}'')$$

$$= \epsilon P^{00'0''} \sum_{s_1 s_2} \int L^{ss_1 s_2}_{\mathbf{k}\mathbf{k}_1\mathbf{k}_2} Q^{s's''s_1 s_2}(\mathbf{k}', \mathbf{k}'', \mathbf{k}_1, \mathbf{k}_2)$$

$$\times \delta(\mathbf{k}_1 + \mathbf{k}_2 - \mathbf{k}) d\mathbf{k}_1 d\mathbf{k}_2$$

$$+ 2\epsilon P^{00'0''} \sum_{s_1 s_2} L^{ss_1 s_2}_{\mathbf{k}-\mathbf{k}'-\mathbf{k}''} Q^{s_1 s'}(\mathbf{k}') Q^{s_2 s''}(\mathbf{k}''). \qquad (1.4)$$

In (1.3) and (1.4), we observe that the time derivatives of cumulants depend on cumulants of higher order. In order to obtain the first closure, the first two equations in the hierarchy will be sufficient. We have also omitted writing down the terms arising from the cubic product of amplitudes in (1.1) but will restore them in the final answer.

The strategy for analyzing the cumulant hierarchy. So what is the strategy for solving (1.3), (1.4) and the other members of the cumulant hierarchy? On the surface, the equation hierarchy has the closure problem of fully developed turbulence. Equations for cumulants of order r involve equal, higher and lower order cumulants. However, when we solve the hierarchy iteratively by expanding each cumulant as an asymptotic expansion in powers of ϵ,

$$Q^{ss'\ldots s^{N-1}}(\mathbf{k}, \mathbf{k}', \ldots \mathbf{k}^{(N-1)}, t)$$
$$= q_0^{ss'\cdots}(0)e^{i(s\omega + s'\omega' + \cdots)t} + \epsilon Q_1^{ss'\cdots} + \epsilon^2 Q_2^{ss'\cdots}, \qquad (1.5)$$

we find that a remarkable simplification occurs. In examining the long time behavior of the iterates Q_r, $r = 1, 2, \ldots$, we find some terms are bounded in t and some grow linearly in t. If the iterates of the Fourier space cumulants should contain generalized functions, then we simply ask that the corresponding expansions of the physical space cumulants are uniformly ordered in the limit of large t. The reason one gains closure is that all the unbounded terms arising in the iterates of the rth-order cumulant involve only zeroth-order cumulants $q_0^{ss'\cdots}$ of order r or less.

For example, simple calculations reveal that there are no secular terms in $Q_1^{ss'}(\mathbf{k}, \mathbf{k}')$ and that the secular terms in $Q_2^{ss'}(\mathbf{k}, \mathbf{k}')$, $s' = -s$ ($Q^{-ss}(\mathbf{k})$ is the waveaction density) are simply ($q_0^{-ss}(\mathbf{k})$ we call $n_\mathbf{k}$ and, for convenience, assume it is s independent)

$$4\pi\mathrm{sgn}(t) \sum_{s_1 s_2} \int L^{ss_1 s_2}_{\mathbf{k}\mathbf{k}_1\mathbf{k}_2} n_{\mathbf{k}} n_{\mathbf{k}_1} n_{\mathbf{k}_2} \left(\frac{L^{-s-s_1-s_2}_{-\mathbf{k}-\mathbf{k}_1-\mathbf{k}_2}}{n_{\mathbf{k}}} + \frac{L^{s_1 s-s_2}_{\mathbf{k}_1\mathbf{k}-\mathbf{k}_2}}{n_{\mathbf{k}_1}} + \frac{L^{s_2 s-s_1}_{\mathbf{k}_2\mathbf{k}-\mathbf{k}_1}}{n_{\mathbf{k}_2}} \right)$$

$$\times \delta(s_1\omega_1 + s_2\omega_2 - s\omega)\delta(\mathbf{k}_1 + \mathbf{k}_2 - \mathbf{k})d\mathbf{k}_1 d\mathbf{k}_2. \tag{1.6}$$

The secular terms arising in $Q_2^{ss'\cdots}$ for all other cumulants of order r (that includes order 2 when $s' \neq -s$) take a remarkably simple form,

$$itq_0^{ss'\cdots}(\mathbf{k}, \mathbf{k}', \ldots)(\Omega_2^s[n_{\mathbf{k}}] + \Omega_2^{s'}[n_{\mathbf{k}'}] + \cdots), \tag{1.7}$$

where

$$\Omega_2^s[n_{\mathbf{k}}] = \sum_{s_2} \int \left(-3iL^{ss s_2 -s_2}_{\mathbf{k}\mathbf{k}\mathbf{k}_2-\mathbf{k}_2} - 4\sum_{s_1} \int L^{ss_1 s_2}_{\mathbf{k}\mathbf{k}_1\mathbf{k}_2} L^{s_1 s-s_2}_{\mathbf{k}_1\mathbf{k}-\mathbf{k}_2} \right)$$

$$\times \left(\hat{P}\frac{1}{s_1\omega_1 + s_2\omega_2 - s\omega} + i\pi\mathrm{sgn}(t)\delta(s_1\omega_1 + s_2\omega_2 - s\omega) \right)$$

$$\times \delta(\mathbf{k}_1 + \mathbf{k}_2 - \mathbf{k})n_{\mathbf{k}_2}d\mathbf{k}_2. \tag{1.8}$$

The symbol $\hat{P}$, not to be confused with the permutation symbol $P^{00'}$, represents the Cauchy principal value. To be meaningful, we have to assume that the quantity $h = s_1\omega_1 + s_2\omega_2 - s\omega$ has only simple zeros. If it has multiple zeros, as it does for acoustic waves, the asymptotics are a bit more complicated (Newell and Aucoin, 1971). In writing (1.8), we have reincluded the terms arising from the cubic interaction with coefficient $L^{ss_1 s_2 s_3}_{\mathbf{k}\mathbf{k}_1\mathbf{k}_2\mathbf{k}_3}$. We also note that the last two terms in (1.6) can be written as $2n_{\mathbf{k}}\mathrm{Im}\Omega_2^s$.

The connection with the random phase approximation: Now, we are in a position to discuss the connection with the so-called random phase approximation which certain authors employ in the derivation of the kinetic equation. In calculating $Q^{-ss}(\mathbf{k})$, we found that the secular terms in the various iterates depend only on $n_{\mathbf{k}}$. All terms involving q_0^N, $N \geq 3$, were bounded. Although we did not, we could have ignored them for this part of the calculations. They have no long time cumulative effect. This means that, had we initially expanded the Fourier amplitudes $A_{\mathbf{k}}^s$ as $A_{\mathbf{k}0}^s + \epsilon A_{\mathbf{k}1}^s + \cdots$, then all product averages $\langle A_{\mathbf{k}0}^s A_{\mathbf{k}'0}^{s'} A_{\mathbf{k}''0}^{s''} \ldots \rangle$ could be decomposed as if the zeroth-order amplitudes had random phases or as if they were joint Gaussian so that only products of two point functions survive (Wick's theorem). But, let us emphasize this point: Only the zeroth-order products of the amplitudes can be expanded as if they had random phases. We stress that averages of products of the complex amplitudes $A^s_{\mathbf{k}}$ cannot be expanded as if they had random phases or as if they were joint Gaussian. We will shortly see that moments such as $P^{00'0''}\langle A^s_{\mathbf{k}1} A^{s'}_{\mathbf{k}'0} A^{s''}_{\mathbf{k}''0}\rangle$ have asymptotic survivors which play central roles in both producing a nontrivial closure and inducing, over long times, weakly decaying long distance correlations. A practical consequence of these facts is that, when deriving the kinetic equation for $n_{\mathbf{k}}$, one can take the following shortcut. After expanding the iterates of the cumulants as products of averages of products of zeroth-order Fourier amplitudes, we can expand the latter as if Wick's theorem obtains and include only the two point averages. But, to evaluate the frequency renormalization, the rule of thumb is a bit more complicated. Therefore, the message is that there is no need to make these approximations in the first place.

The asymptotic closure: In order to remove the secular terms and render the asymptotic expansion (1.5) for the cumulants uniformly valid in time, we allow the zeroth-order cumulants $q^{ss's''}_0(\mathbf{k}, \mathbf{k}', \ldots)$ of order r to vary slowly in time according to the asymptotic expansion

$$\frac{\partial q^{ss'\cdots}_0}{\partial t} = \epsilon^2 F^{(r)}_2 + \epsilon^4 F^{(r)}_4 + \cdots \tag{1.9}$$

and express the initially constant $q^{ss'\cdots}_0$ of (1.5) in terms of the slowly varying $q^{ss'\cdots}_0$ by a reverse Taylor series (e.g. $f(0) = f(T) - Tf'(T)\ldots$). We then choose the members of the sequence $\{F^{(r)}_{2N}\}_{N=1}$ in order to remove the secular terms. This gives us, to leading order,

$$\frac{\partial n_{\mathbf{k}}}{\partial t} = \epsilon^2 \text{sgn}(t) T_2[n_{\mathbf{k}}]$$

$$= \epsilon^2 4\pi \text{sgn}(t) \sum_{s_1 s_2} \int L^{s s_1 s_2}_{\mathbf{k}\mathbf{k}_1\mathbf{k}_2} n_{\mathbf{k}} n_{\mathbf{k}_1} n_{\mathbf{k}_2}$$

$$\times \left(\frac{L^{-s-s_1-s_2}_{-\mathbf{k}-\mathbf{k}_1-\mathbf{k}_2}}{n_{\mathbf{k}}} + \frac{L^{s_1 s-s_2}_{\mathbf{k}_1\mathbf{k}-\mathbf{k}_2}}{n_{\mathbf{k}_1}} + \frac{L^{s_2 s-s_1}_{\mathbf{k}_2\mathbf{k}-\mathbf{k}_1}}{n_{\mathbf{k}_2}} \right)$$

$$\times \delta(s_1\omega_1 + s_2\omega_2 - s\omega)\delta(\mathbf{k}_1 + \mathbf{k}_2 - \mathbf{k})d\mathbf{k}_1 d\mathbf{k}_2. \tag{1.10}$$

$$\frac{\partial q^{ss'\cdots}_0}{\partial t} = i\epsilon^2 q^{ss'\cdots}_0 (\Omega^s_2[n_{\mathbf{k}}] + \Omega^{s'}_2[n_{\mathbf{k}'}] + \cdots), \tag{1.11}$$

where $\Omega^s_2[n_{\mathbf{k}}]$ is given by (1.8). Equation (1.10) is what we call the kinetic equation. Equation (1.11) can be jointly solved for all $q^{ss'\cdots}_0$ simply by renormalizing the frequency

$$s\omega_k \to s\omega_k + \epsilon^2 \Omega^s_2[n_{\mathbf{k}}] + \cdots \tag{1.12}$$

The result is that, over long times, we have a natural closure which we call the asymptotic closure of wave turbulence theory.

Three key features: First, one can see from the nature of the kinetic equation that energy transfer is mediated by resonant wave interactions. This suggests that the relevant "eddies" in wave turbulence (the eddies in fully developed three-dimensional turbulence are loosely identified as vortices of a certain scale) can be identified as wavetrains. It is, thus, truly a weakly nonlinear theory in that the fundamental objects are solutions of the linearized equations. Moreover, in the following section, we shall meet exact solutions $n_0(k)$ of the stationary kinetic equation

$$T_2[n_0(k)] = 0. \tag{1.13}$$

They fall into two categories, the thermodynamic solutions for which the flux of energy density e_k ($\approx \omega_k n_k$) is zero and the KZ solution for which the energy flux is finite but constant everywhere. The latter is the solution most relevant for nonisolated systems and represents situations in which a finite and constant energy flux P is supplied at $k = 0$ and dissipated (the energy dissipation rate) at $k = \infty$. If the coupling coefficient $L^{ss_1s_2}_{\mathbf{k}\mathbf{k}_1\mathbf{k}_2}$ has the homogeneity property that $L^{ss_1s_2}_{\lambda\mathbf{k}\lambda\mathbf{k}_1\lambda\mathbf{k}_2} = \lambda^{\gamma_2} L^{ss_1s_2}_{\mathbf{k}\mathbf{k}_1\mathbf{k}_2}$ and the dispersion relation is $\omega = ck^\alpha$ where c is a dimensional constant, then Zakharov and Filoneko (1967) have shown that the stationary solution is

$$n_0(k) = c_2 P^{1/2} k^{-\gamma_2 - d},$$

where $d = \dim(\mathbf{k})$. It is on these solutions that it is important to test premise P3, the separation of time scales.

The second feature is that one can calculate the deviation of the physical space cumulants from the state of joint Gaussianity. For example, the third-order cumulant in Fourier space is

$$Q_1^{ss's'}(\mathbf{k}, \mathbf{k}', \mathbf{k}'') = \exp(i(s\omega + s'\omega' + s''\omega'')t)$$

$$\times P^{00'0''} \left\{ \sum_{s_1s_2} \int L^{ss_1s_2}_{\mathbf{k}\mathbf{k}_1\mathbf{k}_2} q_0^{s's''s_1s_2}(\mathbf{k}', \mathbf{k}'', \mathbf{k}', \mathbf{k}_1, \mathbf{k}_2) \right.$$

$$\times \Delta(s_1\omega_1 + s_2\omega_2 - s\omega)\delta(\mathbf{k}_1 + \mathbf{k}_2 - \mathbf{k})d\mathbf{k}_1 d\mathbf{k}_2$$

$$+ 2\sum_{s_3s_4} \int L^{ss_3s_4}_{\mathbf{k}-\mathbf{k}'-\mathbf{k}''} q_0^{s_3s'}(\mathbf{k}') q_0^{s_4s''}(\mathbf{k}'')$$

$$\left. \times \Delta(s_3\omega' + s_4\omega'' - s\omega) \right\}. \tag{1.14}$$

In (1.14), $\Delta(x) = \frac{e^{ixt}-1}{ix}$, and we note $\Delta(x)e^{-ixt} = \Delta(-x)$. The long time behavior of the physical space counterpart will lose memory of all terms which are multiplied by a fast oscillation. The reason for this is the RL lemma. For $f(x)$ measurable and L^1 integrable, there are two main results central for our analysis. The first is RL and the second is heavily used in calculating the secular terms which gave rise to (1.6), (1.8) and (1.11); namely,

$$\int f(x)e^{ixt}dx \sim 0$$

$$\int f(x)\frac{e^{ixt}-1}{ix}dx \sim \pi f(0)\text{sgn}(t) + i\hat{P}\int \frac{f(x)}{x}dx. \qquad (1.15)$$

In (1.15), the symbol $\sim$ means the long time limit $t \to \pm\infty$ and $\hat{P}$ is the Cauchy principal value. The only possible asymptotic survivor is

$$R^{ss's''}(\mathbf{r},\mathbf{r}') \sim 2\epsilon P^{00'0''}\int L^{s-s'-s''}_{\mathbf{k}-\mathbf{k}'-\mathbf{k}''}n(\mathbf{k}')n(\mathbf{k}'')$$

$$\times \tilde{\Delta}(s\omega + s'\omega' + s''\omega'')\delta(\mathbf{k}+\mathbf{k}'+\mathbf{k}'')dkdk'dk'', \qquad (1.16)$$

where $\tilde{\Delta}(x) = \pi\delta(x)\text{sgn}(t) + i\hat{P}(1/x)$. For wave turbulence theory, the degree of non-Gaussianity in the third-order cumulant is given by the expression (1.16) containing a quadratic product of waveaction densities. Analysis will show that the asymptotic survivor for the fourth-order cumulant is order ϵ^2 and given by a similar expression to (1.16) with a cubic product of waveaction densities. Such relations can be checked in observations but, to date they have been seen only in a handful of cases that we will talk about later. They lead to a weak and algebraic decay (not L^1 but rather like $|\mathbf{r}|^{-1/2}$) in the physical space cumulants and corresponding structure functions $\langle(u^s(\mathbf{x}+\mathbf{r})-u^s(\mathbf{x}))^N\rangle$. In particular, one can see that for long times, the Fourier transform of the third- (and fourth-) order cumulant is no longer smooth. It has picked up a Dirac delta function behavior. This means that if one were to redo the statistical initial value problem from $t = t_1 = \mathcal{O}(\epsilon^{-2})$, one would find (Benney and Newell, 1969) additional secular terms because of this nonsmoothness. But their manifestation in the kinetic equation is simple and pleasing. They simply add to the $\text{sgn}(t-t_1)T_2[n_\mathbf{k}]$ term an additional contribution $(\text{sgn}(t)-\text{sgn}(t-t_1))T_2[n_\mathbf{k}]$. The sum is $\text{sgn}(t)T_2[n_\mathbf{k}]$. This shows that the kinetic equation is consistent no matter what starting time one uses and tends to relax towards a statistically steady state as $t \to \pm\infty$, with either sign, with time measured from that initial point at which statistical correlations of distance points decay rapidly enough to give smooth Fourier transforms. The physical reason for the development of order ϵ^r ($r = 1$ for third order, $r = 2$ for fourth order) weakly decaying spatial correlations is that the three and four wave resonant interactions couple the field over large distances. In testing the validity of wave turbulence theory, it is important for experimentalists to measure not only the statistically steady spectral attractor $n_0(k)$, the power spectrum, to which solutions of the kinetic equations decay, but also the long distance behavior of structure functions (Biven *et al.*, 2003; Newell *et al.*, 2001).

Third, it is also important to understand that in the derivation of the kinetic equation (1.10) and the frequency modulation (1.11), we have used the *de facto* assumption that the linear $t_L = 1/\omega_k$ and the nonlinear $t_{\rm NL} = \frac{1}{n_{\bf k}} \frac{dn_{\bf k}}{dt}$ time scales are well separated for all k. This assumption is used when considering long time limits of expressions such as, for example, $\sum_{s_1 s_2} \int L_{{\bf k}{\bf k}_1 {\bf k}_2}^{s s_1 s_2} L_{-{\bf k}-{\bf k}_1-{\bf k}_2}^{-s-s_1-s_2} \Delta(s_1\omega_1 + s_2\omega_2 - s\omega)\Delta(s\omega - s_1\omega_1 - s_2\omega_2)n_{{\bf k}_1}n_{{\bf k}_2}\delta({\bf k}_1 + {\bf k}_2 - {\bf k})d{\bf k}_1 d{\bf k}_2$ which leads to the first terms in (1.10). In taking the limit $t \to \infty$, we assume that we keep the time $\epsilon^2 t$, over which $n_{\bf k}$ varies, fixed. We then obtain $\int f(x)\Delta(x)\Delta(-x)dx \sim 2\pi t{\rm sgn}(t)f(0) + 2\hat{P}\int \frac{f(x)}{x}dx$. But, there is a rub! The ratio $t_L/t_{\rm NL}$ depends on k and we will see that, except on a very special spectrum, the generalized Phillips' spectrum (GPS), the ratio always diverges at very small or very large k on all the stationary solutions of the kinetic equation. In particular, it will diverge (along with other measures such as $\frac{\Omega_2^s[n_{\bf k}]}{\omega_k}$ and the deviation of the long time statistics as measured by the structure functions (Biven *et al.*, 2003) from a state close to joint Gaussianity) on the KZ spectrum. This feature does not mean that the wave turbulence closure is wrong. It does say that for either $k < k_1$ or for $k > k_2$, the long and short breakdown wavenumbers for which $t_L/t_{\rm NL} = \mathcal{O}(1)$, we must find another asymptotic state which continuously extends the KZ spectrum that is valid in the regions for which $t_L/t_{\rm NL}$ is small. We have called this phenomenon the breakdown of the theory at high and at low wavenumbers (Newell *et al.*, 2001). Additional physics may have to be added for the regions $k > k_2$ or $k < k_1$ because the new stationary spectrum will not be a solution of $T_2[n_{\bf k}] = 0$ but rather the balance between $T_2[n_{\bf k}]$ and some kind of dissipation. We will discuss two examples taken from contexts of optical waves of diffraction in nonlinear media and ocean gravity waves at the beginning of Sec. 1.5.

Higher order closures: One can consistently continue the wave turbulence closure to higher orders and obtain successive contributions to $dn_{\bf k}/dt$ and the frequency renormalization

$$\frac{dn_{\bf k}}{dt} = \epsilon^2 T_2[n_{\bf k}] + \epsilon^4 T_4[n_{\bf k}] + \cdots , \qquad (1.17)$$

$$s\omega_k \to s\omega_k + \epsilon^2 \Omega_2^s[n_{\bf k}] + \epsilon^4 \Omega_4^s[n_{\bf k}] + \cdots \qquad (1.18)$$

$T_4[n_{\mathbf{k}}]$ will again depend only on $n_{\mathbf{k}}$ and will contain four wave resonances and terms involving three wave resonances mediated by modal interactions. The integrands in such cases will contain products as $\delta(s_1\omega_1 + s_2\omega_2 - s\omega)\hat{P}\frac{1}{s_3\omega_3+s_4\omega_4-s_1\omega_1}$. If triad resonances are impossible (e.g. gravity waves), then the situation is greatly simplified. $T_2[n_{\mathbf{k}}] = 0$ and

$$T_4[n_{\mathbf{k}}] = 12\pi\,\mathrm{sgn}(t) \sum_{s_1 s_2 s_3} \int G^{ss_1 s_2 s_3}_{\mathbf{k}\mathbf{k}_1\mathbf{k}_2\mathbf{k}_3}\, n_{\mathbf{k}} n_{\mathbf{k}_1} n_{\mathbf{k}_2} n_{\mathbf{k}_3}$$

$$\times \left(\frac{G^{-s-s_1-s_2-s_3}_{-\mathbf{k}-\mathbf{k}_1-\mathbf{k}_2-\mathbf{k}_3}}{n_{\mathbf{k}}} + P_{123}\frac{G^{s_1 s-s_2-s_3}_{\mathbf{k}_1\mathbf{k}-\mathbf{k}_2-\mathbf{k}_3}}{n_{\mathbf{k}_1}} \right)$$

$$\times \delta(s_1\omega_1 + s_2\omega_2 + s_3\omega_3 - s\omega)$$

$$\times \delta(\mathbf{k}_1 + \mathbf{k}_2 + \mathbf{k}_3 - \mathbf{k})d\mathbf{k}_1 d\mathbf{k}_2 d\mathbf{k}_3, \qquad (1.19)$$

where

$$G^{ss_1 s_2 s_3}_{\mathbf{k}\mathbf{k}_1\mathbf{k}_2\mathbf{k}_3} = L^{ss_1 s_2 s_3}_{\mathbf{k}\mathbf{k}_1\mathbf{k}_2\mathbf{k}_3} - \frac{2i}{3}P_{123}$$

$$\times \sum_{s_4} L^{s-s_4 s_1}_{\mathbf{k}\mathbf{k}_2+\mathbf{k}_3\mathbf{k}_1} L^{-s_4 s_2 s_3}_{\mathbf{k}_2+\mathbf{k}_3\mathbf{k}_2\mathbf{k}_3}/(s_2\omega_2 + s_3\omega_3 + s_4\omega(\mathbf{k}_2 + \mathbf{k}_3)).$$

For surface gravity waves, there are no resonances for which all the sign parameters are equal. As a consequence, to this order, the total wavenumber $\int n_{\mathbf{k}}d\mathbf{k}$ is formally conserved and (1.19) becomes, for $t > 0$,

$$T_4[n_{\mathbf{k}}] = 12\pi \int |G^{ss_1 s_2 s_3}_{\mathbf{k}\mathbf{k}_1\mathbf{k}_2\mathbf{k}_3}|^2 n_{\mathbf{k}} n_{\mathbf{k}_1} n_{\mathbf{k}_2} n_{\mathbf{k}_3} \left(\frac{1}{n_{\mathbf{k}}} + \frac{1}{n_{\mathbf{k}_1}} - \frac{1}{n_{\mathbf{k}_2}} - \frac{1}{n_{\mathbf{k}_3}} \right)$$

$$\times \delta(\omega_1 + \omega_2 - \omega_3 - \omega)\delta(\mathbf{k}_1 + \mathbf{k}_2 + \mathbf{k}_3 - \mathbf{k})d\mathbf{k}_1 d\mathbf{k}_2 d\mathbf{k}_3. \quad (1.20)$$

Assuming isotropy and angle averaging so that (Ω_0 is the solid angle in d dimensions) $N_\omega = \Omega_0 k^{d-1}\frac{dk}{d\omega}n_\omega(k(\omega))$, the kinetic equation for four wave resonances can be further simplified as

$$\frac{dN_\omega}{dt} = S(\omega) = \int_\Delta S_{\omega\omega_1\omega_2\omega_3} n_\omega n_{\omega_1} n_{\omega_2} n_{\omega_3} \left(\frac{1}{n_\omega} + \frac{1}{n_{\omega_1}} - \frac{1}{n_{\omega_2}} - \frac{1}{n_{\omega_3}} \right)$$

$$\times \delta(\omega_1 + \omega_2 - \omega_3 - \omega)d\omega_1 d\omega_2 d\omega_3, \qquad (1.21)$$

where Δ is the region $\omega_2,\ \omega_3,\ \omega_2 + \omega_3 > 0$ in the $\omega_2,\ \omega_3$ plane. If $G^{ss_1 s_2 s_3}_{\mathbf{k}\mathbf{k}_1\mathbf{k}_2\mathbf{k}_3}$ has homogeneity degree γ_3, $S_{\omega\omega_1\omega_2\omega_3}$ has homogeneity degree $\sigma = (2\gamma_3 + 3d)/\alpha - 4$. Before we discuss the solutions, we make several further remarks.

1. The kinetic equation (1.17) is solved for times $t = \mathcal{O}(1/\epsilon^{2r})$, $r = 1, 2, \ldots$ by successive truncation. For the theory to remain valid, it is important to check that solutions of the truncated equations keep (1.9), (1.17) and

(1.18) uniformly asymptotic in **k**, and that the ratio t_L/t_{NL} is uniformly small in **k**.

2. With forcing and damping, the first truncation leads to a statistically steady state except in cases such as acoustic waves or Alfvén waves where the resonant manifolds foliate wavevector space. See the first challenge in Sec. 1.6 and the shape of the KZ spectrum in Galtier *et al.* (2000).

3. The kinetic equation $dn_{\mathbf{k}}/dt = \epsilon^4 T_4[n_{\mathbf{k}}]$ is independent of the sign of the coefficient $G_{\mathbf{kk_1k_2k_3}}^{ss_1s_2s_3}$, but the first frequency correction is not. This has important ramifications for the equations of nonlinear Schrödinger (NLS) type (see discussion in Sec. 1.5, Q2d).

4. The energy density $e_{\mathbf{k}}$ is related to $n_{\mathbf{k}}$ to order ϵ^2 by $e_{\mathbf{k}} = (\omega_k + \epsilon^2 s^{-1}\Omega_2^s[n_{\mathbf{k}}])n_{\mathbf{k}}$. Formal energy conservation follows from the properties of $L_{\mathbf{k_1k-k_2}...-\mathbf{k_r}}^{s_1s-s_2...-s_r} = (s_1/s)L_{\mathbf{kk_1}...\mathbf{k_r}}^{ss_1...s_r}$ given earlier.

5. As we have already explained for the three wave resonance cases, if one redoes the initial value problem from $t = t_1 = \mathcal{O}(\epsilon^{-2}) > 0$, the coefficients in (1.6) and (1.19) are still sgn(t) not sgn($t - t_1$) because, in addition to (1.6) with the sign factor sgn($t - t_1$), there is an extra term with sign factor (sgn(t) − sgn($t - t_1$)) arising from the fact that the third- and fourth-order cumulants now have nonsmooth, orders ϵ and ϵ^2 respectively, initial values. Thus, the isolated system will always relax to its thermodynamic state for large positive or negative time measured from $t = 0$.

Finally, let us emphasize what we mean by a valid wave turbulence theory. So far, we have only used the conservative part of the underlying dynamical system. We will append, phenomenologically, to the kinetic equation (1.21) two terms S_{IN} and S_{OUT} representing forcing and dissipation. Only if input and output can be represented as $\gamma(\omega)A_{\mathbf{k}}^s$ in (1.1), (1.21) will be a natural closure. In that special case, $S_{\text{IN}} = 2\gamma_{\text{IN}}(\omega)N(\omega)$ and $S_{\text{OUT}} = 2\gamma_{\text{OUT}}(\omega)N(\omega)$. But, the input and output is often too complicated for this to be so. Therefore, in general, we look upon (1.21) with S_{IN} and S_{OUT} added as being a valid approximation because each is much smaller than the first (which feeds $n_{\mathbf{k}}$) and second (proportional to $n_{\mathbf{k}}$) terms in (1.19). We can also take account of weak nonspatial homogeneity by writing $\partial n_{\mathbf{k}}/\partial t$ as $\partial n_{\mathbf{k}}/\partial t + \nabla_{\mathbf{k}}\omega\nabla_{\mathbf{x}}n_{\mathbf{k}} - \nabla_{\mathbf{x}}\omega\nabla_{\mathbf{k}}n_{\mathbf{k}}$, where ω is the renormalized frequency (18). We shall say that wave turbulence theory is valid if solutions of the original or extended equations match what is observed.

1.3. Solutions of the Kinetic Equation

We shall now discuss solutions of the kinetic equation in the context of situations (gravity ocean waves, optical waves; but not vibrations on a thin plate) for which, formally at least, there are two conserved quantities, waveaction or particle number $N = \int n_{\mathbf{k}} d\mathbf{k}$ and energy $E = \int \omega_k n_{\mathbf{k}} d\mathbf{k}$. For these situations, there is a two parameter family of thermodynamic solutions and two pure KZ solutions corresponding to the constant fluxes of energy and waveaction densities. Because there are many sources where the reader can access detailed discussions, here, we emphasize and summarize the main outcomes.

Thermodynamic vs. Kolmogorov solutions: In systems where an ultraviolet-cutoff avoids leakage of energy to infinite wavenumber, zero-flux thermodynamic solutions (Rumpf, 2008) are the statistically steady states. But, most wave turbulent systems of interest are not isolated and have sources and/or sinks which often are widely separated in wavenumber space. In the windows of transparency between sources and sinks, one expects the statistically steady state to be of Kolmogorov finite flux type. It is quite remarkable that, of all the pioneers in establishing the kinetic equation as the centerpiece of wave turbulence theory, only Zakharov saw that it was the finite flux KZ solutions which were more important for nonisolated systems. For over 20 years, very few of his western colleagues took much notice of them and his was a lone voice crying in the wilderness. For this work and his joint discovery with Kraichnan of inverse cascades, he was deservedly awarded the Dirac Medal in 2003.

The pure thermodynamic and Kolmogorov solutions: We begin with the very special pure power-law solutions of (1.21), valid when $L_{\mathbf{k}\mathbf{k}_1 \mathbf{k}_2 \mathbf{k}_3}^{ss_1 s_2 s_3}$ ($L_{\mathbf{k}\mathbf{k}_1 \mathbf{k}_2}^{ss_1 s_2}$) is homogeneous with degree γ_3 (γ_2; $\gamma_3 + \alpha = 2\gamma_2$) and $S_{\omega\omega_1\omega_2\omega_3}$ thereby is homogeneous with degree $\sigma = (2\gamma_3 + 3d)/\alpha - 4$. These solutions are $n_{\mathbf{k}}(\omega) = c_1,\ c_2\omega^{-1},\ c_3 P^{1/3}\omega^{\frac{-2\gamma_3}{3\alpha} - \frac{d}{\alpha}},\ c_4 Q^{1/3}\omega^{\frac{-2\gamma_3}{3\alpha} - \frac{d}{\alpha} + \frac{1}{6}}$, corresponding to equipartition of number density, energy density, finite energy flux P (with zero number flux Q) and finite number flux Q (with zero energy flux P). The first two solutions follow by inspection of (1.21). The last two are obtained by dividing the region of integration Δ into four subregions, Δ_1, $0 < \omega_2, \omega_3 < \omega, \omega_2 + \omega_3 > \omega$; Δ_2, $0 < \omega_2 < \omega, \omega_3 > \omega$; Δ_3, $\omega_2, \omega_3 > \omega$ and Δ_4, $0 < \omega_3 < \omega, \omega_2 > \omega$; and mapping each of $\Delta_2, \Delta_3, \Delta_4$ conformally onto Δ_1, called the Zakharov transformation. Setting $\omega_j = \omega\zeta_j, j = 1, 2, 3$, and $n = c\omega^{-x}$ allows us to write $S(\omega)$ as $c^3\omega^{-y-1} I(x, y(x))$ where

$$I(x, y) = \int_{\Delta_1'} S_{1\zeta_1\zeta_2\zeta_3} (\zeta_1\zeta_2\zeta_3)^{-x} \delta(1 + \zeta_1 - \zeta_2 - \zeta_3)$$

$$\times (1 + \zeta_1^x - \zeta_2^x - \zeta_3^x)(1 + \zeta_1^y - \zeta_2^y - \zeta_3^y) d\zeta_1 d\zeta_2 d\zeta_3 \quad (1.22)$$

and Δ_1' is $0 < \zeta_2, \zeta_3 < 1, \zeta_2 + \zeta_3 > 1$, and $y = 3x + 1 - 2\gamma_3/\alpha - 3d/\alpha$. The choices $x = 0, 1, y = 0, 1$ are the pure equipartition solutions, and finite flux solutions, respectively. Depending on the behavior of $S_{\omega\omega_1\omega_2\omega_3}$ near $\omega_1 = \omega_2 + \omega_3 - \omega = 0$, $I(x,y)$ converges in the ranges including the finite flux solutions. For gravity waves, the KZ finite flux solutions are $x = 23/3$ and $x = 8$, respectively, and $I(x, y(x))$ converges for $5 < x < 19/2$ (see remark 2 below). The Kolmogorov constants c_3 and c_4 can be found by setting $S(\omega) = \partial Q/\partial\omega$ and $\omega S(\omega) = -\partial P/\partial\omega$, respectively, whereupon $Q = -\lim_{y\to 0} c_4^3 Q y^{-1}\omega^{-y} I(x,y) = c_4^3 Q(dI/dy)_{y=0}$ and $P = +\lim_{y\to 0} c_3^3 P(y-1)^{-1} I(x,y) = c_3^3 P(dI/dy)_{y=1}$, respectively. The slopes of $I(x,y)$ at $y = 0, 1$ are negative and positive, respectively. As pointed out in the previous section, a similar analysis for the kinetic equation $dn_{\mathbf{k}}/dt = \epsilon^2 T_2[n_{\mathbf{k}}]$ dominated by three wave resonances gives two special stationary solutions $n_{\mathbf{k}} = T/\omega$ and $n_{\mathbf{k}} = cP^{1/2}\omega^{-(\gamma_2/\alpha+d/\alpha)}$ corresponding to energy equipartition and finite energy flux P.

An alternative derivation: For reasons of pedagogy, it is helpful to rederive the stationary solutions $T_4[n_{\mathbf{k}}] = 0$ in another way. If the coupling coefficient $S_{\omega\omega_1\omega_2\omega_3}$ is localized and supported only near $\omega = \omega_1 = \omega_2 = \omega_3$, one can replace $S(\omega)$ by a differential representation $\partial^2 K/\partial\omega^2$ where $K = S_0\omega^{3x_0+2}n^4 d^2 n^{-1}/d\omega^2$, S_0 is a well-defined integral and $x_0 = 2\gamma_3/(3\alpha) + d/\alpha$. We can identify the particle flux as $Q = \partial K/\partial\omega$ (Q is positive when particles flow from high to low wavenumbers) and P, the direct energy flux, is $K - \omega\partial K/\partial\omega$. The stationary solutions are clearly $K = A\omega + B$, where we can identify the constants A and B with Q and P. The solutions of $K = S_0\omega^{3x_0+2}n^4(d^2(1/n))/(d\omega^2) = Q\omega + P$ will, in general, be a four parameter family with two additional constants T and μ (temperature and chemical potential) arising from the double integration of $K = Q\omega + P$. If $P = Q = 0$, then $n = T/(\omega - \mu)$, the Rayleigh–Jeans solution for classical waves. (Similar solutions for bosons and fermions generalizing the Bose–Einstein and Fermi–Dirac distributions by the addition of finite fluxes can be found in L'vov *et al.* (1998).) The pure KZ solutions occur when we take, in turn, $Q = 0$ and $P = 0$ and look for power-law solutions $n = c_3\omega^{-x}$, $n = c_4\omega^{-x}$ for $K = \omega Q + P$. It is easy to show that these two solutions are $n = c_4 Q^{1/3}\omega^{-2\gamma_3/(3\alpha)-d/\alpha+1/6}$ and $n = c_3 P^{1/3}\omega^{-2\gamma_3/(3\alpha)-d/\alpha}$, respectively. In many practical applications, however, the actual steady state solution may be a more complicated combination of these special solutions. We now make several further points.

1. The pure finite energy (zero particle) flux solution is strictly relevant only when the energy is inserted at the boundary $k = 0$ and removed at $k = \infty$, equivalent to solving (1.21) with boundary conditions $K = P$, $\partial K/\partial\omega = 0$ at $\omega = 0$ and $\omega = \infty$. Likewise, the pure particle flux KZ inverse cascade (zero energy flux) obtains when particles are added to the system at $k = \infty$ and removed at $k = 0$.

2. Note that $T_4[n_{\mathbf{k}}]$ in (1.19) (and $S(\omega)$ in (1.21)) can be written as two integrals, a feed $F_{\mathbf{k}}$ involving integrals over the product $n_{\mathbf{k}_1} n_{\mathbf{k}_2} n_{\mathbf{k}_3}$ and a term $-n_{\mathbf{k}} \Gamma_{\mathbf{k}}$ proportional to $n_{\mathbf{k}}$. Indeed, $\Gamma_{\mathbf{k}}$ is proportional to $\mathrm{Im}\Omega_4[n_{\mathbf{k}}]$. $F_{\mathbf{k}}$ and $\Gamma_{\mathbf{k}}$ are often divergent on the KZ solution but the singularity cancels because of the combination. The first consequence is that when compared to any input or dissipation terms, the contributions $F_{\mathbf{k}}$ and $n_{\mathbf{k}} \Gamma_{\mathbf{k}}$ dominate the phenomenologically added S_{IN} and S_{OUT}. The second is that the breakdown criteria developed in remark 5 can curtail the range of validity of the KZ spectrum even further.

3. Finite and infinite capacity. Both energy and particle number conservation may break down after a finite time. The idea goes back to Onsager. If we compute the amount of energy, $\int_{\omega_0}^{\infty} \omega N_\omega d\omega$, say in the range (ω_0, ∞), under the direct energy flux KZ spectrum $n_{\mathbf{k}} = c_3 P^{1/3} \omega^{-(2\gamma_3)/(3\alpha)-d/\alpha}$, or $\omega N_\omega = (\Omega_0/\alpha) c_3 P^{1/3} \omega^{-(2\gamma_3)/(3\alpha)}$, the integral converges (the spectrum supports finite energy) if $\gamma_3 > 3\alpha/2$ and diverges otherwise (infinite capacity). In the finite capacity case, if energy is added at $\omega = \omega_0$ at a constant rate and is not confined to finite frequencies, it must escape to the dissipation sink at $\omega = \infty$ in a finite time t^* after which energy conservation no longer holds. Likewise, there is an equivalent notion of finite capacity for the particle flux. Ocean gravity waves have finite energy capacity at large wavenumbers and infinite waveaction capacity at low wavenumbers so that the inverse cascade does not build (in infinite wave tanks) condensates. On the other hand, in optical turbulence for which $\alpha = 2$, $\gamma_3 = 0$, the direct cascade of the Hamiltonian density has infinite capacity whereas the power density has finite capacity and, as a result, condensates can form. For three wave resonances, the direct energy flux cascade $n_{\mathbf{k}} = cP^{1/2} \omega^{-(\gamma_2+d)/\alpha}$ has finite capacity if $\gamma_2 > \alpha$ (e.g. capillary waves).

4. How KZ spectra are realized. Pure KZ spectra are generally realized by self-similar solutions $n_{\mathbf{k}} = t^{-b} n(kt^{-a})$ of (1.21) (Falkovich and Shafarenko, 1991). In the infinite capacity case, both exponents a and b are found by scaling arguments and the spectrum consists of a front which joins an exponentially small precursor to a wake spectrum through a front $k_f \propto t^a$, $a > 0$. The wake spectrum behind the front relaxes to the pure KZ spectrum. In the finite capacity case, an anomaly occurs. The front now travels as $k_f \propto (t^* - t)^a$, $a < 0$ and leaves in its wake a spectrum k^{-x} where the exponent $x = b/|a|$ is greater than the KZ value and can only be determined by solving a nonlinear eigenvalue problem. Only after the front hits the dissipation range, does, the KZ spectrum build up backwards from $k = \infty$. It is not known whether such anomalies are present for fully turbulent systems (Connaughton *et al.*, 2003a; Connaughton and Newell, 2010).

5. Breakdown of KZ spectra. To test the validity of P3, let us compute $t_L/t_{NL} = (n_{\mathbf{k}}\omega_k)^{-1}dn_{\mathbf{k}}/dt$ on the spectrum $n_k = ck^{-\alpha x}$ for the case where $T_2[n_{\mathbf{k}}] \equiv 0$. Setting each k_j in T_4 to $k\zeta_j$, we will obtain that

$$\frac{t_L}{t_{NL}} = \frac{k^{2\gamma_3}c^3k^{-3\alpha x}k^{-\alpha}k^{2d}}{ck^{-\alpha x}k^{\alpha}}I, \qquad (1.23)$$

where I is some k-independent integral which we will assume, for the moment, converges. We find

$$\frac{t_L}{t_{NL}} \propto c^2k^{2(\gamma_3+d-\alpha-\alpha x)}. \qquad (1.24)$$

Only on the GPS for which $\alpha x = \gamma_3 + d - \alpha$ is (1.24) $\mathbf{k}$ independent ("critical" balance between linear and nonlinear terms). The ratio then depends on the nondimensional size of c. On the direct energy cascade KZ spectrum, $\alpha x = 2\gamma_3/3 + d$, $c = CP^{1/3}$, we find $t_L/t_{NL} = C^2P^{2/3}k^{2(\gamma_3/3-\alpha)}$. On the inverse number density (waveaction, power) cascade, $\alpha x = 2\gamma_3/3 + d - \alpha/3$, $c = CQ^{1/3}$, we find $t_L/t_{NL} = C^2Q^{2/3}k^{2(\gamma_3-2\alpha)/3}$. We have absorbed the small parameter ϵ into c and thereby both into P and Q. For $\gamma_3 > 3\alpha$, as is the case for ocean gravity waves, the ratio is of order unity or greater and P3 is violated for wavenumbers $k > k_U$ when $P^{2/3}k_U^{2(\gamma_3/3-\alpha)} \sim 1$. For gravity waves, $P^{2/3}k_U/g \sim 1$. For $\gamma_3 < 2\alpha$, as is the case for the NLS equation, the ratio becomes unity at small wavenumbers $k < k_I$ when $Q^{2/3}k_I^{2(\gamma_3-2\alpha)/3} \sim 1$. On the direct energy cascade for three wave resonances, breakdown occurs for $k > k_U$, $P^{1/2}k_U^{\gamma_2-2\alpha} \sim 1$ when $\gamma_2 > 2\alpha$. For capillary waves, $\gamma_2 = 9/4 < 2\alpha = 3$ and so the wave turbulence requirement that $t_L/t_{NL} \ll 1$ gets better the larger k is. As long as the relevant integral I in (1.23) converges, breakdown occurs at the same point for all measures of wave turbulence validity, t_L/t_{NL}, T_4/T_2, Ω_2/ω, Ω_4/ω, $(S_4 - 3S_2^2)/S_2^2$ (with k, ω replaced by r^{-1}, τ^{-1}). But, as we indicated in an earlier remark, it turns out that, for gravity waves, the ratio Ω_4/ω does not converge and this adds a factor $(k_U/k_p)^{1/2}$ to $P^{2/3}k_U/g$, k_p the peak wavenumber in n_k. The breakdown range becomes even larger in this case.

6. Structure functions (Biven *et al.*, 2003). Given that $n_{\mathbf{k}}$ has power-law behavior over an inertial range, one can ask what is the corresponding universal behavior of the structure functions $S_N(\mathbf{r}, \tau) = \langle(\eta(\mathbf{x} + \mathbf{r}, t + \tau) - \eta(\mathbf{x}, t))^N\rangle$, where $\eta(\mathbf{x}, t)$ is, for example, the surface elevation. Since measurements usually involve the time signal at a given point, we have to settle for $S_N(0, \tau)$. For surface waves, the second-order structure function is $S_2(\tau) = 2\int_0^\infty I(\omega)(1 - \cos(\omega\tau))d\omega$, where $I(\omega) = (\pi\omega/2\nu^2)(dk^2/d\omega)n_k(\omega)$, $\omega^2 = k\nu^2 = gk + (S/\rho)k^3$ is the Fourier transform of the two point correlation $\langle\eta(\mathbf{x}, t)\eta(\mathbf{x}, t + \tau)\rangle$. For capillary waves, $I(\omega) = cP^{1/2}(S/\rho)^{1/6}\omega^{-17/6}$ is the direct energy flux spectrum. If this shape was valid for the entire

range, then $S_2(\tau) = 2cP^{1/2}(S/\rho)^{1/6}\tau^{11/6}\int_0^\infty \xi^{-17/6}(1 - \cos\xi)d\xi$. But, of course, in practice, the universal power-law holds at best over a finite range (ω_1, ω_2), where ω_1 (ω_2) denotes the upper (lower) boundary of the forcing (dissipation) range. We can generally take $\omega_2 = \infty$ so that $S_2(\tau) = 2\int_0^{\omega_1} I(\omega)(1 - \cos(\omega\tau))d\omega + 2cP^{1/2}(S/\rho)^{1/6}\tau^{11/6}J(\omega_1\tau)$ where $J(\omega_1\tau) = \int_{\omega_1\tau}^\infty \xi^{-17/6}(1-\cos\xi)d\xi$. The nonuniversal part of $S_2(\tau)$ behaves, for small τ, as τ^2. The universal part behaves as $\tau^{11/6}$. It is very difficult to distinguish the two. To gain separation, it is better to work with averages of higher time differences such as $S'_N(\tau) = \langle(\eta(\mathbf{x}, t + \tau) - 2\eta(\mathbf{x}, t) + \eta(\mathbf{x}, t - \tau))^N\rangle$. In this case, the nonuniversal behavior of S_2 now is τ^4 for small τ whereas the universal behavior remains at $\tau^{11/6}$.

7. Finite box effects. Size matters. If the spectrum is not continuous but quantized, then it is more difficult to satisfy the resonance condition. Some help, however, is gained by the fact that the nonlinear correction to the frequency effectively replaces the Dirac delta function $\delta(\omega_1 + \omega_2 - \omega)$ with a Lorentzian, $\text{Im}(1/(\omega'_1 + \omega'_2 - \omega'))$, $s\omega'_1 = s\omega + \epsilon^2\Omega_2^s$. This means that the resonance manifold is broadened by an amount proportional to ϵ^2 (ϵ^4 in the four wave interaction case). In order that the quantized spectrum allows lots of resonances within the band, we require that the ratio of the box size l to the typical wave length λ participating in triad resonances is greater than ϵ^{-2} times a calculable factor. For gravity waves when $\epsilon \sim .1$, to resolve resonances involving waves of $60\,\text{m}$ would require a tank of approximately $60\,\text{km}$. In finite boxes, it is important to check that boundary reflections also lead to joint Gaussian behavior in the linear limit. In the open ocean, the fields at a set of points in some finite region consists of sums of surface displacements carried by uncorrelated waves which have traveled from afar. In a finite box, one cannot *a priori* assume that the contributions from the incoming waves will be uncorrelated (cf. nonergodic behavior of billards in certain shaped boxes).

1.4. Experimental Evidence

We examine the available experimental evidence. While there is some evidence of consistency between theory and experiment, much remains to be done. In particular, it would be valuable to measure the joint space–time power spectrum $\Phi(\mathbf{k}, \omega)$, the Fourier transform of $\langle\eta(\mathbf{x}, t)\eta(\mathbf{x} + \mathbf{r}, t + \tau)\rangle$. A necessary condition for a valid theory is that its support be concentrated upon the modified dispersion relation.

1.4.1. *Capillary wave turbulence*

Weakly nonlinear capillary waves support three wave resonances. Two groups, based in Paris (Falcon *et al.*, 2007a, 2007b, 2008a, 2008b, 2009)

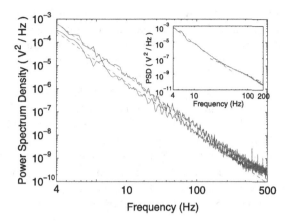

Fig. 1.1. Capillary wave turbulence: power spectrum density of surface wave height on the surface of a fluid layer in low gravity. Lower curve: random forcing 0–6 Hz. Upper curve: sinusoidal forcing at 3 Hz. Dashed lines had slopes of -3.1 (lower) and -3.2 (upper). Inset: same with gravity. Slopes of dashed lines are -5 (upper) and -3 (lower) corresponding, respectively, to gravity and capillary wave turbulence regimes. Figure by courtesy of Eric Falcon. [Falcon *et al.* 2009. *EPL* 86: 14002.]

and Chernogolovka (Kolmakov *et al.*, 2004, 2006) have carried out a series of experiments on the surface response to broad and narrowband forcing at wavenumbers $k_f < k_0 = \sqrt{\rho g / S}$. To increase the range where pure capillary influences are dominant, both groups have sought to decrease viscosity and the effective gravity. The Paris group used shallow layers of mercury ($\lambda_0 = 2\pi/k_0 \sim 1$ cm) and layers of ethanol and water in very low gravity situations (kudos to their courage in flying loop the loops) where $\lambda_0 \sim 10$ cm. The forcing in both the cases was sinusoidal (via subharmonic generation) and low pass filtered, broadband and random in the 0 to 6 Hz range. There is clear evidence of the theoretically predicted pure KZ energy flux frequency spectrum $I(\omega) = (2\pi)^{-1} \int \langle \eta(x,t)\eta(x,t+\tau)\rangle \exp(-i\omega\tau)d\tau = cP^{1/2}(S/\rho)^{1/6}\omega^{-17/6}$ over at least a decade of frequencies in the first experiment and over two decades in the second. The observed power-law $I(\omega) \sim \omega^{-s}$ found $s = 3.1$ for broadband and $s = 3.2$ for narrowband input and is shown in Fig. 1.1. The pdf for the surface elevation η (Falcon *et al.*, 2007b) is almost Gaussian with the usual Tayfun correction expected from second harmonics excited by quadratic interactions. In the normal gravity experiments, the spectrum in the gravity wave range is much steeper than the KZ spectrum predictions, consistent with what is seen for gravity waves in finite tanks (Nazarenko *et al.*, 2010). However, the dependence of $I(\omega)$ on P, the energy flux, is neither $P^{1/2}$ (nor $P^{1/3}$) as predicted but seems to be proportional to P. The reason for this is unclear but one might argue that P, the constant flux in the inertial range and dissipation rate, is not measured simply by the mean of a very widely distributed input flux (as measured by forces on the driving paddles) which has fluctuations much larger than the mean itself and can take on both positive and negative values. In addition, the structure function measurements do not corroborate the theory but that may be attributable to comment 6 in Sec. 1.3.

The Russian group studied capillary turbulence on quantum fluids, liquid hydrogen and helium in both its normal and superfluid states. For broadband forcing, they found $I(\omega) \sim \omega^{-s}$ where $s = 2.8 \pm 0.2$. The result for narrowband forcing was steeper, $s = 3.7$, and the group studied the transition between this and the KZ regime. We do not understand why the difference between narrow (not expected to follow wave turbulence theory) and broadband forcing in the Russian experiments should be so much greater than that observed by the Paris group.

1.4.2. *Gravity wave turbulence*

The views expressed in this section have been informed by experimental results of Donelan *et al.* (1992, 2005, 2006) and Toba (1972, 1973a, 1973b, 1997), by the books of Young (1999) and Phillips (1977), by the numerical simulations of both the forced and damped kinetic and water wave equations by Badulin *et al.* (2005, 2007), Korotkevich (2008), and by recent review articles of Zakharov (2005). While most measurements have involved time signals at a fixed location, Hwang *et al.* (2000, 2004, 2006) (see Fig. 1.2) measured spatial correlations by flying precision parallel courses over the ocean surface.

Despite the fact that Hasselmann derived the kinetic equation in 1962 and Zakharov and Filonenko found the finite flux solutions in 1968, it took a long time for the oceanographic community to accept the fact that, over the largest range, the observed spectra $E(k) = 2\pi k \omega_k n_k$ and $I(\omega)$ had more connections with the "pure" KZ spectra $n_k = c_3 P^{1/3} k^{-4}$ ($E(k) = 2\pi c_3 g^{1/2} P^{1/3} k^{-5/2}$, $I(\omega) = d_3 P^{1/3} g\omega^{-4}$) and $n_k = c_4 Q^{1/3} k^{-23/6}$ ($E(k) = 2\pi c_4 g^{1/2} Q^{1/3} k^{-7/3}$, $I(\omega) = d_4 Q^{1/3} g\omega^{-11/3}$) than with the flux-independent spectrum $I(\omega) \sim g^2 \omega^{-5}$ which Phillips had proposed. Eventually in the 1980's, Donelan *et al.* (1985) argued (Young, 1999, p. 119) that the data did not support the earlier JONSWAP spectrum proportional to ω^{-5}, but rather the one given by an $I(\omega)$ with a tail frequency of ω^{-4}. Further, the spectrum $I(\omega) = d_3 P^{1/3} \omega^{-4}$ is consistent with the observationally deduced Toba's law. The total energy per square meter of ocean surface is proportional to $I = \langle \eta^2 \rangle \approx \int_{\omega_p}^{\infty} I(\omega) d\omega = (1/3) d_3 P^{1/3} g\omega_p^{-3}$ which translates into the law that the mean square height $\sqrt{\langle \eta^2 \rangle}$ is proportional to $T_p^{3/2}$, the 3/2 power of the period of the peak wave. This is analogous to the widely accepted result that $\epsilon = \langle \eta^2 \rangle$ satisfies $\epsilon \omega_p^4 / g^2 = \alpha((d\epsilon/dt)\omega_p^3 / g^2)^{1/3}$

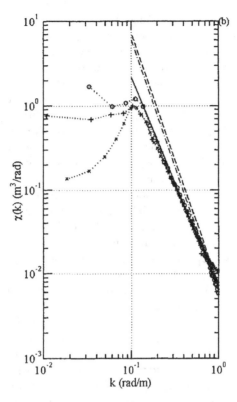

Fig. 1.2. Surface elevation spectra proportional to $E(k)/g$ with three comparison slopes. Solid line $k^{-5/2}$, dashed and dashed dotted curves k^{-3} (Phillips' spectrum) with different normalizations. Note near the peak, the spectrum is slightly less steep (waveaction flux has slope $7/3$), whereas for meter length scales the spectrum is steeper and closer to Phillips. Figure by courtesy of Paul A. Hwang. [Hwang *et al.* 2000. *J. Phys. Oceanogr.* 30: 2753–2767.]

where the "flux" P is replaced by $g d\epsilon/dt$ and $d\epsilon/dt$ is interpreted as a total dissipation rate. Note that neither this nor Toba's law is much changed if we were to use the pure inverse waveaction flux KZ spectrum instead of the pure direct energy flux KZ spectrum.

While the evidence would seem to favor the wave turbulence prediction, there is a rub. Although, little is known about the precise form of input and dissipation, it is generally agreed that the main input occurs at small ($<1\,\mathrm{m}$) scales by a variant of the Miles (1967) instability. (An aside: The Kelvin–Helmholz (KH) instability which amplifies perturbations of wavenumber k on the surface between two layers of fluid (e.g. air, water) with constant tangential velocities (the ultra simple model) U and U_w with densities ρ_a and ρ, $(\rho_a U^2 \simeq \rho U_w^2)$ has a growth rate $(\rho + \rho_a)\sigma = -ik(\rho_a U + \rho U_w) + \nu$, $\nu^2 = -g(\rho^2 - \rho_a^2)k(1 - k\rho\rho_a(U - U_w)^2/(g(\rho^2 - \rho_a^2)) + Sk^2/(g(\rho - \rho_a)))$. It would require a U of at least $5\,\mathrm{m/s}$ ($\rho/\rho_a \sim 10^3$, $S/(\rho g) \sim 7 \cdot 10^{-2}\,\mathrm{cm}^2$) in order to excite waves of wavelengths more than cms. The fact that wave generation occurs at these centimeter scales for much lower wind speeds would seem to rule out the KH instability as the primary mechanism. However, as we argue in Sec. 1.5, it may play a role in initiating whitecap events.) The generation of longer waves is primarily thought to be the result of an inverse cascade where both waveaction and energy are carried to long waves. But the pure KZ ω^{-4} spectrum is predicated on an energy flux from long to short waves. Numerical simulations (Badulin *et al.*, 2007) suggest that the observed spectrum, an evolving spectral shape $I(\omega)$ with a front at ω_f moving from high towards low frequencies rising quickly to a peak at ω_p and then decaying algebraically as ω^{-s}, $11/3 \le s \le 4$ in the wake $\omega > \omega_p$, is a combination of an almost constant waveaction flux (especially near ω_p) and a nonconstant energy flux which changes sign so as to provide a net direct energy flux. The change of sign can be attributed to the deposition of energy by the dual inverse fluxes that conserve both energy and waveaction. The conclusion we draw is that the observed spectrum is consistent with a wave turbulence solution but one which is more complicated than that of pure waveaction or energy fluxes. An additional observation is that at small scales, in strongly driven seas, the spectrum seems to be steeper and more aligned with the Phillips' prediction. The appearance of the Phillips' spectrum in strongly driven situations is consistent with the findings by Korotkevich (2009) in direct numerical simulations and with our own attempts to amend the KZ spectrum at wavenumbers where it is no longer consistent with the wave turbulence closure (Newell and Zakharov, 2008). We discuss this further in Sec. 1.5.

1.4.3. *Vibrating plate turbulence: can one hear the Kolmogorov spectrum?*

During *et al.* (2006) derived and analyzed the wave turbulence of vibrations $\omega_\mathbf{k} \propto |\mathbf{k}|^2$ on large, thin, elastic plates with normal deformations $\eta(\mathbf{x}, t)$ governed by the von Karman–Donnell equations. They found a kinetic equation similar to (1.21), but with the distinct feature that, unlike gravity waves and optical waves of diffraction, the restriction that the four wave interaction preserves wavenumbers is no longer required. As a result, there is only one equipartition solution $n_\mathbf{k} = T/\omega_\mathbf{k} \propto k^{-2}$ and the KZ direct energy flux solution $n_\mathbf{k} \propto k^{-2}$. Because of the degeneracy, the KZ solution requires a log correction and one can show $\langle |\eta_k|^2 \rangle = d(P^{1/3}/k^4)\ln^{1/3}(k^*/k)$ where k^* is some cutoff wavenumber. Numerical simulations appear to corroborate these findings to within a surprising (even the log power fit!) accuracy. However, subsequent experiments by Mordant (2008), Boudaoud *et al.* (2006) have failed to observe these spectra. In frequency space, the corresponding KZ spectrum is $\ln^{1/2}(\omega^*/\omega)$ but the experiments show a definite power-law decay ω^{-s} where $s \approx 0.6$. Space–time spectra confirm that, except for very large waves, the joint power spectrum $\Phi(\mathbf{k},\omega)$ is supported on a very thin curve which closely follows the linear dispersion relation. The real (nonlinear) frequency correction (Ω_2 in (1.8)) is found, however, to be proportional to P^s where $s \geq 1/2$ rather than $s = 1/3$.

Three reasons might be given to explain the steeper spectrum. The first has been suggested by Josserand (private communication) who included broadband damping and found that the spectrum indeed did decay as ω^{-s}, $s > 0$ where s could be tuned by the choice of damping. The second is that the spectrum is steepened by finite size effects. The third, an explanation more in tune with the discussion in Sec. 5.2 is that the plate deformation dynamics is dominated by d-cone singularities (Cerda and Mahadevan, 1998). This is currently being investigated by the authors and L. Mahadevan.

1.4.4. *Condensates of classical light waves*

In a recent article Sun *et al.* (2011) have demonstrated experimentally that classical light waves can, via resonant nonlinear wave interactions, condense into a $k = 0$ plane wave mode. They use a defocusing photorefractive crystal $Sr_{0.75}Ba_{0.25}Nb_2O_6$ whose coupling strength is controlled by applying a voltage. They show clearly that as the nonlinearity increases, there is a strong condensation. The system is isolated and indeed at high wavenumbers, they observe the familiar k^{-2} Rayleigh–Jeans spectrum with zero chemical potential. These observations are also consistent with entropy arguments. Entropy is increased as the system cascades energy to high wavenumber modes and explores more and more of the phase space. The statistically steady state consists of a condensed wave immersed in a sea of small scale fluctuations whose fine details, in the absence of dissipation, contains all the information necessary to effect reversibility.

1.5. Two Open Questions

We have seen that the KZ solutions of the kinetic equation are almost never uniformly valid over all wavenumbers as, on the solutions, premise P3 is violated. Indeed, the demands that the ratio $t_l/t_{NL} \ll 1$, that the asymptotic expansions (1.9), (1.17), (1.18) and the asymptotic expansions for the structure functions are uniformly valid in k and r, respectively, almost always fail at very high or very low wavenumbers (or small or large r scales). While it is satisfying to note that the theory is self-regulating in that it allows one test its validity, the failure does raise two important questions.

1. If premise P3 fails on the KZ spectrum, can we always find other solutions of the stationary kinetic equation, perhaps with the addition of dissipation, which would both meld smoothly with the KZ solution and capture the observed behavior in those wavenumber ranges where the KZ solution fails?

   And another more worrying possibility.

2. Are there situations for which the wave turbulence closure is not relevant at all, at any scale? If this were to be the case, and it is, can we identify what additional premises are needed, premises whose validity can be checked beforehand, which, together with P1, P2 and P3, would be sufficient to guarantee the self-consistency of the wave turbulence closure or to establish that transfer is achieved by processes entirely different from resonant wave interactions?

The answers to these questions are:

1. Sometimes. We illustrate how with two very relevant cases, (i) ocean gravity waves in the case where the energy flux is sufficiently great that the breakdown scale is much longer than the capillary scale, and (ii) optical waves where the medium has only a finite capacity to absorb the inverse cascade of power and some additional means must be found to effect dissipation.

2. Yes. There are examples, thankfully fairly rare and usually one dimensional, for which conclusions drawn from the wave turbulence closure are not relevant at all. And no, to date there are no known *a priori* conditions beyond P1, P2, P3 which would either guarantee that weakly nonlinear wave processes dominate the dynamics or indicate if other fully nonlinear structures are important. We address this question by first carrying out an exercise which shows how the same methods used to obtain the wave turbulence closure can fail if the fields are L_2 and smooth, second by discussing three examples of systems where initial weak nonlinearity can become strong, and third by offering two candidates for additional premises.

We begin with question 1.

Q1a. *Ocean gravity waves*

Given, over some range, a direct flux KZ spectrum for gravity waves, we have shown that it will break premise P3 for wavenumbers $k > k_U$ when $P^{2/3}k_U/g$ is of order unity. It is then natural to ask if there is another physical process, in this case capillary wave action, which may come into play for wavenumbers (k_U, ∞) if $k_0 < k_U$, $k_0 = (\rho g/S)^{1/2}$, which can regularize the breakdown, or, if $k_0 > k_U$, what new spectral state can we find for $n_{\mathbf{k}}$ in (k_U, ∞) (we take $k_0 = \infty$ in this case) which provides for dissipation and can be legitimately attached to the KZ direct flux spectrum at k_U? For ocean waves, $k_0 < k_U$ if $P < (gS/\rho)^{3/4}$ or, because $P \sim (\rho_a/\rho)^{3/2}U^3$, at wind speeds U of less than 5 m/s. In this case, the direct three wave resonance energy transfer carries the energy flux from the four wave KZ spectrum to millimeter scales where viscosity acts to absorb the energy. However, for $P > (gS/\rho)^{3/4}$ or $U > (\rho/\rho_a)^{1/2}(gS/\rho)^{1/4}$ or for wind speeds much greater than 5 m/s, a new spectrum must be appended for $k > k_U$. We note that this is precisely the criterion for which there is a range of wavenumbers for which the Kelvin–Helmholz instability is active. Such an instability leads to wavebreaking and may be responsible for whitecapping events. It has been our suggestion (Newell and Zakharov, 2008) that the new spectrum is the Phillips' spectrum $n_{\mathbf{k}} = cg^{1/2}k^{-9/2}$. It has exactly the right properties to satisfy the amended kinetic equations $S(\omega) + S_{\text{OUT}} = 0$ in the integral sense so that, ignoring surface tension, all energy crossing k_U with flux P is absorbed in (k_U, ∞) by whitecapping events. Moreover, since the nondimensional constant c is very small (the estimate is 0.2), the wave turbulence approximation (see (1.24)) is still valid.

Our suggestion is that the GPS for which

$$n_{\mathbf{k}} = ck^{-(\gamma_3 + d - \alpha)} \tag{1.25}$$

or $ck^{-(2\gamma_2 + d - 2\alpha)}$ in the case for which three wave resonances dominate, plays a central role in complementing the KZ spectrum in those situations where the latter is nonuniformly valid in k. It has four remarkable properties:

1. It is the only spectrum for which the statistical properties, all derived from $n_{\mathbf{k}}$, inherit the symmetry of the governing equation (1.1). To see this, let $\mathbf{K} = \lambda \mathbf{k}$, $T = \lambda^{-\alpha}t$, $A_{\mathbf{k}}^s(t) = \lambda^b B_{\mathbf{K}}^s(T)$, insert in (1.1) and use the homogeneity properties of the coefficients $L_{\mathbf{k}\mathbf{k}_1...\mathbf{k}_r}^{ss_1...s_r}$. In order that $B_{\mathbf{K}}^s(T)$ satisfies the same equation as $A_{\mathbf{k}}^s(t)$, we require $b = d + \frac{\gamma_r - \alpha}{r - 1}$ for all r. (Note that for both gravity and capillary waves $\gamma_3 + \alpha = 2\gamma_2$.) Now suppose that $\langle A_{\mathbf{k}}^s A_{\mathbf{k}'}^{-s} \rangle = \delta(\mathbf{k} + \mathbf{k}')ck^{-\alpha x}$. Then, $\lambda^{2b} \langle B_{\mathbf{K}}^s B_{\mathbf{K}'}^{-s} \rangle = \lambda^{d+\alpha x}\delta(\mathbf{K} + \mathbf{K}')cK^{-\alpha x}$ inherits this symmetry if $2b = d + \alpha x$ or $\alpha x = d + 2\frac{\gamma_r - \alpha}{r - 1}$, the GPS. Since the behavior of the waveaction density $n_{\mathbf{k}}$ governs all other statistical quantities, the statistics inherits the symmetry of the governing equations. We suggest in such cases that there will be no anomalies.

2. As is shown in (1.24), it is the only spectrum for which the ratio $\frac{t_L}{t_{\text{NL}}}$ is independent of k. Thus, it has no low k high k breakdowns.

3. As we have shown (Newell and Zakharov, 2008), it is the spectrum that allows us to balance the nonlinear transfer $S(\omega)$ or $T_4[n_\mathbf{k}]$ with dissipation. To see this, write the kinetic equation as $\frac{\partial E_k}{\partial t} = -\frac{\partial p}{\partial k} - \gamma_k$ where $\int \omega_k n_\mathbf{k} d\mathbf{k} = \int E_k dk$, p is the angle average flux and γ_k is dissipation. Assume a stationary spectrum, $\frac{\partial E_k}{\partial t} = 0$, $n_\mathbf{k} = Ck^{-\alpha x}$, for which $p(k) = C^3 I(x)k^{3\alpha(x_{KZ}-x)}$ where $\alpha x_{KZ} = \frac{2\gamma_3}{3} + d$, the KZ spectrum. Integrate $\frac{\partial p}{\partial k} = -\gamma_k$ from the breakdown number k_1 to $k = \infty$. Find $p(k_1) = \int_{k_1}^{\infty} \gamma_k dk$. If the dissipation is such that all the flux P passing through k_1 is absorbed in (k_1, ∞), we find $C^3 I(x)k_1^{3\alpha(x_{KZ}-x)} = P$. But, the breakdown relation for KZ is $P^{2/3}k_1^{2(\gamma_3/3-\alpha)}$ in order one or $P \sim k_1^{-(\gamma_3-3\alpha)}$. Thus, $3\alpha(x_{KZ} - x) = -\gamma_3 + 3\alpha$ or $2\gamma_3 + 3d - 3\alpha x = -\gamma_3 + 3\alpha$ or $\alpha x = \gamma_3 + d - \alpha$, the GPS.

4. Another interesting property is that, on the GPS, the entropy production divided by the frequency ω, is constant.

Finally, in Newell and Zakharov (2008), we have shown that the GPS for ocean gravity waves is entirely consistent with a picture of sea of finite length crests with derivative discontinuities when one averages over both angle and over the family of crest shapes $\eta(x, y, t) = \frac{1}{2\sigma}e^{-\sigma|x|}e^{-y^2/L^2}$, parametrized by σ, where x and y measure distances perpendicular and parallel to the crest line. The parameter σ approximately captures the whitecap shape at different stages of its evolution. Admittedly, this is a crude approximation but, to date, no exact expression for the shape of an emerging whitecap is known.

Q1b. *Collapses and the cycle of intermittency.*

The model, applicable for optical waves of diffraction in a Kerr nonlinear medium, is the NLS equation with Hamiltonian $\frac{1}{2}\int(|\nabla u|^2 + \lambda|u|^4)d\mathbf{x}$. For power input Q_0 and energy input P_0 (here energy is the spectral density of the Hamiltonian) put in at middle scales ω_0, the power cascades to low wavenumbers. The KZ inverse cascade suffers breakdown at low wavenumbers. If there is no dissipation sink there (as there might be in semiconductor laser context where lasing occurs at the minimum bandgap, namely low k), then a condensate will build. We will discuss here what happens in the self-focusing case for which $\lambda = -1$. The condensate is Benjamin–Feir unstable, and the resulting structures which emerge are fully nonlinear coherent pulses which collapse by becoming singular (infinite amplitudes, infinitesimal width) in a finite time t^*. The collapses are the mechanisms through which power is carried from low to high wavenumbers at which dissipation acts. But in two dimensions, the collapse burnout is incomplete because each collapse carries the minimum amount of power required to sustain it. Thus, a collapse begins to be arrested as soon as it begins to lose power. A fraction f (roughly 20%) of the power is dissipated. The remaining power (and corresponding energy; the energy of a collapsing pulse is zero; the failing collapse puts energy back into the waves) becomes a new source of (waveaction) power and the cascade rate increases from Q_0 until eventually it reaches Q_0/f at which point the collapsing pulses dissipate exactly the same power fQ_0/f as the input Q_0 and the statistics becomes stationary. We have coined the phrase "cycle of intermittency" to describe this scenario because if one measures the signal of power dissipation it is intermittent, consisting principally of spikes with each one corresponding to a collapse event (Dyachenko *et al.*, 1992). In such cases, the steady state statistics arise from the interplay of two species, waves and collapses. If the driving is mild, the number of collapses required to effect dissipation is small, intermittency is low and the spectral shapes associated with wave turbulence predictions will be seen. If the driving is moderate to strong, the gas of waves and collapses may not behave so that their individual signatures are apparent, although the mechanics and physics of the situation involve both wave resonances and collapses. However, collapses are so fast that collapses and waves do not interact at mid wavenumbers so one may very well see the signature of a strong inverse cascade. They will only feel each other at low (where the collapses refuel the finite capacity inverse cascade) and high (when collapses dissipate and provide a new source of wave power) wavenumbers. It is not yet clear whether one can derive a theory which contains both waves and collapses in this two species gas. A similar behavior is found in the discrete NLS equation (Rumpf and Newell, 2004), where discrete breathers take the place of collapses. This system is somewhat simpler to analyze because of the periodic wavenumber space of lattices.

We now turn to question 2.

Q2a. *An exercise.*

Imagine that, instead of starting with an ensemble of random fields bounded at infinity, we begin with deterministic fields (say both L_2 and smooth, think hyperbolic secant) so that they have ordinary Fourier transforms $A_{\mathbf{k}}^s$. Since the amplitudes are ordinary functions, we can directly work with the governing equation. Set

$$A_{\mathbf{k}}^s = A_{\mathbf{k}0}^s + \epsilon A_{\mathbf{k}1}^s + \cdots, \tag{1.26}$$

compute the iterates just as we did for the cumulant equations and then analyze their long time behaviors. All expansions for the $A_{\mathbf{k}j}^s$ are integrals with time-dependent kernels, $\Delta(x)$, $E(x,y) = \int_0^t \Delta(x-y)\exp(iyt)dt$. But, without any Dirac delta functions arising from moment decompositions, the arguments $x, y, \ldots$ e.g. $x = s_1\omega_1 + s_2\omega_2 - s\omega$, $y = s_3\omega_3 + s_4\omega_4 - s_1\omega_1$, remain independent. As a result, no secular terms arise from resonances such as $\mathbf{k}_1 + \mathbf{k}_2 - \mathbf{k} = \mathbf{0}$, $s_1\omega_1 + s_2\omega_2 - s\omega = 0$. The reason is that unlike the case of infinite wavetrains, the wavepackets of the deterministic problem are finite in length and do not interact for long enough to create long time secular behaviors. Using the same logic as we did in effecting the wave turbulence closure, and not taking any account of convergence, we might conclude that the deterministic problem is asymptotically linear. But we know this conclusion to be incorrect without further constraints on the coefficients $L_{\mathbf{k}\mathbf{k}_1\ldots\mathbf{k}_r}^{ss_1\ldots s_r}$. In general, the influence of nonlinearities, even starting from initial conditions for which the local field amplitude is small, can be such that, over time, strong nonlinear structures such as shocks or solitary waves are created. The Fourier basis is inadequate to help us to capture such behaviors. Before we discuss briefly three different examples where this occurs, let us define a new premise P4. This premise says that one must test the deterministic theory first. If the field remains asymptotically linear (which might be tested by numerical simulations), we might surmise that this would rule out the appearance of coherent structures also dominating the long time behavior of the random system. The thinking here is that the deterministic problem would rule out resonances creating secular behavior (because the wavepackets are finite in length and so resonances do not produce long time cumulative effects) but not the appearance of coherent structures. If the latter do not appear in the deterministic system, the argument is that they will play no role in statistical ensembles either. The following three examples bear out this idea.

Q2b. *The lesson from Fermi–Pasta–Ulam (FPU).*

The Korteweg deVries (KdV) equation derived by Zabusky and Kruskal, (1965) as a long wave model of the FPU mass-spring chain and used to understand the negative and surprising outcomes of that experiment in which a system of weakly coupled oscillators did not thermalize, provides an illustrative lesson. Imagine an initial field with small amplitudes which in the far field has a tail which behaves as $Ae^{-\Theta}$, where $\Theta = 2kx - 8k^3t > 0$ and which satisfies the linearized KdV equation. Iterate this solution (the exercise is best done by converting the equation into Hirota form) and we find that when $A = 8k^2$, we can sum the iterates to find the exact convergent expansion $8k^2 e^{-\Theta}/(1 + e^{-\Theta})^2$ or $2k^2 \text{sech}^2 k(x - 4k^2t)$, which is bounded for all x. In taking as our basis only wavetrain solutions, we have overlooked these bounded coherent objects. So, in order to conclude that the field is eventually dominated by statistically-independent wavetrains $A_{0k}^s \exp(kx - is\omega_k t)$ and not by coherent solitary waves for which the Fourier amplitudes are highly correlated, we have to know beforehand that such coherent objects are either not present or sufficiently rare so as not to contaminate the statistics.

Q2c. *Self-induced Transparency (SIT).*

Another well-known one-dimensional example in which coherent structures can play a central role in the long-term dynamics arises in optics and is connected with the phenomenon of self-induced transparency. In that experiment, an envelope pulse centered on a carrier frequency ω close to $\omega_{12} = (E_1 - E_2)/\hbar$ is propagated into a predominantly two energy (E_1, E_2) level medium. The medium is called inhomogeneously broadened because of a random mismatch $\omega - \omega_{12}$ of the input and medium frequencies induced by Doppler shifts of the thermally excited atoms. Some parts of the initial field behave as wavetrains made incoherent by interactions with the random medium. Because of their random phases, these parts are trapped (cf. Anderson localization). For them the medium is opaque. But another part of the initial pulse, the soliton component, can synchronize the ensemble of random oscillators and cause them to oscillate in lock step with each other and with the central frequency of the incoming pulse. For them, the medium is transparent. There is no Anderson localization. The only observed effect is that, in the medium, the soliton pulse is slowed down. A Fourier decomposition of the initial pulse into wavetrains would have been misleading if one further assumed the amplitudes were uncorrelated. Because the system turns out to be exactly integrable, the inverse scattering transform decomposition into solitons (for which the partaking Fourier amplitudes are highly correlated) and radiation (for which the Fourier amplitudes rapidly become uncorrelated) is the more enlightened one. And so, it would be even as the envelope became infinitely long. The dynamical evolution of the "soliton" (coherent structure) and "radiation" (uncorrelated wavetrains) is distinctly different even if by some amplitude measure, the original field is locally small. The existence of a soliton component depends on an integral and not a local measure. So if we had converted SIT equations into the form (1.1) and then iterated on the small local amplitude measure, we would have got the wrong answer.

Prelude to Q2d: The MMT Model

The model by Majda, McLaughlin, Tabak (MMT) (Majda *et al.*, 1997)

$$
\begin{aligned}
iu_t &= Lu + \lambda u^2 u^*, \\
Le^{ikx} &= |k|^{1/2} e^{ikx}, \\
\lambda &= \pm 1,
\end{aligned}
\tag{1.27}
$$

was invented as a means to check carefully by numerical simulations the predictions of wave turbulence. It uses the linear dispersion relation for ocean gravity waves and a simple cubic nonlinearity. The Fourier space representation of the dispersion is $\omega = |k|^{1/2}$. Its Hamiltonian $H = \int (|\partial^{1/4} u / \partial x^{1/4}|^2 + \lambda |u|^4 / 2) dx$ includes fractional derivatives. For comparison, the NLS equation in one dimension $i u_t = -u_{xx}/2 + \lambda u^2 u^*$ is obtained for $L e^{ikx} = k^2 e^{ikx}/2$ corresponding to the dispersion $\omega = k^2/2$. The NLS Hamiltonian is $H = \int (|\partial u / \partial x|^2 + \lambda |u|^4)/2 dx$. Statistically, stationary nonequilibrium states are obtained if an external driving force injects wave action and energy, and these quantities are dissipated by a damping term. Damping may be applied at low $|k|$ in order to avoid the formation of a condensate of long waves, and at high k corresponding to viscosity or hyperviscosity. The driving force may be applied at some intermediate wavenumber.

The MMT model exhibits collapses for $\lambda = -1$ (Cai *et al.*, 2001). For this choice of λ, the coupling energy $\int \omega_k n_k d\mathbf{k}$ and the quartic part of the Hamiltonian $-\int |u|^4 d\mathbf{x}/2$ have opposite signs, so that a matching growth of these two contributions can take place on a phase space shell with fixed H during a collapse. It has been claimed (e.g. Cai *et al.*, 1999, 2001) that collapses emerge from a modulational instability of waves. However, we want to point out that there is no long wave modulational instability for $\lambda = -1$. Collapses depend on the signs of ω and λ. In contrast, the Benjamin–Feir instability depends on the signs of the second derivative ω'' and of λ. The sign of ω'' matches the sign of ω for NLS, while the signs are opposite for MMT. Waves of the focusing ($\lambda = -1$) NLS equation $i u_t = -u_{xx}/2 - u^2 u^*$ are Benjamin–Feir unstable as ω'' is positive for $\omega = k^2/2 > 0$. For the MMT equation with $\omega = \sqrt{|k|} > 0$ it follows from the negative second derivative $\omega'' = -|k|^{-3/2}/4 < 0$ that waves are not Benjamin–Feir unstable for $\lambda = -1$ (but still this equation produces collapses). The envelope equation of this MMT equation is a defocusing NLS equation. For these reasons, one must take care in using the expressions "focusing" and "defocusing" in the context of the MMT equation.

There are instabilities of waves in the MMT that are different from those in the NLS equation. We consider a wave with wavenumber k and a modulation with wavenumber q and introduce the function $D(k,q) = \sqrt{|k+q|} + \sqrt{|k-q|} - 2\sqrt{|k|}$ which is negative for $0 < q/k < 5/4$ and positive for $q/k > 5/4$. An instability occurs when $\lambda D(k,q)$ is small and negative. The MMT equation with $\lambda = 1$ has a modulational instability at small q/k, and an instability for $0 < 5/4 - q/k \ll 1$ so that the modulation has a wavelength that is shorter than the carrier wave. It is the latter's instability which leads to the coherent structures responsible for the anomalous spectrum we discuss in Q2d. An instability of this kind occurs also for $\lambda = -1$ at $0 < q/k - 5/4 \ll 1$ which initiates collapses.

The collapses for MMT with $\lambda = -1$ are observed when the strength of the driving force exceeds a critical threshold. As a result, the collapses lead to fluxes of energy and waveaction in amplitude space in addition to fluxes in the wavenumber space that are due to wave turbulence (Rumpf and Biven, 2005). Pure wave turbulence without strongly nonlinear events has been observed when the strength of the driving force is below a certain critical threshold. It is necessary to apply a sufficiently strong dissipation at small $|k|$ to avoid the formation of a condensate of long waves. In this case, most of the wave action that is injected by the driving force moves towards small $|k|$. A fraction of the wave action moves towards high $|k|$, where it is dissipated by the the high-wavenumber damping.

In Q2d, we will discuss the highly important case of the MMT equation with $\lambda = 1$, which is Benjamin–Feir unstable, but has no collapses.

Q2d. *The Majda-McLaughlin-Tabak conundrum. MMT for $\lambda = 1$.*

This brings us the conundrum of the one-dimensional MMT equation (1.27) for a complex field $u(x,t)$ for which we have recently (Rumpf *et al.*, 2009) obtained a new understanding of anomalous results presented by Majda *et al.* (1997). Numerical simulations of this equation showed that for $\lambda = -1$, with sufficiently small initial conditions, one would recover the KZ spectrum $n_k \sim k^{-1}$ spectrum as long as collapse events were rare. However, for the case $\lambda = 1$, MMT found a statistical steady state with the spectrum $n_k \sim k^{-5/4}$, distinctly different from KZ over the whole range of wavenumbers. We now know (Rumpf *et al.*, 2009) that the reason for the MMT spectrum is that, even for the case of small initial amplitudes, the field becomes dominated by an ensemble, not of resonantly interacting wavetrains, but of coherent objects. They are created from a Benjamin–Feir instability with one marked difference from that found for the focusing NLS equation. For NLS, wavetrains with a dominant carrier wavenumber k_0 became unstable to close sideband modes k_0+q, $q/k_0 \ll 1$. For the MMT, as we have discussed in the prelude to this subsection, there is an instability for which $q \approx 5k_0/4$. They look like short envelope pulses with only a few oscillations under the envelope. As such they are rather reminiscent of "rogue" or "freak" waves, found in oceans and in nonlinear optical media. These coherent objects evolve in shape and in speed because they generate radiating tails (cf. Cerenkov radiation) which drain their power. As a result, their central wavenumber becomes larger and they evolve so as eventually to resemble quasisolitons whose q/k_0 is small. At this stage, the feed to the radiating tail becomes exponentially small. If one posits that the field is dominated by an ensemble of such objects at different stages of their lifetimes, then the MMT spectrum is recovered. Moreover, the observed features of the inverse cascade, created by the long wavelength radiating tails, can also be explained (Rumpf *et al.*, 2009).

Possible Answers to Q2

These lessons from Q2a,b,c,d beg for an answer to question 2. How might we have known, beforehand, whether the long time dynamics of an initially weakly nonlinear system is dominated by wavetrains, or coherent objects, or a mixture of both? We have suggested one possible condition as premise P4; namely, that the governing equation (1.1) is asymptotically linear if $A_{\mathbf{k}}^s$ is an ordinary function. And now, we suggest another condition based on a new and truly remarkable result. We have found that, in certain rare circumstances, the KZ solution can be unstable to disturbances which spontaneously break the symmetry of spatial homogeneity (Newell *et al.* 2012), namely break P1!

What we have found is that the KZ solution $n_k \sim k^{-1}$ for MMT is unstable when $\lambda = +1$ and neutrally stable when $\lambda = -1$ when perturbations which assume that the two point correlations such as $\langle u^s(\mathbf{x})u^{-s}(\mathbf{x} + \mathbf{r})\rangle$ are allowed to depend weakly on the base coordinate $\mathbf{x}$. The analysis uses the Vlasov form of the kinetic equation described at the end of Sec. 1.2. The reader might ask how can it be that for one sign of λ the KZ solution is stable and for the other unstable since the kinetic equations for both cases are the same, depending only on λ^2. But whereas the kinetic equations are the same, the first nonlinear frequency corrections depend on the sign of λ and in the nonlinear dependence of ω_k on $n_{\mathbf{k}}$ which determines stability or instability. In addition, we find that even when $\lambda = 1$ but the dimension of the system is two or greater, then again the KZ spectrum is neutrally stable against perturbations which break spatial homogeneity.

In a numerical experiment, we created the k^{-1} spectrum by allowing an ensemble of solutions of (1.27) reach a statistically steady state for the choice of sign $\lambda = -1$. Then, we switched the sign of λ and watched as coherent structures resembling ultra short pulse waves ("rogue waves") came to dominate the dynamics and the statistics. The instability is slow in the sense that it does not occur on a time scale $(\epsilon^2\omega_0)^{-1}$, ω_0 the frequency of the spectral peak associated with the frequency renormalization, but rather on the same time scale $(\epsilon^4\omega_0)^{-1}$ over which the four wave resonances act.

So, we introduce premise P5. All KZ solutions are stable against perturbations which spontaneously break the spatial homogeneity symmetry. It would be very interesting to know if, along with P1, P2, P3, either or both of these two conditions P4, P5 give sufficient conditions for wave turbulence closure to be in effect and if, and in what way, they may be connected.

In the introduction, we made the point that the wave turbulence closure depends on very mild statistical assumptions. Certainly if P5 holds, one can be sure that the validity of P1 is not in doubt. P2, the fact that the initial fields are uncorrelated at widely separated points is also very mild. P3 is a serious precondition but its validity can always be tested. P4, which is not *a priori* condition on the statistics but on the dynamics of individual solutions, may be stronger than one requires to have the long time behavior of the random system dominated by resonantly interacting waves. But, at the moment, we know of no good substitute.

1.6. Open Challenges

To conclude, we list some additional challenges.

Acoustic Turbulence, Isotropy or Shocks?

The resonant manifolds for the dispersion relation $\omega = c|\mathbf{k}|$ are rays in wavevector space. The first closure transfers spectral energy along but not between the rays. Given an initial anisotropic energy distribution, do the nonlinear interactions of the next closure lead to an isotropic distribution or to condensation along particular rays which would likely produce fully nonlinear shocks (L'vov *et al.*, 1997)? Or, to use a more colorful vernacular: Were the dinosaurs frozen or fried? The asteroid collision of 65 million years ago created a huge disturbance which must have propagated throughout the atmospheric sheath. If the energy were to have condensed along certain propagation directions, the resulting huge shock waves would have led to enormous temperature fluctuations. Initial extinctions would have been very quick (fried!) followed by a much slower death by nuclear winter (frozen!).

Energy Exchange Times

For a discrete set of interacting triads, the nonlinear energy exchange time is ϵ^{-1}. For a continuum set of such triads, "cancellations" cause this time to be extended to ϵ^{-2}. Why? In a recent review article, Kartashova (2010) partially addresses this question.

Condensate Formation

Modeled by the defocussing ($\lambda = -1$) NLS equation, this is an open and hot topic. Three questions: (i) Given an input Q_0, P_0 of number and energy flux at finite wavenumber k_0, can one follow, using wave turbulence theory, the creation of a condensate by inverse flux action and the subsequent relaxation of waves and vortices on the condensate (Dyachenko *et al.*, 1992; Lacaze *et al.*, 2001)? (ii) Given a finite total energy and number of particles (and an ultraviolet cutoff k_c), can one find in the subcritical temperature range $T < T_c$, (at $T = T_c$, μ, the chemical potential is zero; for $T < T_c$, μ would be positive and the Rayleigh–Jeans equilibrium $n_k = T/(\omega_k - \mu)$ singular), a wave turbulence description with a gradual transition from free waves ($\omega_k = k^2$) to Bogoliubov waves ($\omega = \pm\sqrt{2|\lambda|n_0 k^2 + k^4}$) where n_0 is the number of particles in the condensate? (iii) How is the second-order phase transition in (ii) affected if one initiates the dynamics with nonzero fluxes?

Wave Turbulence in Astrophysics

Magnetized plasmas, found in the solar corona, solar wind and earth's magnetosphere support waves and, like ocean waves, have a continuum of scales (up to 18 decades!) and are a natural playground for wave turbulence (Boldyrev and Perez, 2009; Galtier *et al.*, 2000, 2003, 2006, 2009; Goldreich and Sridhar, 1995, 1997; Goldstein and Roberts, 1999; Kuznetsov, 1972, 2001; Ng *et al.*, 1996, 2003; Sridhar and Goldreich, 1994; Sahraoui *et al.*, 2003, 2007). To date, however, only the signatures of strong turbulence (Kolmogorov, rather than Iroshnikov–Kraichnan) have been found experimentally although wave turbulence behavior has been clearly seen in direct numerical simulations by Bigot *et al.* (2008). Given present satellite capabilities, what are the best hopes for observing wave turbulence spectra such as the $E(\mathbf{k}_\perp, k_\parallel) \sim f(k_\parallel)\mathbf{k}_\perp^{-2}$, $k_\parallel = \mathbf{k} \cdot \mathbf{b}$, $\mathbf{k}_\perp = \mathbf{k} - k_\parallel\mathbf{b}$, $f(k_\parallel)$ nonuniversal, $\mathbf{b}$ unit vector in the magnetic field direction, behavior for a sea of oppositely traveling Alfvén waves? For a review, see Galtier (2009).

Continuum Limit of Finite Dimensional Wave Turbulence

In a box L^d, one can define a natural probability measure on Fourier amplitudes avoiding the difficulties of such measures in infinite dimensions. The resulting Liouville hierarchy for the Fourier amplitudes pdf leads, via the Brout–Prigogine equation for its vacuum component, to a kinetic equation if one assumes that the vacuum pdf can be factored into a product of its marginals (a closure assumption!). Can one show that a natural closure occurs in L^d or, if not, how the natural closure arises in taking the $L \to \infty$ limit? (Jakobsen and Newell, 2004).

A Priori Conditions for Wave Turbulence

Can one find mathematically rigorous *a priori* conditions on the governing equation (1.1) or its statistical hierarchy which guarantees that the wave turbulence theory will hold? (See earlier comments in Sec. 1.5.)

Homogeneity

Is broken spatial homogeneity (P1) a potential problem for all turbulence theories?

Anomalous Exponents

Are all finite capacity Kolmogorov solutions reached with anomalous exponents? Do they have anything to do with positive entropy production?

Appendix 1. Derivation of the Governing Equation for Gravity-Capillary Waves

We define the Fourier transform (d is dimension, for surface waves we have $d = 2$ and $\mathbf{x} = \binom{x}{y}$)

$$
v(\mathbf{x}) = \int V(\mathbf{k})e^{i\mathbf{k}\cdot\mathbf{x}}d\mathbf{k},
$$
$$
V(\mathbf{k}) = \frac{1}{(2\pi)^d} \int v(\mathbf{x})e^{-i\mathbf{k}\cdot\mathbf{x}}d\mathbf{x}.
$$
(A.1.1)

If the Fourier transforms of $v(\mathbf{x})$ and $w(\mathbf{x})$ are $V(\mathbf{k})$ and $W(\mathbf{k})$, then the Fourier transform of $v(\mathbf{x})w(\mathbf{x})$ is

$$
\int V(\mathbf{k}_1)W(\mathbf{k}_2)\delta(\mathbf{k}_1 + \mathbf{k}_2 - \mathbf{k})d\mathbf{k}_1 d\mathbf{k}_2.
$$
(A.1.2)

The equations for the free surface $\eta(\mathbf{x}, t)$ and velocity potential ϕ $(\mathbf{x}, z, t)$ are

$$
\nabla^2\phi = 0
$$
(A.1.3)

$$
\frac{\partial\phi}{\partial z} \to 0 \quad \text{as } z \to -\infty
$$
(A.1.4)

$$
\frac{\partial\eta}{\partial t} + \frac{\partial\phi}{\partial x}\frac{\partial\eta}{\partial x} + \frac{\partial\phi}{\partial y}\frac{\partial\eta}{\partial y} = \frac{\partial\phi}{\partial z},
$$
(A.1.5)

$$
\frac{\partial\phi}{\partial t} + \frac{1}{2}\left(\frac{\partial\phi}{\partial x}\right)^2 + \frac{1}{2}\left(\frac{\partial\phi}{\partial y}\right)^2 + \frac{1}{2}\left(\frac{\partial\phi}{\partial z}\right)^2
$$
$$
+ g\eta - \sigma\nabla^2\eta/(1 + (\nabla\eta)^2)^{3/2} = 0.
$$
(A.1.6)

The boundary conditions (A.1.5) and (1.6) are applied at the surface $z = \eta(\mathbf{x}, t)$. Here g is gravity and $\sigma = S/\rho$, the surface tension divided by density. We expand (A.1.5) and (1.6) about $z = \eta$ and keep only terms to quadratic order.

$$
\frac{\partial\eta}{\partial t} + \frac{\partial\phi}{\partial x}\frac{\partial\eta}{\partial x} + \frac{\partial\phi}{\partial y}\frac{\partial\eta}{\partial y} = \frac{\partial\phi}{\partial z} + \eta\frac{\partial^2\phi}{\partial z^2} \quad \text{at } z = 0,
$$
(A.1.7)

$$
\frac{\partial\phi}{\partial t} + \eta\frac{\partial^2\phi}{\partial t\partial z} + \frac{1}{2}\left(\frac{\partial\phi}{\partial x}\right)^2 + \frac{1}{2}\left(\frac{\partial\phi}{\partial y}\right)^2
$$
$$
+ \frac{1}{2}\left(\frac{\partial\phi}{\partial z}\right)^2 + g\eta - \sigma\nabla^2\eta = 0 \quad \text{at } z = 0.
$$
(A.1.8)

Let $N_\mathbf{k}$ and $F_\mathbf{k}$ be the Fourier transforms of $\eta(\mathbf{x}, t)$ and $\phi(\mathbf{x}, 0, t)$. We can thus represent $\eta(\mathbf{x}, t)$, $\phi(\mathbf{x}, z, t)$ and solve (A.1.3) and satisfy (A.1.4) by

$$\eta(\mathbf{x}, t) = \int N_\mathbf{k} e^{i\mathbf{k} \cdot \mathbf{x}} d\mathbf{k},$$

$$\phi(\mathbf{x}, z, t) = \int F_\mathbf{k} e^{i\mathbf{k} \cdot \mathbf{x}} e^{kz} d\mathbf{k},$$

$k = |\mathbf{k}|$. Substituting into (A.1.7) and (A.1.8), we obtain

$$\frac{\partial N_\mathbf{k}}{\partial t} - kF_\mathbf{k} = R_1, \tag{A.1.9}$$

$$\frac{\partial F_\mathbf{k}}{\partial t} + \nu_k^2 N_\mathbf{k} = R_2, \tag{A.1.10}$$

where $\nu_k^2 = g + \sigma k^2$ and

$$R_1 = \int \mathbf{k}_1 \cdot \mathbf{k}_2 F_{\mathbf{k}_1} N_{\mathbf{k}_2} \delta_{12,0} d\mathbf{k}_{12} + \int k_2^2 N_{\mathbf{k}_1} F_{\mathbf{k}_2} \delta_{12,0} d\mathbf{k}_{12}, \tag{A.1.11}$$

$$R_2 = \frac{1}{2} \int (\mathbf{k}_1 \cdot \mathbf{k}_2 - k_1 k_2) F_{\mathbf{k}_1} F_{\mathbf{k}_2} \delta_{12,0} d\mathbf{k}_{12} - \int N_{\mathbf{k}_1} k_2 \frac{\partial F_{\mathbf{k}_2}}{\partial t} \delta_{12,0} d\mathbf{k}_{12}. \tag{A.1.12}$$

We use the notations $\delta_{12,0} = \delta(\mathbf{k}_1 + \mathbf{k}_2 - \mathbf{k})$, $d\mathbf{k}_{12} = d\mathbf{k}_1 d\mathbf{k}_2$.

We diagonalize the linear part of the system (A.1.9) and (A.1.10) by setting

$$N_\mathbf{k} = c_k(A_\mathbf{k}^+ + A_\mathbf{k}^-) = \sum_s c_k A_\mathbf{k}^s, \tag{A.1.13}$$

$$F_\mathbf{k} = \frac{i\nu_k^2}{\omega_k}(A_\mathbf{k}^+ - A_\mathbf{k}^-) = \sum_s \frac{is\nu_k^2}{\omega_k} c_k A_\mathbf{k}^s. \tag{A.1.14}$$

as the eigenvalues of $\begin{pmatrix} 0 & +k \\ -\nu_k^2 & 0 \end{pmatrix}$ are $\pm i\omega_k$ when $\omega_k^2 = k\nu_k^2$ and the corresponding eigenvectors are $\begin{pmatrix} 1 \\ \pm i\nu_k^2/\omega_k \end{pmatrix}$. The sum is over all s values. The parameter s, here equal to ± 1, counts the number of frequencies for each wavevector. For gravity-capillary waves, there are two. For magnetohydrodynamics there are six, bidirectional fast, slow and Alfvén waves. The constant c_k, depending only on $k = |\mathbf{k}|$, is chosen for convenience to ensure that the average energy per unit area of surface is $\int \omega_k n_\mathbf{k} d\mathbf{k}$ to leading order. We will see it turns out to be $\frac{1}{\sqrt{2}} \frac{\sqrt{\omega_k}}{\nu_k}$ so that we rewrite

(A.1.13) and (A.1.14) as

$$N_{\mathbf{k}} = \sum_s \frac{\sqrt{\omega_k}}{\sqrt{2}\nu_k} A_{\mathbf{k}}^s,$$

$$F_{\mathbf{k}} = \sum_s \frac{is\nu_k}{\sqrt{2}\omega_k} A_{\mathbf{k}}^s. \qquad (\text{A.1.15})$$

Substituting (A.1.15) into (A.1.9) and (A.1.10), adding and subtracting, we find

$$\frac{dA_{\mathbf{k}}^s}{dt} - is\omega_k A_{\mathbf{k}}^s = \frac{1}{\sqrt{2}} \frac{\nu_k}{\sqrt{\omega_k}} R_1 - \frac{is}{\sqrt{2}} \frac{\sqrt{\omega_k}}{\nu_k} R_2. \qquad (\text{A.1.16})$$

To leading order, $\frac{\partial F_{\mathbf{k}_2}}{\partial t}$ in R_2 is replaced by $-\nu_{k_2}^2 N_{\mathbf{k}_2}$. If one wishes to go to the cubic order, the R_2 component of $\frac{\partial F_{\mathbf{k}}}{\partial t}$ in (A.1.10) must be accounted for. A little calculation lets us write (A.1.16) as

$$\frac{dA_{\mathbf{k}}^s}{dt} - is\omega_k A_{\mathbf{k}}^s = \sum \int L_{\mathbf{k}\mathbf{k}_1\mathbf{k}_2}^{ss_1s_2} A_{\mathbf{k}_1}^{s_1} A_{\mathbf{k}_2}^{s_2} \delta_{12,0} d\mathbf{k}_{12}, \qquad (\text{A.1.17})$$

where

$$L_{\mathbf{k}\mathbf{k}_1\mathbf{k}_2}^{ss_1s_2} = \frac{i}{4\sqrt{2}} \frac{\nu_k}{\sqrt{\omega_k}}$$

$$\times \left\{ (\mathbf{k}_1 \cdot \mathbf{k}_2 + k_1^2) \frac{s_1\nu_{k_1}}{\nu_{k_2}} \sqrt{\frac{\omega_{k_2}}{\omega_{k_1}}} + (\mathbf{k}_1 \cdot \mathbf{k}_2 + k_2^2) \frac{s_2\nu_{k_2}}{\nu_{k_1}} \sqrt{\frac{\omega_{k_1}}{\omega_{k_2}}} \right\}$$

$$+ \frac{is}{4\sqrt{2}} \frac{\sqrt{\omega_k}}{\nu_k} \left\{ (\mathbf{k}_1 \cdot \mathbf{k}_2 - k_1 k_2) \frac{s_1\nu_{k_1} s_2\nu_{k_2}}{\sqrt{\omega_{k_1}\omega_{k_2}}} + (\omega_{k_1}^2 + \omega_{k_2}^2) \frac{\sqrt{\omega_{k_1}\omega_{k_2}}}{\nu_{k_1}\nu_{k_2}} \right\}.$$

$$(\text{A.1.18})$$

Using the fact that $\omega_k^2 = k\nu_k^2$, and writing ω_{k_j}, ν_{k_j} as ω_j, ν_j, we can rewrite (A.1.18) as

$$L_{\mathbf{k}\mathbf{k}_1\mathbf{k}_2}^{ss_1s_2} = \frac{i}{4\sqrt{2}} \left\{ (\mathbf{k}_1 \cdot \mathbf{k}_2 + k_1^2)s_1 \left(\frac{k_2\omega\omega_1}{kk_1\omega_2} \right)^{1/2} + (\mathbf{k}_1 \cdot \mathbf{k}_2 + k_2^2)s_2 \left(\frac{k_1\omega\omega_2}{\omega_1 kk_2} \right)^{1/2} \right.$$

$$\left. + (\mathbf{k}_1 \cdot \mathbf{k}_2 - k_1 k_2)ss_1s_2 \left(\frac{\omega_1\omega_2 k}{\omega k_1 k_2} \right)^{1/2} + s(\omega_1^2 + \omega_2^2) \left(\frac{kk_1 k_2}{\omega\omega_1\omega_2} \right)^{1/2} \right\}.$$

$$(\text{A.1.19})$$

One can check that $L_{\mathbf{k}\mathbf{k}_1\mathbf{k}_2}^{ss_1s_2}$ has the following properties:

(1) It is pure imaginary and symmetric under interchange of the indices 1 and 2.

(2) $L^{ss_1s_2}_{\mathbf{k}\mathbf{k}_1\mathbf{k}_2} = 0$ when $\mathbf{k} - \mathbf{k}_1 + \mathbf{k}_2 = 0$. This guaranties that if $\langle A^s_{\mathbf{k}} \rangle = 0$ initially, it remains so.

(3) $L^{s_1s-s_2}_{\mathbf{k}_1\mathbf{k}-\mathbf{k}_2} = \frac{s_1}{s} L^{ss_1s_2}_{\mathbf{k}\mathbf{k}_1\mathbf{k}_2}$ when $\mathbf{k}_1 + \mathbf{k}_2 = \mathbf{k}$ and $s_1\omega_1 + s_2\omega_2 = s\omega$. This is slightly more difficult to prove. It allows us to show that the energy $\int \omega_k n_k dk$ is formally conserved. As we have already learned, however, in Sec. 1.3, energy conservation fails after a finite time t^* if the corresponding KZ solution has finite capacity.

If (A.1.17) is in the Hamiltonian form, the canonically conjugate variables are $A^s_{\mathbf{k}}$ and $A^{-s}_{-\mathbf{k}}$. Finally, for pure gravity waves $L^{ss_1s_2}_{\lambda\mathbf{k}\lambda\mathbf{k}_1\lambda\mathbf{k}_2} = \lambda^{7/4} L^{ss_1s_2}_{\mathbf{k}\mathbf{k}_1\mathbf{k}_2}$ and for capillary waves $L^{ss_1s_2}_{\lambda\mathbf{k}\lambda\mathbf{k}_1\lambda\mathbf{k}_2} = \lambda^{9/4} L^{ss_1s_2}_{\mathbf{k}\mathbf{k}_1\mathbf{k}_2}$. Thus for gravity waves $\gamma_2 = 7/4$ (and $\gamma_3 = 3$). For capillary waves, $\gamma_2 = 9/4$ (and $\gamma_3 = 3$).

Appendix 2. Asymptotic Analysis

Because the results required to carry out the asymptotic analysis almost never appear in the literature (the exception is Benney and Newell, 1969), here we provide, albeit briefly, details. Using these results, the reader can continue wave turbulence theory beyond the first closure. Triad and quartet resonances appear as combinations such as $\delta(\sum_i s_i\omega_i - s\omega)\delta(\sum_i \mathbf{k}_i - \mathbf{k})$ in the transfer integrals appearing in the kinetic equation where the summations go from 1 to 2 (triad) or 1 to 3 (quartet), respectively. Terms such as $\delta(s_1\omega_1 + s_2\omega_2 - s\omega)\hat{P}\frac{1}{s_3\omega_3+s_4\omega_4-s_1\omega_1}\delta(\mathbf{k}_1 + \mathbf{k}_2 - \mathbf{k})\delta(\mathbf{k}_3 + \mathbf{k}_4 - \mathbf{k}_1)$ indicate triad resonances coupled with modal interactions. They correspond to higher order loops in the Feynman diagram techniques.

The starting result is: If $f(x)$ is absolutely integrable and measurable, the RL lemma tells us that

$$\lim_{t\to\infty} \int_{-\infty}^{\infty} f(x)e^{ixt} dx = 0 \qquad (A.2.1)$$

which we write symbolically as $e^{ixt} \sim 0$. The next important result is:

$$\lim_{t\to\pm\infty} \int_{-\infty}^{\infty} f(x)\Delta(x,t)dx = \pi\,\text{sgn}(t)f(0) + i\hat{P}\int_{-\infty}^{\infty} \frac{f(x)}{x} dx, \qquad (A.2.2)$$

where (omitting the t in the argument)

$$\Delta(x) = \frac{e^{ixt} - 1}{ix} = \int_0^t e^{ixt} dt \qquad (A.2.3)$$

and $\hat{P}$ denotes the Cauchy principal value. Symbolically, we write

$$\Delta(x) \sim \tilde{\Delta}(x) = \pi\,\text{sgn}(t)\delta(x) + i\hat{P}\left(\frac{1}{x}\right), \qquad (A.2.4)$$

(A.2.2) is easily proved by splitting the integral from $(-\infty, -\epsilon)$, $(-\epsilon, \epsilon)$, (ϵ, ∞). Since $\int_{-\epsilon}^{\epsilon} f(x)\frac{e^{ixt}-1}{ix}dx$ tends to zero as $\epsilon \to 0$,

$$\int_{-\infty}^{\infty} f(x)\Delta(x)dx = \hat{P}\int_{-\infty}^{\infty} f(x)\frac{e^{ixt}}{ix}dx + i\hat{P}\int_{-\infty}^{\infty} \frac{f(x)}{x}dx.$$

We can carry out the first integral by contour integration if $f(x)$ admits analytic continuation. Otherwise, we write $f(x)$ as $f(x) - f(0)e^{-x^2} + f(0)e^{-x^2}$. The first part will vanish by Riemann–Lebesgue. The second can be calculated by differentiating with respect to t, evaluating the resulting integral, integrating to obtain $\pi f(0)\mathrm{sgn}(t)$ in the limit $t \to \pm\infty$.

The next objects we meet are $\Delta(x)\Delta(y)$ and

$$E(x,y) = \int_0^t \Delta(x-y)e^{iyt}dy = \frac{\Delta(x) - \Delta(y)}{i(x-y)}. \qquad (A.2.5)$$

It is easy to show that (for $t \to +\infty$)

$$\Delta(x)\Delta(y) \sim \tilde{\Delta}(x)\tilde{\Delta}(y) = \pi^2\delta(x)\delta(y) + i\pi\delta(x)\hat{P}_y\left(\frac{1}{y}\right)$$

$$+ i\pi\delta(y)\hat{P}_x\left(\frac{1}{x}\right) - \hat{P}\left(\frac{1}{xy}\right). \qquad (A.2.6)$$

We now consider a very important result. From the identity

$$\Delta(x)\Delta(y) = i\left(\frac{\Delta(x)}{y} + \frac{\Delta(y)}{x} - \left(\frac{1}{x} + \frac{1}{y}\right)\Delta(x+y)\right), \qquad (A.2.7)$$

the left-hand side should be asymptotic to

$$i\pi\delta(x)\hat{P}_y\left(\frac{1}{y}\right) + i\pi\delta(y)\hat{P}_x\left(\frac{1}{x}\right) - 2\hat{P}\left(\frac{1}{xy}\right)$$

$$- i\left(\frac{1}{x} + \frac{1}{y}\right)\left(\pi\delta(x+y) + i\hat{P}\left(\frac{1}{x+y}\right)\right). \qquad (A.2.8)$$

Equating (A.2.6) and (A.2.8) gives

$$\hat{P}_x\left(\frac{1}{x}\right)\hat{P}_y\left(\frac{1}{x+y}\right) + \hat{P}_y\left(\frac{1}{y}\right)\hat{P}_x\left(\frac{1}{x+y}\right) = \pi^2\delta(x)\delta(y) + \hat{P}_{xy}\left(\frac{1}{xy}\right),$$

$$\qquad (A.2.9)$$

where the subscripts indicate the required order of integration. (A.2.9) is a nontrivial result. It is known as the Poincaré–Bertrand lemma. By subtracting the second term on the right-hand side from the second term on the left we can rewrite it as

$$\int \frac{dx}{x} \int \frac{f(x,y)dy}{x+y} - \int dy \int \frac{f(x,y)dx}{x(x+y)} = \pi^2 f(0,0), \qquad (A.2.10)$$

the usual form of this lemma (see Muskhelishvili (1953)). It tells us that there is a price to be paid for changing the order of integration in singular integrals. From (A.2.5),

$$E(x,y) \sim -i\pi\delta(x)\hat{P}_y\left(\frac{1}{x-y}\right) + \hat{P}_x\left(\frac{1}{x}\right)\hat{P}_y\left(\frac{1}{x-y}\right)$$

$$+ i\pi\delta(y)\hat{P}_x\left(\frac{1}{x-y}\right) - \hat{P}_y\left(\frac{1}{y}\right)\hat{P}_x\left(\frac{1}{x-y}\right)$$

$$= +i\pi\delta(x)\hat{P}_y\left(\frac{1}{y}\right) + i\pi\delta(y)\hat{P}_x\left(\frac{1}{x}\right) + \pi^2\delta(x)\delta(y) - \hat{P}_{xy}\left(\frac{1}{xy}\right)$$

from (A.2.9) (by changing $y \to -y$). Thus

$$E(x,y) \sim \tilde{\Delta}(x)\tilde{\Delta}(y). \qquad (A.2.11)$$

It is convenient to introduce

$$\Delta_n(x) = \int_0^t \Delta_{n-1}(x)dt,$$

$$\Delta_0(x) = \Delta(x), \qquad (A.2.12)$$

$$\Delta_n(0) = \frac{t^{n+1}}{(n+1)!}.$$

By an inductive argument, it is easy to show that

$$\Delta_n(x) \sim \frac{1}{n!}\tilde{\Delta}(x)\left(t - i\frac{\partial}{\partial x}\right)^n. \qquad (A.2.13)$$

The reductions of $\Delta(x)\Delta(y)$ and $E(x,y)$ have the following asymptotic behavior

$$\Delta(x)\Delta(x) = \Delta_1(x) + \Delta_1(-x) \sim 2\pi t\delta(x) + 2\hat{P}\left(\frac{1}{x}\right)\frac{\partial}{\partial x}, \qquad (A.2.14)$$

$$E(x,0) \sim \tilde{\Delta}(x)\left(t - i\frac{\partial}{\partial x}\right), \qquad (A.2.15)$$

$$E(x, -x) = \frac{1}{2}\Delta(x)\Delta(-x) \sim \pi t \delta(x) + \hat{P}\left(\frac{1}{x}\right)\frac{\partial}{\partial x}, \qquad \text{(A.2.16)}$$

$$E(x, x) \sim i\tilde{\Delta}(x)\frac{\partial}{\partial x}, \qquad \text{(A.2.17)}$$

$$E(0, 0) = t^2/2. \qquad \text{(A.2.18)}$$

We illustrate how to prove the first result (A.2.14). By direct calculation, $\Delta(x)\Delta(-x) = \frac{2(1-\cos(xt))}{x^2}$. Consider $\int f(x)\Delta(x)\Delta(-x)dx = -2\int f(x)(1 - \cos(xt))d(\frac{1}{x}) = 2t\int f(x)\frac{\sin(xt)}{x}dx + 2\int \frac{f(x)}{x}(1 - \cos(xt))dx$ assuming that $f(x)$ vanishes at the end points. The first integral tends to $2\pi t \text{sgn}(t)f(0)$ as $t \to \pm\infty$. The second can be written as the sum of two principal value integrals by dividing the region of integration from $(-\infty, -\epsilon)$, $(-\epsilon, \epsilon)$, (ϵ, ∞) and taking the limit as $\epsilon \to 0$. By the Riemann–Lebesgue $\hat{P}\int f(x)\frac{\cos(xt)}{x}dx \sim 0$ and the result (A.2.14) follows.

This calculation is important. Many well-known experts in the field of wave turbulence have declared that it is impossible to go beyond the first closure if triad resonances are present because one is unable to handle integrals where there is a term such as $(s_1\omega_1 + s_2\omega_2 - s\omega)^2$ in the denominator. That statement is nonsense. Integration by parts of well-defined integrals such as $\int f(x)\frac{1-\cos xt}{x^2}dx$ present no problems. It is certainly true that beyond the first closure one will encounter terms which grow as t^2 but these will cancel and only terms which grow like t will remain. It is the latter which provide the next term in the asymptotic expansion for $\frac{\partial n_k}{\partial t}$.

A complete list of the time-dependent functions which arise in going beyond the first closure and their asymptotic behaviors is given in Benney and Newell (1969). In Newell (1968), a generalization of the Poincaré–Bertrand lemma for n iterated integrals is given. It would be a fun exercise to correlate all these results with Feynman diagram techniques used in field theories.

Acknowledgment

This review is partly based on our recent article (Newell and Rumpf, 2011). One of the authors (ACN) would like to express his appreciation for NSF support under grant DMS 0809189.

Bibliography

Badulin SI, Pushkarev AN, Resio D, Zakharov VE. 2005. Self-similarity of wind-driven seas. *Nonlinear Proc. Geophys.* 12: 891–945.

Badulin SI, Babanin AV, Zakharov V, Resio D. 2007. Weakly turbulent laws of wind-wave growth. *J. Fluid Mech.* 591: 339–378.

Balk AM, Nazarenko SV, Zakharov VE. 1990. Non-local turbulence of drift waves. *Sov. Phys. JETP* 71: 249–260.

Balk AM, Nazarenko SV, Zakharov VE. 1991. New invariant for drift turbulence. *Phys. Lett. A* 152: 276–280.

Battacharjee A, Ng CS. 2001. Random scattering and anisotropic turbulence of shear Alfvén wave packets. *Astrophys. J.* 548: 318–322.

Benney DJ, Saffman PG. 1966. Nonlinear interactions of random waves in a dispersive medium. *Proc. R. Soc. Lond. A* 289: 301–320.

Benney DJ, Newell AC. 1967. Sequential time closures of interacting random waves. *J. Math. Phys.* 46: 363–393.

Benney DJ, Newell AC. 1969. Random wave closures. *Stud. Appl. Math.* 48: 29–53.

Bigot B, Galtier S, Politano H. 2008. Energy decay laws in strongly anisotropic magnetohydrodynamic turbulence. *Phys. Rev. Lett.* 100: 074502.

Biven L, Connaughton C, Newell AC. 2003. Structure functions and breakdown criteria for wave turbulence. *Physica D* 184: 98–113.

Boldyrev S, Perez JC. 2009. Spectrum of weak magneohydrodynamic turbulence. *Phys. Rev. Lett.* 103: 225001.

Boudaoud A, Cadot O, Odille B, Touzé C. 2008. Observation of wave turbulence in vibrating plates. *Phys. Rev. Lett.* 100: 234504.

Cai D, Majda AJ, McLaughlin DW, Tabak EG. 1999. Spectral bifurcations in dispersive wave turbulence. *Proc. Natl. Acad. Sci. U. S. A.* 96: 14216–14221.

Cai D, Majda AJ, McLaughlin DW, Tabak EG. 2001. *Physica D* 152–153: 551–572.

Cerda, E, Mahadevan L. 1998. Conical surfaces and crescent singularities in crumpled sheets. *Phys. Rev. Lett.* 80: 2358–2361.

Connaughton C, Newell AC, Pomeau Y. 2003a. Non-stationary spectra of local wave turbulence. *Physica D* 184: 64–85.

Connaughton C et al. 2003b. Dimensional analysis and weak turbulence. *Physica D* 184: 86–97.

Connaughton C et al. 2005. Condensation of classical nonlinear waves. *Phys. Rev. Lett.* 95: 263901.

Connaughton C, Newell AC. 2010. Dynamical scaling and the finite capacity anomaly in 3-wave turbulence. *Phys. Rev. E* 81: 036303.

Donelan MA, Hamilton J, Hui WH. 1985. Directional spectra of wind-generated waves. *Phil. Trans. R. Soc. London A* 315: 509–562.

Donelan MA, Pierson Jr WJ. 1987. Radar scattering and equilibrium ranges in wind-generated waves with application to scatterometry. *J. Geophys. Res.* 92: 4971–5029.

Donelan MA, Skafel M, Graber H, Liu P, Schwab D, Venkatesh S. 1992. On the growth rate of wind-generated waves. *Atmos. Ocean* 30: 457–478.

Donelan MA, Babanin AV, Young IR, Banner ML, McCormick C. 2005. Wave follower field measurements of wind-input spectral function. Part I. Measurements and calibrations. *J. Atmos. Ocean Technol.* 22: 799–813.

Donelan MA, Babanin AV, Young IR, Banner ML. 2006. Wave follower field measurements of wind-input spectral function. Part II. Parametrization of the wind input. *J. Phys. Oceanogr.* 36: 1672–1689.

During G, Josserand C, Rica S. 2006a. Weak turbulence for a vibrating plate: Can you hear a Kolmogorov spectrum. *Phys. Rev. Lett.* 97: 025503.

During G, Picozzi A, Rica S. 2006b. Breakdown of weak turbulence and nonlinear wave condensation. *Physica D* 238: 1524–1549.

Dyachenko A, Falkovich G. 1996. Condensate turbulence in two dimensions. *Phys. Rev. E* 54: 5095–5099.

Dyachenko S, Newell AC, Pushkarev A, Zakharov VE. 1992. Optical turbulence: Weak turbulence, condensates and collapsing filaments in the nonlinear Schrodinger equation. *Physica D* 57: 96–160.

Falcon E, Laroche C, Fauve S. 2007a. Observation of gravity-capillary wave turbulence. *Phys. Rev. Lett.* 98: 094503.

Falcon E, Fauve S, Laroche C. 2007b. Observation of intermittency in wave turbulence. *Phys. Rev. Lett.* 98: 154501.

Falcon E, Aumaitre S, Falcon C, Laroche C, Fauve S. 2008. Fluctuations of energy flux in wave turbulence. *Phys. Rev. Lett.* 100: 064503.

Falcon C, Falcon E, Bortolozzo U, Fauve S. 2009. Capillary wave turbulence on a spherical fluid surface in zero gravity. *EPL* 86: 14002.

Falkovich GE, Shafarenko AV. 1991. Non-stationary wave turbulence. *J. Nonlinear Sci.* 1: 457–480.

Frisch U. 1996. *Turbulence*. Cambridge: Cambridge University Press.

Galtier S, Nazarenko SV, Newell AC, Pouquet A. 2000. A weak turbulence theory for incompressible MHD. *J. Plasma Phys.* 63: 447–488.

Galtier S. 2003. Weak inertial-wave turbulence theory. *Phys. Rev. E* 68: 015301.

Galtier S, Bhattachardjee A. 2003. Anisotropic weak whistler wave turbulence in electron magnetohydrodynamics. *Phys. Plasmas* 10: 3065–3076.

Galtier S. 2006. Wave turbulence in incompressible Hall magnetohydrodynamics. *J. Plasma Phys.* 72: 721–769.

Galtier S. 2009. Wave turbulence in magnetized plasmas. *Nonlin. Processes Geophys.* 16: 83–98.

Goldreich P, Sridhar S. 1995. Toward a theory of interstellar turbulence II. Strong alfénic turbulence. *Astrophys. J.* 438: 763–775.

Goldreich P, Sridhar S. 1997. Magnetohydrodynamic turbulence revisited. *Astrophys. J.* 485: 680–688.

Goldstein ML, Roberts DA. 1999. Magnetohydrodynamics turbulence in the solar wind. *Phys. Plasmas* 6: 4154–4160.

Hasselmann K. 1962. On the non-linear energy transfer in a gravity-wave spectrum. Part 1. General theory. *J. Fluid Mech.* 12: 481–500.

Hasselmann K. 1963a. On the non-linear energy transfer in a gravity-wave spectrum. Part 2. Conservation theorems — wave-particle analogy — irreversibility. *J. Fluid Mech.* 15: 273–281.

Hasselmann K. 1963b. On the non-linear energy transfer in a gravity-wave spectrum. Part 3. Evaluation of the energy flux and swell-sea interaction for a Neumann spectrum. *J. Fluid Mech.* 15: 385–398.

Hwang PA, Wang DW, Walsh EJ, Krabill WB, Swift RN. 2000. Airborne measurements of the wavenumber spectra of ocean surface waves. Part 1. Spectral slope and dimensionless spectral coefficient. *J. Phys. Oceanogr.* 30: 2753–2767.

Hwang PA, Wang DW. 2004. Field measurements of duration-limited growth of wind-generated ocean surface waves at young stage of development. *J. Phys. Oceanogr.* 34: 2316–2326.

Hwang PA. 2006. Duration- and fetch-limited growth functions of wind-generated waves parametrized with three different scaling wind velocities. *J. Geophys. Res.* 111: C02005.

Iroshnikov PS. 1964. Turbulence of a conducting fluid in a strong magnetic field. *Soviet Astron.* 7: 566–571.

Jakobsen P, Newell AC. 2004. Invariant measures and entropy production in wave turbulence. *J. Stat. Mech.: Theor. Exp.* L10002.

Kartashova E. 2010. *Nonlinear Resonance Analysis*. Cambridge: Cambridge University Press.

Kitaigorodskii SA. 1983. On the theory of equilibrium range spectrum of wind-generated gravity-waves. *J. Phys. Oceanogr.* 13: 816–827.

Kolmakov GV, Levchenko AA, Brazhnikov MYu, Mezhov-Deglin LP, Silchenko AN, McClintock PVE. 2004. Quasiadiabatic decay of capillary turbulence on the charged surface of liquid hydrogen. *Phys. Rev. Lett.* 93: 074501.

Kolmakov GV, Brazhnikov MY, Levchenko AA, Silchenko AN, McClintock PVE, Mezhov-Deglin LP. 2006. Nonstationary nonlinear phenomena on the charged surface of liquid hydrogen. *J. Low Temp. Phys.* 145: 311–335.

Komen G, Cavaleri L, Donelan M, Hasselmann S, Janssen P. 1996. *Dynamics and Modelling of Ocean Waves*. Cambridge: Cambridge University Press.

Korotkevich AO. 2008. Simultaneous numerical simulation of direct and inverse cascades in wave turbulence. *Phys. Rev. Lett* 101: 074504.

Kraichnan RH. 1965. Inertial-range spectrum in hydromagnetic turbulence. *Phys. Fluids* 8: 1385–1387.

Kraichnan RH. 1967. Inertial ranges in two-dimensional turbulence. *Phys. Fluids* 10: 1417–1423.

Kuznetsov EA. 1972. On turbulence of ion sound in plasma in a magnetic field. *Sov. Phys. JETP* 35: 310–314.

Kuznetsov EA. 2001. Weak magnetohydrodynamic turbulence of a magnetized plasma. *Sov. Phys. JETP* 93: 1052–1064.

Lacaze R, Lallemand P, Pomeau Y, Rica S. 2001. Dynamical formation of a Bose-Einstein condensate. *Physica D* 152–153: 779–786.

Longuet-Higgins MS, Gill AE, Kenyon K. 1967. Resonant interactions between planetary waves (and discussion). *Proc. R. Soc. Lond. A* 299: 120–144.

L'vov VS, L'vov Y, Newell AC, Zakharov V. 1997. Statistical description of acoustic turbulence. *Phys. Rev. E* 56: 390–405.

Lvov YV, Newell AC. 1997. Semiconductor lasers and Kolmogorov spectra. *Phys. Lett. A* 235: 499–503.

Lvov YV, Binder R, Newell AC. 1998. Quantum weak turbulence with applications to semiconductor lasers. *Physica D* 121: 317–343.

Lvov YV, Nazarenko SV, West R. 2003. Wave turbulence in Bose-Einstein condensates. *Physica D* 184: 333–351.

L'vov VS, Nazarenko SV, Rudenko O. 2007. Bottleneck crossover between classical and quantum superfluid turbulence. *Phys. Rev. B* 76: 024520.

Majda AJ, McLaughlin DW, Tabak EG. 1997. One-dimensional model for dispersive wave turbulence. *J. Nonlinear Sci.* 7: 9–44.

Maron J, Goldreich P. 2001. Simulations of incompressible magnetohydrodynamic turbulence. *Astrophys. J.* 554: 1175–1196.

Miles JW. 1967. On the generation of surface waves by shear flows. *J. Fluid Mech.* 30: 163–175.

Mordant N. 2008. Are there waves in elastic wave turbulence? *Phys. Rev. Lett.* 100: 234505.

Muskhelishvili NI. 1953. *Singular Integral Equations.* Groningen: Noordhoff.

Nazarenko SV, Newell AC, Galtier S. 2001. Non-local MHD turbulence. *Physica D* 152–153: 646–652.

Nazarenko S, Onorato M. 2006. Wave turbulence and vortices in Bose-Einstein condensation. *Physica D* 219: 1–12.

Nazarenko SV, Lukaschuk S, McLelland S, Denissenko P. 2010. Statistics of surface gravity wave turbulence in the space and time domains. *J. Fluid Mech.* 642: 395–420.

Newell AC. 1968a. The closure problem in a system of random gravity waves. *Rev. Geophys.* 6: 1–31.

Newell AC. 1968b. An alternative proof of the Poincaré-Bertrand lemma and its generalization. *J. Math. Anal. Appl.* 24: 149–155.

Newell AC, Aucoin PJ. 1971. Semidispersive wave systems. *J. Fluid Mech.* 49: 593–609.

Newell AC, Zakharov VE. 1992. Rough sea foam. *Phys. Rev. Lett.* 69: 1149–1151.

Newell AC, Zakharov VE. 1995. Optical turbulence. In *Turbulence: A Tentative Dictionary*, ed. P Tabeling, O Cardoso, pp. 59–66. New York: Plenum Press.

Newell AC, Nazarenko SV, Biven L. 2001. Wave turbulence and intermittency. *Physica D* 152–153: 520–550.

Newell AC, Zakharov VE. 2008. The role of the generalized Phillips' spectrum in wave turbulence. *Phys. Lett. A* 372: 4230–4233.

Newell AC, Rumpf B. 2011. Wave Turbulence. *Annu. Rev. Fluid Mech.* 43: 59–78.

Newell AC, Rumpf B, Zakharov VE. 2012. Spontaneous breaking of the spatial homogeneity symmetry of wave turbulence. *Phys. Rev. Lett.* 108 194502.

Ng CS, Bhattacharjee A. 1996. Interaction of shear-Alfvén wave packets: Implication for weak magnetohydrodynamic turbulence in astrophysical plasmas. *Astrophys. J.* 465: 845–854.

Ng CS, Bhattacharjee A, Germaschewski K, Galtier S. 2003. Anisotropic fluid turbulence in the interstellar medium and solar wind. *Phys. Plasmas* 10: 1954–1962.

Nordheim LW. 1928. On the kinetic method in the new statistics and its application in the electron theory of conductivity. *Proc. R. Soc. Lond. A* 119: 689–698.

Peierls R. 1929. Zur kinetischen Theorie der Wärmeleitung in Kristallen. *Annalen der Physik* 395: 1055–1101.

Phillips OM. 1977. *The Dynamics of the Upper Ocean.* Cambridge: Cambridge University Press.

Phillips OM. 1985. Spectral and statistical properties of the equilibrium range in wind-generated gravity-waves. *J. Fluid Mech.* 156: 505–531.

Pushkarev AN, Zakharov VE. 1996. Turbulence of capillary waves. *Phys. Rev. Lett.* 76: 3320–3323.

Pushkarev AN, Resio D, Zakharov V. 2003. Weak turbulence approach to wind-generated gravity sea waves. *Physica D* 184: 29–63.

Rumpf B, Newell AC. 2004. Intermittency as a consequence of turbulent transport in nonlinear systems. *Phys. Rev. E* 69: 026306.

Rumpf B, Biven L. 2005. Weak turbulence and collapses in the Majda-McLaughlin-Tabak equation: Fluxes in wavenumber and in amplitude space. *Physica D* 204: 188–203.

Rumpf B. 2008. Transition behavior of the discrete nonlinear Schrödinger equation. *Phys. Rev. E* 77: 036606.

Rumpf B, Newell AC, Zakharov VE. 2009. Turbulent transfer of energy by radiating pulses. *Phys. Rev. Lett.* 103: 074502.

Sahraoui F, Belmont G, Rezeau L. 2003. Hamiltonian canonical formulation of Hall MHD: Toward an application to weak turbulence. *Phys. Plasmas* 10: 1325–1337.

Sahraoui F, Galtier S, Belmont G. 2007. Incompressible Hall MHD waves. *J. Plasma Phys.* 73: 723–730.

Sridhar S, Goldreich P. 1994. Toward a theory of interstellar turbulence I. Weak Alfvénic turbulence. *Astrophys. J.* 432: 612–621.

Sun C, Jia S, Barsi C, Rica S, Picozzi A, Fleischer J. 2011. Observation of the condensation of classical waves. To appear in *Nature*.

Toba Y. 1972. Local balance in air-sea boundary processes. I. On the growth process of wind waves. *J. Oceanogr. Soc. Jpn.* 28: 109–120.

Toba Y. 1973a. Local balance in air-sea boundary processes. II. Partition of wind stress to waves and current. *J. Oceanogr. Soc. Jpn.* 29: 70–75.

Toba Y. 1973b. Local balance in air-sea boundary processes. On the spectrum of wind waves. *J. Oceanogr. Soc. Jpn.* 29: 209–220.

Toba Y. 1997. Wind-wave strong wave interactions and quasi-local equilibrium between wind and wind sea with the friction velocity proportionality. In *Nonlinear Ocean Waves*, ed. W Perrie, *Advances in Fluid Mechanics* 17: 1–59.

Young IR. 1999. *Wind Generated Ocean Waves.* Elsevier Ocean Eng. Book Series, Vol. 2. Amsterdam: Elsevier.

Zabusky NJ, Kruskal MD. 1965. Interaction of "solitons" in a collisionless plasma and the recurrence of initial states. *Phys. Rev. Lett.* 15: 240–243.

Zakharov VE. 1965. Weak turbulence in media with a decay spectrum. *J. Appl. Mech. Tech. Phys.* 6: 22–24.

Zakharov VE, Filonenko NN. 1967a. Weak turbulence of capillary waves. *J. Appl. Mech. Tech. Phys.* 8: 37–42.

Zakharov VE, Filonenko NN. 1967b. Energy spectrum for stochastic oscillations of the surface of a liquid. *Sov. Phys. Dokl.* 11: 881–884.

Zakharov VE, Zaslavsky MM. 1982. The kinetic equation and Kolmogorov spectra in the weak-turbulence theory of wind waves. *Izv. Atmos. Ocean. Phys.* 18: 747–753.

Zakharov VE, L'vov V, Falkovich G. 1992. *Kolmogorov Spectra of Turbulence I.* Berlin: Springer-Verlag.

Zakharov VE. 1998. *Nonlinear Waves and Weak Turbulence*, AMS Translations Series 2, Vol. 182. Providence, RI: American Mathematical Society.

Zakharov VE, Korotkevich AO, Pushkarev AN, Dyachenko AI. 2000. Mesoscopic wave turbulence. *JETP Lett.* 82: 487–491.

Zakharov VE. 2005. Theoretical interpretation of fetch limited wind-driven sea observations. *Nonlinear Proc. Geophys.* 12: 1011–1020.

Zakharov VE, Nazarenko SV, 2005. Dynamics of Bose-Einstein condensation. *Physica D* 201: 203–211.

Chapter 2

Fluctuations of the Energy Flux in Wave Turbulence

S. Aumaître*, E. Falcon[†,§] and S. Fauve[‡]

*SPEC, DSM, CEA-Saclay, CNRS,
91 191 Gif-sur-Yvette, France

†Université Paris Diderot, Sorbonne Paris Cité,
MSC, UMR 7057, CNRS, F-75 013 Paris, France

‡LPS, ENS Paris, CNRS, 24 rue Lhomond, 75 005 Paris, France
§eric.falcon@univ-paris-diderot.fr

The key governing parameter of wave turbulence is the energy flux that drives the waves and cascades to small scales through nonlinear interactions. In the inertial range, the energy flux is conserved across the scales, and is assumed to be constant in most theoretical approaches. It is only recently that measurements of the injected power into wave turbulence have been performed at the scale of the wave maker (integral scale). In this review, we focus on the statistical properties of the injected power fluctuations in gravity-capillary wave turbulence in a stationary regime. Fluctuations of the injected power have been found much larger than their mean value. In addition, events related to a negative injected power, i.e. an instantaneous reversed energy flux, occur with a fairly large probability. Both features are well described using a Langevin type equation. Finally, we consider the experimental dependence of the scaling law of the wave spectrum with the mean injected power and discuss possible reasons for the discrepancy with weak turbulence theory.

Contents

2.1. Introduction

A fundamental problem of wave turbulence consists of describing the transfer of energy among different scales through the weak interaction between waves. It was understood first by Zakharov (1967) that the kinetic equations obtained in the weak turbulence limit have stationary solutions that involve a finite mean energy flux per unit surface and density ϵ across the scales. In the case of a n-wave process, the energy spectrum of the wave heights $E(k)$ is proportional to $\epsilon^{1/(n-1)}$. Once this is taken into account, the power-law dependence of $E(k)$ on the wavenumber k often follows from dimensional analysis (Connaughton *et al.*, 2003),

$$E(k) \propto \epsilon^{1/(n-1)} k^{\alpha}, \qquad (2.1)$$

where the proportionality constant involves parameters of the dispersion relation. Since measurements are often performed in time at a fixed point in space, it is useful to consider an equation similar to (2.1) in the frequency domain

$$E(\omega) \propto \epsilon^{1/(n-1)} \omega^{\beta}. \qquad (2.2)$$

In weak turbulence theory, Eqs. (2.1) and (2.2) are related through the dispersion relation $\omega = W(k)$. This has been checked only recently by comparison of the spatial and temporal spectra determined in experiments of waves on elastic plates (Cobelli *et al.*, 2009) and capillary-gravity surface waves (Snouck *et al.*, 2009; Herbert *et al.*, 2010). A fair agreement has been found in the former case whereas a more complex structure in the spatio-temporal domain exists in the latter ones. Most of the experiments performed so far have dealt with the determination of the exponents α and β related to the spatial (respectively temporal) spectrum. Although a fair agreement with theory has been often claimed, detailed measurements show disagreement in many cases: for elastic waves on plates (see the chapter by N. Mordant) or gravity waves for which the spectrum has been shown to depend on the amplitude and frequency content of the forcing (Falcon *et al.*, 2007a; Denissenko *et al.*, 2007; Nazarenko *et al.*, 2010). The spectrum of capillary waves looks more robust, an exponent in rough agreement with theory being found in different configurations, high frequency part of surface waves (Falcon *et al.*, 2007a), pure capillary waves observed in microgravity (Falcon *et al.*, 2009), and capillary waves at the interface between two fluids with similar density (Düring and Falcón, 2009).

It is thus of primary interest to test other predictions of the weak turbulence theory in order to determine its range of validity. Measuring the dependence of the spatial (respectively temporal) spectrum on the mean energy flux ϵ is the next step. No direct measurement of the energy flux within the inertial range exists so far, but its mean value is equal to the mean injected power of the force driving the system. That quantity can be measured. We will discuss these measurements in Sec. 2.3.

A second aspect concerns the fluctuations of the energy flux. It has been known for a long time that the power needed to maintain a dissipative system in a statistically stationary regime is generally a fluctuating quantity. This has been shown in the case of spatio-temporally chaotic waves generated by the Faraday instability (Ciliberto *et al.*, 1991), hydrodynamic turbulence (Labbé *et al.*, 1996) and many other dissipative systems such as granular gases (Aumaître *et al.*, 2001, 2004). However, very few general results have been obtained so far about the properties of the fluctuations of global quantities in dissipative systems driven far from equilibrium. In all these systems, dissipative processes damp out any initial motion in the absence of an external forcing. In order to reach a statistically stationary regime, an external operator should provide an injected power $I(t)$ that, on average, compensates the dissipation $D(t)$. The equation for the energy budget takes the form

$$\frac{dE(t)}{dt} = I(t) - D(t), \qquad (2.3)$$

where E is an energy, for instance the kinetic energy of a turbulent flow or of a granular gas. In majority of the situations, the external operator never controls one of the global quantities, I, D or E. It usually drives the system by imposing locally a given force or velocity, for instance, the velocity or the torque applied to a propeller generating a turbulent flow, or the vibration velocity of a piston driving waves in a fluid or motions of the particles in a granular gas. The injected power $I(t)$ is thus generally a fluctuating quantity, that can often take instantaneous negative values depending on the phase of the response of the system to the driving (Aumaître *et al.*, 2001). It should be positive on average, since we have in a statistically stationary regime

$$\langle I \rangle = \langle D \rangle, \qquad (2.4)$$

with $D(t) > 0$ in a macroscopic description of a dissipative system. Equation (2.3) is probably one of the most common equations of physics since it only states that the time variation of some quantity results from the difference between input and output. It is thus surprising that its general properties have not been emphasized more often. In a statistically stationary regime, Eq. (2.4) is of course well known, but to the best of our knowledge, this is not the case for the relations involving higher moments of the fluctuations of $I(t)$ and $D(t)$. For instance, $I(t)$ and $D(t)$ should have the same spectrum at zero frequency (Aumaître *et al.*, 2004; Farago, 2004)

$$|\hat{I}(0)|^2 = |\hat{D}(0)|^2, \tag{2.5}$$

or using Wiener–Kintchine theorem,

$$\int_0^\infty [\langle I(\tau)I(0)\rangle - \langle I\rangle^2]d\tau = \int_0^\infty [\langle D(\tau)D(0)\rangle - \langle D\rangle^2]d\tau. \tag{2.6}$$

If the correlation functions decay fast enough at large τ, the above equation shows that the variances of the injected and dissipated power are related by

$$\sigma_I^2 \tau_I = \sigma_D^2 \tau_D, \tag{2.7}$$

where τ_I (respectively τ_D) is the correlation time of $I(t)$ (respectively $D(t)$). As explained by Aumaître *et al.* (2004) and Farago (2004), the above results trace back to the large deviation functions for I and D that have the same Taylor expansion about $\langle I\rangle = \langle D\rangle$.

This chapter is organized as follows: first, we recall experimental results about gravity-capillary surface waves in a turbulent regime and show how the spectra depend on the injected power in the system. Then, we consider the fluctuations of the injected power and their statistical properties. We show that the experimental observations can be modeled using a Langevin-type equation driven by colored noise.

2.2. Spectra in the Gravity and Capillary Regimes

The experimental setup is shown in Fig. 2.1. It consists of a square plastic vessel, 20 cm side, filled with a fluid up to a depth $h = 1.8$ or 2.3 cm leading to an almost deep water limit ($\lambda \lesssim 2\pi h$ for our range of wavelengths λ). Mercury is chosen as the working fluid because of its low kinematic viscosity (one order of magnitude smaller than that of water), thus reducing wave dissipation. Note, however, that similar qualitative results to the ones reported here are found when changing mercury by water. The properties of mercury are, density $\rho = 13.5 \ 10^3 \ \text{kg/m}^3$, kinematic viscosity $\nu = 1.15 \ 10^{-7} \ \text{m}^2/\text{s}$ and surface tension $\gamma = 0.4 \ \text{N/m}$. Surface waves are generated by the horizontal motion of two rectangular ($10 \times 3.5 \ \text{cm}^2$) plunging Plexiglas wave makers driven by two electromagnetic vibration exciters. The wave makers are driven with a random noise (in amplitude and frequency) band-pass filtered within a frequency bandwidth between 1 Hz and f_p (f_p being typically from $f_p = 4$ to 6 Hz). This corresponds to wavelengths of surface waves larger than 4 cm. The height η of the surface waves is measured at a given location (7 cm away from the wave makers) by a capacitive wire gauge plunging perpendicularly to the fluid surface at rest. The latter is made of an

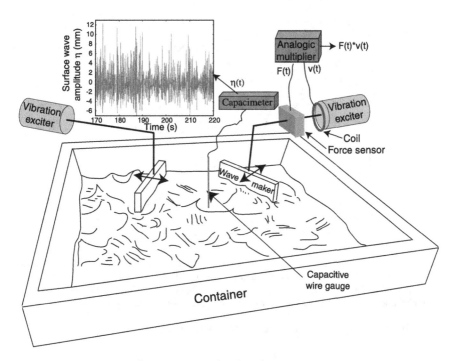

Fig. 2.1. Schematic view of the experimental setup showing a typical time recording of the surface wave height, $\eta(t)$, at a given location during 50 s. $\langle \eta \rangle \simeq 0$.

insulated copper wire, 0.1 mm in diameter, the insulation (a varnish) being the dielectric of an annular capacitor with the wire as the inner conductor and mercury as the outer one. The capacitance is thus proportional to the fluid level. A low-cost home-made analogic multivibrator with a response time 0.1 ms is used as a capacitance meter in the range 0–200 pF. The linear sensing range of the sensor allows wave height measurements from $10\,\mu$m up to 2 cm with a 20 mm/V sensitivity. We have checked that the exact location of the capacitive sensor has almost no influence on the wave spectrum. Note that the maximum forcing amplitude is less than the onset of the water drop ejection or wave breaking.

A typical recording of the surface wave amplitude, $\eta(t)$, at a given location is displayed in the inset of Fig. 2.1 as a function of time. The wave amplitude is very erratic with a large distribution of amplitudes. The largest values of the amplitude are of the order of the fluid depth, whereas the mean value of the amplitude is close to zero. The asymmetry of the fluctuations (larger crests than troughs) is stronger when the height of the layer is smaller but it persists in the deep layer limit (Falcon *et al.*, 2007a).

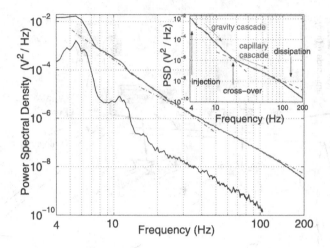

Fig. 2.2.   Power spectra of the surface wave height for two different injected powers $\langle I \rangle = 1.6$ and 32.4 mW (from bottom to top). Dashed-dotted lines have slopes -4.3 and -3.2. Random forcing within a 1–6 Hz frequency bandwidth. Inset: Same with $\langle I \rangle = 32.4$ mW and a 1–4 Hz bandwidth. Dashed lines have slopes -6.1 and -2.8.

The power spectrum density of the wave height,

$$S_\eta(f) \equiv \int \langle \eta(t + \tau)\eta(t)\rangle_t e^{-i2\pi f\tau} d\tau,$$

is recorded using a spectrum analyzer from 4 Hz up to 200 Hz and averaged during 2000 s. For a small forcing, peaks related to the forcing and its harmonic are visible in the low frequency part of the spectrum in Fig. 2.2. At a higher forcing, those peaks are smeared out and a power-law can be fitted. At higher frequencies, the slope of the spectrum changes, and a crossover is observed near 20 Hz between two regimes. This corresponds to the transition from gravity to capillary wave turbulence. For a narrower frequency band of excitation (1–4 Hz), similar spectra are found but with a broader power-law in the gravity range (see inset of Fig. 2.2).

    For linear waves, the crossover between gravity and capillary regimes corresponds to a wave number k of the order of the inverse of the capillary length $l_c \equiv \sqrt{\gamma/(\rho g)}$, i.e. to a critical frequency, $f_c = \sqrt{g/2l_c}/\pi$, where g is the acceleration of gravity. For mercury, $l_c = 1.74$ mm and $f_c \simeq 17$ Hz corresponding to a wavelength of the order of 1 cm. The inset of Fig. 2.2 shows a correct agreement in the case of a narrow driving frequency band. We also observe that the crossover frequency increases with the driving amplitude and with the width of the driving frequency band (Falcon *et al.*, 2007a). This can be due to the fact that the above estimate of f_c is only valid for linear waves.

Surface wave turbulence is usually described as a continuum of interacting waves governed by kinetic-like equations in the case of small nonlinearity and weak wave interactions. Weak turbulence theory predicts that the surface height spectrum $S_\eta(f)$, i.e. the Fourier transform of the autocorrelation function of $\eta(t)$, is scale invariant with a power-law frequency dependence. Such a Kolmogorov-like spectrum writes

$$S_\eta(f) \propto \epsilon^{\frac{1}{2}} \left(\frac{\gamma}{\rho}\right)^{\frac{1}{6}} f^{-\frac{17}{6}} \quad \text{for capillary waves}$$

(Zakharov and Filonenko, 1967a)

$$S_\eta(f) \propto \epsilon^{\frac{1}{3}} g f^{-4} \quad \text{for gravity waves (Zakharov and Filonenko, 1967b),}$$

$$(2.8)$$

where ϵ is the energy flux per unit surface and density [$S_\eta(f)$ has dimension $L^2 T$ and ϵ has dimension $(L/T)^3$]. In both regimes, these frequency power-law exponents are compared with the slopes of surface height spectra measured for different forcing intensities and bandwidths. The experimental values of the scaling exponent of capillary spectra are close to the expected $f^{-2.8}$ scaling as already shown with one driving frequency (Wright *et al.*, 1996; Lommer and Levinsen, 2002; Brazhnikov, Kolmakov and Levchenko, 2002) or with noise (Brazhnikov, Kolmakov and Levchenko, 2002). However, in these previous works, peaks and their harmonics (related to the parametric forcing) are observed on the spectrum with maximal amplitudes decreasing as a frequency power-law (Snouck *et al.*, 2009; Brazhnikov, Kolmakov and Levchenko 2002; Brazhnikov *et al.*, 2002). The frequency-spectrum exponent estimated in that way is thus not very accurate. A recent study has even found an exponent in disagreement with weak turbulence underlying the difficulty to reach a wave turbulence regime with a parametric forcing (Snouck *et al.*, 2009). In the case of our measurements, we observe that this exponent for the capillary range does not depend on the amplitude and the frequency band of the forcing, within our experimental precision. For the gravity spectrum, no power-law is observed at a small forcing since the turbulence is not strong enough to hide the first harmonic of the forcing (see Fig. 2.2). At high enough forcing, the scaling exponent of gravity spectra is found to increase with the intensity and the frequency band from -6 to -4 such that the predicted f^{-4} scaling of Eq. (2.8) is only observed for the largest forcing intensities and bandwidth (Falcon *et al.*, 2007a). The dependence of the slope of the gravity waves spectrum on the forcing characteristics has been ascribed to finite size effects (Falcon *et al.*, 2007a). Similar results in the gravity range have been found in a much larger tank with sinusoidal forcing

(Denissenko *et al.*, 2007; Nazarenko *et al.*, 2010). This dependence has been also ascribed to the presence of strong nonlinear waves (Cobelli *et al.*, 2011). However, recent experiments have shown that it is related to the anisotropy and inhomogeneity of the forcing (Issenmann and Falcon, 2013).

2.3. Direct Measurement of the Injected Power

The power injected into the fluid by the wave maker is determined as follows. The velocity $V(t)$ of the wave maker is measured using a coil placed on the top of the vibration exciter (see Fig. 2.1). The voltage induced by the moving permanent magnet of the vibration exciter is proportional to $V(t)$. The force $F_A(t)$ applied by the vibration exciter on the wave maker is measured by a piezoresistive force transducer (FGP 10 daN). $F_A(t)$ and $V(t)$ recorded by means of an acquisition card with a 1 kHz sampling rate during 300 s. The time recordings of $V(t)$ and $F_A(t)$ together with their PDFs are displayed in Fig. 2.3. Both $V(t)$ and $F_A(t)$ are Gaussian with zero mean value. For a given excitation bandwidth, the rms value σ_V of

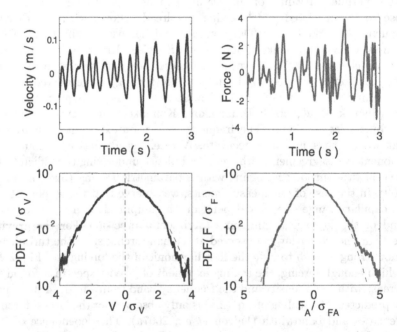

Fig. 2.3.   Time recordings of the velocity of the wave maker and the force applied to the wave maker by the vibration exciter ($\langle F_A \rangle \approx \langle V \rangle \approx 0$). The fluid is mercury, with $h = 23$ mm. Both PDFs are Gaussian (dashed lines) with zero mean value.

the velocity fluctuations of the wave maker is proportional to the driving voltage U applied to the electromagnetic shaker and does not depend on the fluid density ρ. On the contrary, the standard deviation σ_{F_A} of the force applied to the wave maker is decreased by the density ratio (~ 13) when mercury is replaced by water. The rms velocity of the wave maker is thus prescribed in our experiments. We have checked that $\sigma_{F_A} \propto \rho S \sigma_V$ where S is the immersed area of the wave maker. This linear behavior has been measured on one decade up to $\sigma_{F_A} \sim 2$ N and $\sigma_V \sim 0.1$ m/s. The power injected into the fluid by the wave maker is $I(t) = -F_R(t)V(t)$ where $F_R(t)$ is the force applied by the fluid on the wave maker. It generally differs from $F_A(t)V(t)$ which is measured here, because of the piston inertia (see below). However, their time averages are equal, thus $\langle I \rangle = \langle F_A(t)V(t) \rangle$.

The mean energy flux ϵ is estimated by the measurement of $\langle I \rangle/(\rho S)$ where $\langle I \rangle$ is the mean power injected by the wave maker and S is the immersed area of the wave maker. With given σ_V, we have first checked that $\langle I \rangle$ is proportional to S and decreases by a factor 13 when mercury is replaced by water. Our measurements also show that $\langle I \rangle \propto \sigma_V^2$ with a proportionality coefficient of order $10 \, \mathrm{W/(m/s)^2}$ (see the inset of Fig. 2.4). We thus have $\epsilon \propto c\sigma_V^2$ where c has the dimension of a velocity. If we assume that ϵ should involve only large scale quantities, it cannot depend

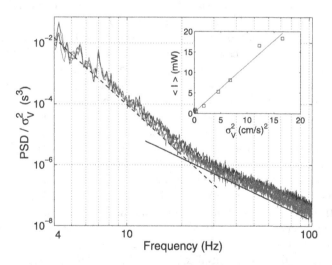

Fig. 2.4. Spectra of the surface wave height divided by the variance σ_V^2 of the velocity of the wave maker for different forcing amplitudes, $\sigma_V = 2.1, 2.6, 3.5$ and 4.1 cm/s. Random forcing with a 1–4 Hz frequency bandwidth. The dashed line has slope -5.5 whereas the full line has slope $-17/6$. The mean injected power is displayed as a function of σ_V^2 in the inset. The best fit gives a slope $11.5 \, \mathrm{W/(m/s)^2}$.

on surface tension or viscosity. Then, we need an additional length scale or time scale besides g in order to be able to determine c. In a deep fluid range, the height of the layer h cannot be used and the additional parameter could be the horizontal size of the layer or equivalently the travel time of the large scale gravity waves. Without this additional parameter, we should get $\epsilon \propto \sigma_V^3$ as usually assumed in weak turbulence theory. The dependence of ϵ on c can be ascribed to finite size effects. The inverse of travel time of a wave within the tank or the frequency difference between the discrete modes of the tank, both scale with c. Note that the peaks are visible at low frequencies in the spectrum of Fig. 2.2 at low forcing amplitude. Discreteness also explains the presence of these peaks that correspond to vessel eigenvalue modes (Deike *et al.*, 2012). However, the large enough values of ϵ required to observe power-laws, are more than one order of magnitude smaller than the critical flux $(\gamma g/\rho)^{3/4} \approx 2200\,(\text{cm/s})^3$ corresponding to the breakdown of weak turbulence (Newell and Zakharov, 1992).

We consider now how the spectra of the gravity and capillary waves scale with the mean energy flux. The best choice in order to collapse our experimental spectra on a single curve for different values of σ_V is displayed in Fig. 2.4 where the power spectral density divided by σ_V^2 is plotted versus f. Surprisingly, spectra are collapsed on both the gravity and capillary ranges by this single scaling. The wave height spectrum is found to scale as $\langle I \rangle^{1 \pm 0.1}$ for both gravity and capillary wave turbulence regimes over almost one decade in $\langle I \rangle$ as shown in Fig. 2.4. This scaling does not depend on the vessel geometry when using a circular one of similar size. A similar spectrum scaling $\sim \langle I \rangle^1$ has been observed for both regimes by horizontally vibrating the whole container (Issenmann and Falcon, 2013) for the same range of $\langle I \rangle$, for the capillary regime with a parametric forcing (Xia *et al.*, 2010), and for the inverse cascade of gravity wave turbulence (Deike *et al.*, 2011). Moreover, it has been checked that $\langle I \rangle \sim \sigma_\eta^2$, where σ_η is the rms value of the wave height. This is coherent with $S_\eta(f) \sim \langle I \rangle^1$ (Issenmann and Falcon, 2013) since $\int_0^\infty S_\eta(f)df = \sigma_\eta^2/(2\pi)$.

Another way to estimate the mean energy flux consists of measuring the wave energy decay rate after switching off the wave maker (Denissenko *et al.*, 2007; Nazarenko *et al.*, 2010). This method gives a good estimate of the mean energy flux provided large scale dissipation is negligible. Measurements performed in Deike *et al.* (2012) show that the decay rate does not depend on the initial wave amplitude, thus giving a capillary spectrum scaling in ϵ^1 instead of the predicted $\epsilon^{1/2}$ one. This shows that the large scale waves lose most of their energy through large scale dissipation and not by transferring energy to smaller scales.

Thus, both the scaling of the mean injected power $\langle I \rangle \sim \sigma_V^2$ with respect to the forcing velocity σ_V and the linear scaling of the spectrum $S_\eta(f) \sim \langle I \rangle^1$ are in disagreement with the weak turbulence theory that predicts a spectrum $\sim \epsilon^{1/3}$ in the gravity regime and $\sim \epsilon^{1/2}$ in the capillary regime (see above). Note also that matching the gravity regime to the capillary one using Eq. (2.8) gives a crossover frequency that slightly decreases when ϵ is increased whereas an increase with ϵ is observed. This also shows a discrepancy with respect to weak turbulence theory that is likely to be related to the previous ones but does not require the measurement of $\langle I \rangle$.

We now discuss some possible reasons for the discrepancy between the experimental measurements presented above and the theory of weak turbulence. We first consider the measurements involving the injected power. A first possibility is that one part of the power is directly provided to the bulk flow and dissipated by viscosity without cascading through the wave field. For instance, the wave maker can generate vortices that would dissipate some part of the injected power. Although this mechanism is certainly present, it is unlikely to be the dominant one. Indeed, as said above, the scaling laws involving the mean injected power do not change when the forcing is made by horizontally vibrating the whole container instead of using wave makers. In addition, $\langle I \rangle \propto \sigma_V^2$ would correspond to the Stokes regime whereas the Reynolds number of the wave maker is larger than 1000 in mercury.[1] Thus, we rather think that most of the injected power is transferred to large scale waves. A second possibility to explain the observed discrepancy is that, although chaotic (a broad band spectrum is observed at large enough forcing), these large scale waves transfer a small amount of energy flux to higher harmonics compared to their direct dissipation by viscosity. This speculation is strengthened by recent experiments of decaying wave turbulence on the surface of a fluid that have shown that only a small part of the initial power injected into the waves feeds the capillary cascade, whereas the major part is dissipated at large scales (Deike *et al.*, 2012). For stationary wave turbulence, it is thus likely that only some fraction of the power injected into waves cascades through the scales, the rest being dissipated at various scales. This unknown dissipated fraction of injected power could be at the origin of the discrepancy with weak turbulence theory for the scaling of the spectrum with the injected power. This can also explain the discrepancies that do not involve the measurement of the mean energy flux. It is likely that finite size

---

[1]For a typical wave maker forcing (0.4 cm amplitude and 4 Hz frequency), its velocity is 0.1 m/s, and its Reynolds number is 4000 in mercury ($\nu = 10^{-7}$ m^2/s).

effects, by inhibiting the energy transfers among large scale waves, prevent the observation of a cascade regime in the gravity range with the scaling predicted by weak turbulence theory (Falcon *et al.*, 2007a). The presence of strong fluctuations of the injected power (see Sec. 2.4) can also affect the predictions of the theory. This issue is still open and deserves more studies. For instance, it will be of primary interest to measure the energy flux at various scales (instead of the injected power by the wave maker) by a fully resolved space–time measurement.

2.4. Fluctuations of the Energy Flux

We now study the fluctuations of the power injected by the wave maker in the fluid. When the wave maker inertia is negligible, the power $I(t)$ injected into the fluid is roughly given by $F_A(t)V(t)$ (see below). The time recording of $I(t)$ is shown in the inset of Fig. 2.5. Contrary to the velocity or the force, the injected power consists of strong intermittent bursts. Although the forcing is statistically stationary, there are quiescent periods with a small amount of injected power interrupted by bursts where $I(t)$ can take both positive and negative values. The PDFs of $I/\langle I \rangle$ are displayed in Fig. 2.5. They show that the most probable value of I is zero and display two asymmetric exponential tails (or stretched exponential in the smaller

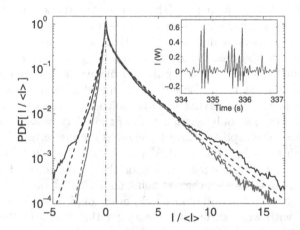

Fig. 2.5.   PDF of $I(t)/\langle I \rangle$ for mercury: $\sigma_V = 5\,\text{cm/s}$, container size $57 \times 50\,\text{cm}^2$ [red (light grey)] and $20 \times 20\,\text{cm}^2$ (black) ($h = 18\,\text{mm}$). Red (light grey) solid line: experiment with $\langle I \rangle = 51\,\text{mW}$ and $\sigma_{F_A} = 1.6\,\text{N}$. Black solid line: experiment with $\langle I \rangle = 2\,\text{mW}$ and $\sigma_{F_A} = 0.73\,\text{N}$. Dashed lines are the related predictions from Eq. (2.12) without fitting parameter, $r = 0.7$ [red (light grey)] and $r = 0.6$ (black). Vertical solid line show the mean injected power. Inset: time recording of $I(t)$.

container). We observe that events with $I(t) < 0$, i.e. for which the wave field gives back energy to the wave maker, occur with a fairly high probability. The standard deviation σ_I of the injected power is much larger than its mean value $\langle I \rangle$ and rare events with amplitude up to $7\sigma_I$ are also detected.

We also observe in Fig. 2.5 that the probability of negative events strongly decreases when the container size is increased whereas the positive fluctuations are less affected. This shows that the backscattering of the energy flux from the wave field to the driving device is related to the waves reflected by the boundary that can, from time to time, drive the wave maker in phase with its motion. We note that we have less statistics for the negative tail of the PDF when the size of the container is increased.

We emphasize the bias that can result from the system inertia when one tries a direct measurement of the fluctuations of injected power. The equation of motion of the wave maker is

$$M\dot{V} = F_A(t) + F_R(t), \tag{2.9}$$

where M is the mass of the wave maker and $F_R(t)$ is the force due to the fluid motion. The power injected into the fluid by the wave maker is $I(t) = -F_R(t)V(t)$. When $M\dot{V}$ is not negligible, $I(t)$ generally differs from $F_A(t)V(t)$ which is experimentally determined. This obviously does not affect the mean value $\langle I \rangle$ but may lead to wrong estimates of the fluctuations. Using an accelerometer, we have checked that $M\dot{V}$ is negligible compared to F_A when the working fluid is mercury. On the contrary, inertia cannot be neglected for experiments in water for which an error as large as one order of magnitude can be made on the probability of rare events if one use $F_A V$ to estimate I. Thus, the correction due to $M\dot{V}$ has been taken into account in water experiments.

The PDF of injected power for the same driving in the same container for mercury displays a larger asymmetry than the one for water. This is related to the larger mean energy flux, i.e. mean dissipation, for mercury, as shown below.

The qualitative features of the PDF of injected power can be described with the following simple model. Guided by our experimental observation of the linearity of σ_{F_A} in σ_V (see Sec. 2.3), we assume that the force F_R due to the fluid can be roughly approximated by a friction force $-M\gamma V$, where γ is a constant (the inverse of the damping time of the wave maker). We are aware that a better approximation to the force due to the fluid should involve both $\dot{V}$ and an integral of $V(t)$ with an appropriate kernel. Thus, we only claim here to give a heuristic understanding of the qualitative properties of the PDF of I. Since in our experiment the forcing is stochastic and low-pass filtered at frequency $1/\beta$, we model the forcing with an

Ornstein–Uhlenbeck process:

$$\dot{V} = -\gamma V + F, \quad \dot{F} = -\beta F + \xi, \tag{2.10}$$

where β is the inverse of the correlation time of the applied force ($F = F_A/M$) and ξ is a Gaussian white noise with $\langle \xi(t)\xi(t')\rangle = \Delta\delta(t - t')$. The PDF $P(V, F)$ is the bivariate normal distribution (Risken, 1996)

$$P(V, F) = \frac{\exp\left[-\frac{1}{2(1-r^2)}\left(\frac{V^2}{\sigma_V^2} - \frac{2rVF}{\sigma_V\sigma_F} + \frac{F^2}{\sigma_F^2}\right)\right]}{2\pi\sigma_V\sigma_F\sqrt{1 - r^2}}, \tag{2.11}$$

with $\sigma_F = \sqrt{\Delta/2\beta}$, $\sigma_V = \sqrt{\Delta/(2\gamma\beta(\gamma + \beta))}$ and $r = \sqrt{\gamma/(\gamma + \beta)}$. Changing variables (V, F) to $(\tilde{I} = FV = I/M, F)$ and integrating over F gives

$$P(\tilde{I}) = \frac{\exp\left[\frac{r\tilde{I}}{(1-r^2)\sigma_V\sigma_F}\right]}{\pi\sigma_V\sigma_F\sqrt{1 - r^2}}K_0\left[\frac{|\tilde{I}|}{(1 - r^2)\sigma_V\sigma_F}\right], \tag{2.12}$$

where $K_0(X)$ is the zeroth-order modified Bessel function of the second kind. Using the method of steepest descent, this predicts exponential tails, $P(X) \sim (1/\sqrt{|X|})\exp(rX - |X|)$ where $X = \tilde{I}/[(1-r^2)\sigma_V\sigma_F]$. In addition, we have $\langle\tilde{I}\rangle = \Delta/[2\beta(\gamma+\beta)] = r\sigma_V\sigma_F$. Thus, (2.12) is determined once $\langle I\rangle$, σ_V and σ_F have been measured and can be compared to the experimental PDF without using any fitting parameter. This is displayed with dashed lines in Fig. 2.5. Taking into account the strong approximation made in the above model, we observe a good agreement in the larger container. More importantly, this model captures the qualitative features of the PDF: its maximum for $I = 0$ and the asymmetry of the tails that is governed by the parameter $r = \sqrt{\gamma/(\gamma + \beta)} = \langle I\rangle/(\sigma_V\sigma_{F_A})$. For given σ_V and σ_{F_A}, the larger is the mean energy flux, i.e. the dissipation, the more asymmetric is the PDF. For mercury, direct determination of r from the measurement of $\langle I\rangle$, σ_V and σ_{F_A} gives $r \sim 0.7$ for the large container and $r \sim 0.6$ for the small one. Smaller values of r are achieved in water for which the dissipation is smaller. The PDFs are more stretched for water, in particular, in the smaller container.

In order to explain this modification of the PDF, we consider the effect of nonlinearities in the model, for instance, in the form of a velocity-dependent damping term, $\gamma(V) = \frac{\gamma_o}{1+aV^2}$, in (2.10). The weaker damping at large velocity does generate stretched exponential tails in the PDF of injected power with more probable large events as shown in Fig. 2.6. However, the PDF of the velocity is no longer Gaussian in that case (inset of Fig. 2.6).

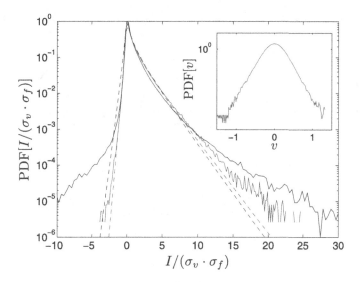

Fig. 2.6. Solid lines: PDF of the injected power from the numerical resolution of Eq. (2.10) with a nonlinear damping coefficient $\gamma(V) = \frac{\gamma_o}{1+aV^2}$ with $\gamma_o = 1$, $\beta = 0.5$, $\sigma_f = 0.22$: $a = 1.0$ (red) and $a = 2.0$ (blue). Dashed lines correspond to Eq. (2.12) for the same values of $\langle I \rangle$, σ_V, and σ_{F_A}. Inset: PDF of the velocity for $a = 1.0$.

One could notice that the PDF of I given by (2.12) respects a symmetry similar to the one of the fluctuation theorem (Evans *et al.*, 1993; Gallavotti and Cohen, 1995), but without any time averaging. Indeed, we get

$$\frac{1}{\tau_F} \log \frac{P(I)}{P(-I)} = \frac{2}{\sigma_V^2} I, \qquad (2.13)$$

where $\tau_F = 1/\beta$ is the correlation time of the force. Thus, $\sigma_V^2/2$ plays the role of an effective temperature, $k_B T = \sigma_V^2/2$.

However, the symmetry related to the fluctuation theorem is not observed with our data if we consider, as we should, the injected power averaged on a time interval τ

$$I_\tau(t) = \frac{1}{\tau} \int_t^{t+\tau} I(t')dt'. \qquad (2.14)$$

The PDFs of I_τ for $\tau/\tau_c = 1, 3, 11$ and 50 where τ_c is the correlation time of $I(t)$, are displayed in Fig. 2.7. They become more and more peaked around $I_\tau \simeq \langle I \rangle$, as they should. However, one needs to average on a rather large time interval ($\tau \sim 50\tau_c$) in order to get a maximum probability $P(I_\tau)$

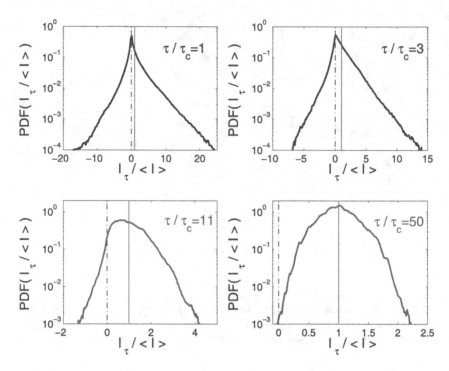

Fig. 2.7.   PDFs of the injected power I_τ averaged on a time interval τ: $\tau = 1, 3, 11$ and $50\tau_c$, where $\tau_c = 0.03\,\mathrm{s}$ is the correlation time of $I(t)$. Solid lines indicate the value of $\langle I \rangle$ (water, $h = 23\,\mathrm{mm}$).

for $I_\tau = \langle I \rangle$ (Fig. 2.7, bottom right). Then, the probability of negative events become so small that almost none can be observed. Figure 2.8 shows that the quantity $\frac{1}{\tau} \log \frac{P(I_\tau/\langle I \rangle)}{P(-I_\tau/\langle I \rangle)}$ for different values of τ that has been predicted to be linear in $I_\tau/\langle I \rangle$ for $\tau \gg \tau_c$ when the hypotheses of the fluctuation theorem (in particular time reversibility) are fulfilled (Evans *et al.*, 1993; Gallavotti and Cohen, 1995; Kurchan, 1998). As we clearly observe in Fig. 2.8, this is not the case in general for macroscopic (i.e. energy $\gg kT$) dissipative systems. As already mentioned (Aumaître *et al.*, 2001) and studied in detail (Puglisi *et al.*, 2005; Visco *et al.*, 2005), the linear behavior reported in several experiments or numerical simulations results from the too small values of $I_\tau/\langle I \rangle$ that are probed when $\tau \gg \tau_c$. Large enough values are obtained in the present experiment and the expected nonlinear behavior is thus reached. The shape of the curve in Fig. 2.8 is found in good agreement with the analytical calculation (Farago, 2002) performed with a Langevin-type equation with white noise.

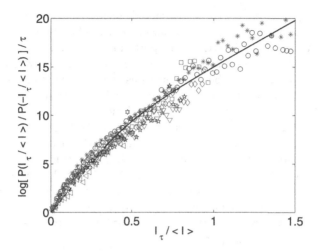

Fig. 2.8. Plot of $\frac{1}{\tau} \log \frac{P(I_\tau/\langle I\rangle)}{P(-I_\tau/\langle I\rangle)}$ for $16 < \tau/\tau_c < 39$ $[\tau/\tau_c = 17$ (*), 19.5 (o), 22 ($\Box$), 25 ($\Diamond$), 28 (pentagram), 30.5 (∇), 33.5 (hexagram), 39 ($\triangleleft$)]. Langevin model of Farago (2002): 4γ for $I_\tau/\langle I\rangle \leq 1/3$ (dashed line) and $7\gamma I_\tau/(4\langle I\rangle) + 3\gamma/2 - \gamma\langle I\rangle/(4I_\tau)$ for $I_\tau/\langle I\rangle \geq 1/3$ (solid line) with $\gamma = 5\,\mathrm{Hz}$.

Finally, we emphasize that a fluctuating injected power implies fluctuations of the energy flux at all wavenumbers in the energy cascade from injection to dissipation. In any system where an energy flux cascades from the injected power at large scales to dissipation at small scales, one has for the energy $E_<$ for wavenumbers smaller than k within the inertial range, $\dot{E}_< = I(t) - \Phi(k,t) \equiv R$, where $\Phi(k,t)$ is the energy flux at k toward large wavenumbers. Thus, $\int_0^\infty \langle R(\tau)R(0)\rangle d\tau = 0$ in order to prevent the divergence of $\langle E_<^2 \rangle$. Dimensionaly, this implies that $\sigma_\Phi^2 \tau_k$ does not depend on k, where σ_Φ is the standard deviation of the energy flux and τ_k is its correlation time. If this dimensional scaling is correct, fluctuations of the energy flux are expected to increase during the cascade from large to small scales since τ_k decreases. Such fluctuations have been found numerically and experimentally in hydrodynamic turbulence (Cerutti and Meneveau, 1998; Tao *et al.*, 2002). To which extent this is related or modified by small scale intermittency (Falcon *et al.*, 2007b) remains an open question.

2.5. Conclusion

The key governing parameter of wave turbulence is the energy flux that drives the waves and cascades to small scales through nonlinear interactions.

This quantity is assumed to be conserved during the cascades across the scales. It is only recently that measurements of the injected power into wave turbulence systems have been performed at the scale of the wave maker (integral scale). Fluctuations of the injected power much larger than the mean value have been observed as well as instantaneous negative events that occur with a fairly large probability. Taking into account these energy flux fluctuations in theoretical models of cascades remains an open problem. Moreover, it is likely that only an unknown fraction of the power injected into the system cascades through the scales, the rest being directly dissipated at various scales. To which extent, this could explain the discrepancy with theory for the observed scaling law of the power spectrum with the injected power, is an open problem that deserves more studies. For instance, an experimental challenge is to measure the evolution of the energy flux at various scales of the cascade using fully resolved space–time measurements of wave amplitude.

Acknowledgment

This work has been supported by ANR Turbulence 12-BSO4-005.

Bibliography

Aumaître S, Fauve S, McNamara S, Poggi P. 2001. Power injected in dissipative systems and the fluctuation theorem. *Eur. Phys. J. B* 19: 449.

Aumaître S, Farago J, Fauve S, McNamara S. 2004. Energy and power fluctuations in vibrated granular gases. *Eur. Phys. J. B* 42: 255.

Brazhnikov MYu, Kolmakov GV, Levchenko AA. 2002. The turbulence of capillary waves on the surface of liquid hydrogen. *Sov. Phys JETP* 95: 447–454.

Brazhnikov MYu, Kolmakov GV, Levchenko AA, Mezhov-Deglin LP. 2002. Observation of capillary turbulence on the water surface in a wide range of frequencies. *Europhys. Lett.* 58: 510–516.

Cerutti S, Meneveau C. 1998. Intermittency and relative scaling of subgrid-scale energy dissipation in isotropic turbulence. *Phys. Fluids* 10: 928.

Ciliberto S, Douady S, Fauve S. 1991. Investigating space-time chaos in Faraday instability by means of the fluctuations of the driving acceleration. *Europhys. Lett.* 15: 23.

Cobelli P, Petitjeans P, Maurel A, Pagneux V, Mordant N. 2009. Space-time resolved wave turbulence in a vibrating plate. *Phys. Rev. Lett.* 103: 204301.

Cobelli P, Przadka A, Petitjeans P, Lagubeau G, Pagneux V, Maurel A. 2011. Different regimes for water wave turbulence. *Phys. Rev. Lett.* 107: 214503.

Connaughton C, Nazarenko S, Newell AC. 2003. Dimensional analysis and weak turbulence. *Physica D* 184: 86–97.

Deike L, Laroche C, Falcon E. 2011. Experimental study of the inverse cascade in gravity wave turbulence. *Europhys. Lett.* 96: 34004.

Deike L, Berhanu M, Falcon E. 2012. Decay of capillary wave turbulence. *Phys. Rev. E.* 85: 066311.

Denissenko P, Lukaschuk S, Nazarenko S. 2007. Gravity wave turbulence in a laboratory flume. *Phys. Rev. Lett.* 99: 014501.

Düring G, Falcón C. 2009. Symmetry induced Four-wave capillary wave turbulence. *Phys. Rev. Lett.* 103: 174503.

Evans DJ, Cohen EGD, Morriss GP. 1993. Probability of second law violations in shearing steady states. *Phys. Rev. Lett.* 71: 2401.

Falcon E, Laroche C, Fauve S. 2007. Observation of gravity-capillary wave turbulence. *Phys. Rev. Lett.* 98: 094503.

Falcon E, Fauve S, Laroche C. 2007. Observation of intermittency in wave turbulence. *Phys. Rev. Lett.* 98: 154501.

Falcón C, Falcon E, Bortolozzo U, Fauve S. 2009. Capillary wave turbulence on a spherical fluid surface in zero gravity. *Europhys. Lett.* 86: 14002.

Farago J. 2002. Injected power fluctuations in Langevin equation. *J. Stat. Phys.* 107: 781.

Farago J. 2004. Power fluctuations in stochastic models of dissipative systems. *Physica A* 331: 69.

Gallavotti G, Cohen EGD. 1995. Dynamical ensembles in nonequilibrium statistical mechanics. *Phys. Rev. Lett.* 74: 2694.

Herbert E, Mordant N, Falcon E. 2010. Observation of the nonlinear dispersion relation and spatial statistics of wave turbulence on the surface of a fluid. *Phys. Rev. Lett.* 105: 144502.

Issenmann B, Falcon E. 2013. Gravity wave turbulence revealed by horizontal vibrations of the container. *Phys. Rev. E* 87:011001(R).

Kurchan J. 1998. Fluctuation theorem for stochastic dynamics. *J. Phys. A* 31: 3719.

Labbé R, Pinton JF, Fauve S. 1996. Power fluctuations in turbulent swirling flows. *J. Phys II France* 6: 1099.

Lommer M, Levinsen MT. 2002. Using laser-induced fluorescence in the study of surface wave turbulence. *J. Fluoresc.* 12: 45–50.

Nazarenko S, Lukaschuk S, McLelland S, Denissenko P. 2010. Statistics of surface gravity wave turbulence in the space and time domains. *J. Fluid Mech.* 642: 6902080.

Newell AC, Zakharov VE. 1992. Rough sea foam. *Phys. Rev. Lett.* 69: 1149–1151.

Puglisi A, *et al.* 2005. Fluctuations of internal energy flow in a vibrated granular gas. *Phys. Rev. Lett.* 95: 110202.

Risken H. 1996. *The Fokker-Planck Equation.* Berlin: Springer-Verlag.

Snouck D, Westra M-T, van de Water W. 2009. Turbulent parametic surface waves. *Phys. Fluids* 21: 025102.

Tao B, Katz J, Meneveau C. 2002. Statistical geometry of subgrid-scale stresses determined from holographic particle image velocimetry measurements, *J. Fluid Mech.* 457: 35.

Visco P, *et al.* 2005. Injected power and entropy flow in a heated granular gas. *Europhys Lett.* 72: 55.

Wright WB, Budakian R, Putterman SJ. 1996. Diffusing light photography of fully developed isotropic ripple turbulence. *Phys. Rev. Lett.* 76: 4528–4531.

Xia H, Shats M, Punzmann H. 2010. Modulation instability and capillary wave turbulence. *Europhys Lett.* 91: 14002.

Zakharov VE. 1967. Weak turbulence in media with a decay spectrum. *J. Appl. Mech. Tech. Phys.* 6(4): 22–24.

Zakharov VE, Filonenko NN. 1967. Weak turbulence of capillary waves. *J. App. Mech. Tech. Phys.* 8: 37–40.

Zakharov VE, Filonenko NN. 1967. Energy spectrum for stochastic oscillations of the surface of a liquid. *Sov. Phys. Dokl.* 11: 881–884.

Chapter 3

Wave Turbulence in Astrophysics

Sebastien Galtier

Institut d'Astrophysique Spatiale (IAS),
Université Paris-sud, Orsay, France
Sebastien.Galtier@ias.u-psud.fr

This chapter reviews the recent progress made mainly during the last two decades in wave turbulence in astrophysical plasmas (MHD, Hall MHD and electron MHD) in the incompressible and compressible cases. The emphasis is on homogeneous and anisotropic turbulence which usually provides the best theoretical framework to investigate space and laboratory plasmas. The interplanetary medium and the solar atmosphere are presented as two examples of media where anisotropic wave turbulence is relevant. The most important results of wave turbulence are reported and discussed in the context of space and simulated magnetized plasmas. Important issues and possible spurious interpretations are eventually discussed.

Contents

3.1. Introduction

Weak wave turbulence is the study of the long-time statistical behavior of a sea of weakly nonlinear dispersive waves (Zakharov *et al.*, 1992; Nazarenko, 2011). The energy transfer between waves occurs mostly among resonant sets of waves and the resulting energy distribution, far from a thermodynamic equilibrium, is characterized by a wide power-law spectrum and a high Reynolds number. This range of wavenumbers — the inertial range — is generally localized between large scales at which energy is injected in to the system (sources) and small scales at which waves break or dissipate (sinks). Pioneering works on wave turbulence[1] date back to the sixties when it was established that the stochastic initial value problem for weakly coupled wave systems has a natural asymptotic closure induced by the dispersive nature of the waves and the large separation of linear and nonlinear time scales (Benney and Saffman, 1966; Benney and Newell, 1967, 1969). In the meantime, Zakharov and Filonenko (1966) showed that the wave kinetic equations derived from the wave turbulence analysis (with a Gaussian Ansatz applied to the four-point correlations of the wave amplitude) have exact equilibrium solutions which are the thermodynamic zero flux solutions but also — and more importantly — finite flux solutions which describe the transfer of conserved quantities between sources and sinks. The solutions, first published for isotropic turbulence (Zakharov, 1965; Zakharov and Filonenko, 1966), were then extended to anisotropic turbulence (Kuznetsov, 1972).

---

[1]For simplicity, we will use "wave turbulence" instead of "weak wave turbulence" but it is an abuse of the language.

Wave turbulence is a very common natural phenomenon with applications, for example, in capillary waves (Kolmakov *et al.*, 2008; Abdurakhimov *et al.*, 2004) gravity waves (Falcon *et al.*, 2007), super-fluid helium and processes of Bose–Einstein condensation (Kolmakov and Pokrovsky, 1995; Lvov *et al.*, 2003), nonlinear optics (Dyachenko *et al.*, 1992), inertial waves (Galtier, 2003; Morize *et al.*, 2005) or Alfvén waves (Galtier *et al.*, 2000; Kuznetsov, 2001; Chandran, 2005). The most important difference between plasmas and incompressible neutral fluids is the plethora of linear waves supported by the former. The direct consequence is that in the weakly nonlinear plasmas, the fundamental entities are waves rather than the eddies of strong turbulence (Kolmogorov, 1941; Krommes, 2002). In the situation discussed here — magnetized plasmas seen as fluids — wave and strong turbulence may coexist and therefore, both waves and eddies have in fact an impact on the nonlinear dynamics which is strongly anisotropic. Anisotropic turbulence is particularly well observed in space plasmas since a magnetic field is often present on the largest scale of the system, like in the inner interplanetary medium where the magnetic field lines form an Archimedean spiral near the equatorial plane (Goldstein and Roberts, 1999), at the solar surface where coronal loops and open magnetic flux tubes are found (Cranmer *et al.*, 2007) or in planetary magnetospheres where shocks and discontinuities are measured (Sahraoui *et al.*, 2006).

In the present chapter, we review the wave turbulence in astrophysical plasmas. A plasma is a gaseous state of matter in which the atoms or molecules are strongly ionized. Mutual electromagnetic forces between the ions and the free electrons are then playing dominant roles which adds to the complexity as compared to the situation in neutral atomic or molecular gases. In the situation discussed here, the plasmas are described by the magnetohydrodynamics (MHD) approximation in the incompressible or compressible case. The role played by the Hall term is discussed through the Hall MHD description as well as its small-scale limit of the electron MHD.

The structure of the chapter is as follows. Physical motivations for developing wave turbulence theories are given in Sec. 3.2, where we first describe multiscale solar wind turbulence, and then present the coronal heating problem. Section 3.3, emphasizes the differences between strong and wave turbulence in MHD and the path followed historically by researches to finally obtain the MHD wave turbulence theories. In Sec. 3.4, the wave turbulence formalism is exposed with the basic ideas to derive the wave kinetic equations. Section 3.5 deals with the results obtained under different approximations (MHD, Hall MHD and electron MHD) in the incompressible or compressible case. Finally, we conclude with a general discussion in the last section.

3.2. Waves and Turbulence in Space Plasmas

Waves and turbulence are ubiquitous in astrophysical plasmas. Their signatures are found in the Earth's magnetosphere (Sahraoui *et al.*, 2006), the solar corona (Chae *et al.*, 1998), the solar wind (Bruno and Carbone, 2005) or the interstellar medium (Elmegreen and Scalo, 2004; Scalo and Elmegreen, 2004). These regions are characterized by extremely large (magnetic) Reynolds numbers, up to 10^{13}, with a range of available scales from 10^{18} m to a few meters.

3.2.1. *Interplanetary medium*

Extensive investigations are made in the interplanetary medium (and in the Earth's magnetosphere which is not the subject discussed here) where many *in situ* spacecrafts measurements are available. The solar wind plasma is found to be in a highly turbulent state with magnetic and velocity fluctuations detected from 10^{-6} Hz up to several hundred Hz (Coleman, 1968; Roberts *et al.*, 1987; Leamon *et al.*, 1998; Smith *et al.*, 2006). The turbulent state of the solar wind was first suggested in 1968 (Coleman, 1968) when a power-law behavior was reported for energy spectra with spectral indices lying between -1 and -2 (with the use of the Taylor "frozen-in flow" hypothesis). More precise measurements revealed that the spectral index at low frequency (<1 Hz) is often about -1.7 which is closer to the Kolmogorov prediction (Kolmogorov, 1941) for neutral fluids ($-5/3$) rather than the Iroshnikov–Kraichnan prediction (Iroshnikov, 1964; Kraichnan, 1965) for magnetized fluids ($-3/2$). Alfvén waves are also well observed since 1971 (Belcher and Davis, 1971) with a strong domination of antisunward propagative waves at short heliocentric distances (less than 1 AU). Since pure (plane) Alfvén waves are exact solutions of the ideal incompressible MHD equations (see e.g. Pouquet, 1993), nonlinear interactions should be suppressed if only one type of waves is present. Therefore, sunward Alfvén waves, although subdominant, play an important role in the internal solar wind dynamics.

The variance analysis of the magnetic field components and of its magnitude shows clearly that the magnetic field vector of the (polar) solar wind has a varying direction but with only a weak variation in magnitude (Forsyth *et al.*, 1996). Typical values give a normalized variance of the field magnitude smaller than 10% whereas for the components it can be as large as 50%. In these respects, the inner interplanetary magnetic field may be seen as a vector lying approximately around an Archimedean spiral direction with only weak magnitude variations (Barnes, 1981). Solar wind anisotropy with more power perpendicular to the mean magnetic field than that parallel to it, is pointed out by data analysis (Klein *et al.*, 1993) that provides a ratio of power up to 30. From single-point spacecraft measurements, it is not possible to specify the exact three-dimensional form of the spectral tensor of the magnetic or velocity fluctuations. However, it is possible to show that the spacecraft-frame spectrum may depend on the angle between the local magnetic field direction and the flow direction (Horbury *et al.*, 2008). In the absence of such data, a quasi two-dimensional model was proposed (Bieber *et al.*, 1996) in which wave vectors are nearly perpendicular to the large-scale magnetic field. It is found that about 85% of solar wind turbulence possesses a dominant 2D component. Additionally, solar wind anisotropies are detected through radio wave scintillations which reveal that density spectra close to the Sun are highly anisotropic with irregularities stretched out mainly along the radial direction (Armstrong *et al.*, 1990).

More recently, for frequencies larger than 1 Hz, a steepening of the magnetic fluctuation power-law spectra is observed over more than two decades (Coroniti *et al.*, 1982; Denskat *et al.*, 1983; Leamon *et al.*, 1998; Bale *et al.*, 2005; Smith *et al.*, 2006) with a spectral index mainly between -2.5 and -2.8. This new inertial range seems to be characterized by a bias of the polarization suggesting that these fluctuations are likely to be right-hand polarized, outward propagating waves (Goldstein *et al.*, 1994). Various indirect lines of evidence indicate that these waves propagate at large angles to the background magnetic field and that the power in fluctuations parallel to the background magnetic field is still less than the perpendicular one (Coroniti *et al.*, 1982; Leamon *et al.*, 1998). For these reasons, it is thought (Stawicki *et al.*, 2001) that Alfvén — left circularly polarized — fluctuations are suppressed by proton cyclotron damping and that the high frequency power-law spectra are likely to consist of Alfvén whistler waves (also called kinetic Alfvén waves) (Rogers *et al.*, 2001; Sahraoui *et al.*, 2012). This scenario is supported by multi-dimensional direct numerical simulations of compressible Hall MHD turbulence in the presence of an ambient field (Ghosh *et al.*, 1996) where a steepening of the spectra was found on a narrow range of wavenumbers, and associated with the appearance of right circularly polarized fluctuations. This result has been confirmed numerically with a turbulent cascade (shell) model based on 3D Hall MHD in which a well-extended steeper power-law spectrum was found at scale smaller than the ion skin depth (Galtier and Buchlin, 2007). (Note that in this cascade model no mean magnetic field is assumed.) However, the exact origin of the change of statistical behavior is still under debate (Markovskii *et al.*, 2008): for example, an origin from compressible effects is possible in the context of Hall MHD (Alexandrova *et al.*, 2008); a gyrokinetic description is also proposed in which kinetic Alfvén waves play a central role (Howes *et al.*, 2008).

The solar wind plasma is currently the subject of a new extensive research around the origin of the spectral break observed in the magnetic fluctuations accompanied sometimes by the absence of intermittency (Kiyani *et al.*, 2009). We will see that wave turbulence may have a central role in the sense that it is a useful point of departure for understanding the detailed physics of solar wind turbulence. In particular, it gives strong results in regards to the possible multiscale behavior of magnetized plasmas as well as the intensity of the anisotropic transfer between modes.

3.2.2. *Solar atmosphere*

Although it is not easy to measure directly the coronal magnetic field, it is now commonly accepted that the structure of the low solar corona is mainly due to the magnetic field (Aschwanden *et al.*, 2001). The high level of coronal activity appears through a perpetual impulsive reorganization of magnetic structures over a large range of scales, from about 10^5 km until the limit of resolution, about one arcsec (<726 km). The origin of the coronal reorganization is currently widely studied in solar physics. Information about the solar corona comes from spacecraft missions like SoHO/ESA or TRACE/NASA launched in the 1990s, or from more recent spacecrafts STEREO/NASA, Hinode/JAXA or SDO/NASA (see Fig. 3.1). The most recent observations reveal that coronal loops are not yet resolved transversely and have to be seen as tubes made of a set of strands which radiate alternatively. In fact, it is very likely that structures at much smaller scales exist but have not yet been detected (see e.g. Warren, 2006).

Observations in UV and X-ray show a solar corona extremely hot with temperatures exceeding 10^6 K — close to hundred times the solar surface temperature. These coronal temperatures are highly inhomogeneous: in the quiet corona much of the plasma lies near 1–2×10^6 K and 1–8×10^6 K in active regions. Then, one of the major questions in solar physics concerns the origin of such high values of coronal temperature. The energy available in the photosphere — characterized by granules — is clearly sufficient to supply the total coronal losses (Priest, 1982) which is about $10^4 \, \mathrm{J\,m^{-2}s^{-1}}$ for active regions and about one or two orders of magnitude smaller for the quiet corona and coronal holes where open magnetic field lines emerge. The main issue is thus to understand how the available photospheric energy is transferred and accumulated in the solar corona, and by what processes it is dissipated.

In active region loops, analyses made by spectrometers show that the plasma velocity can reach values up to 50 km/s (Brekke *et al.*, 1997). The highly dynamical nature of some coronal loops is also pointed out by nonthermal velocities reaching sometimes 50 km/s as it was revealed

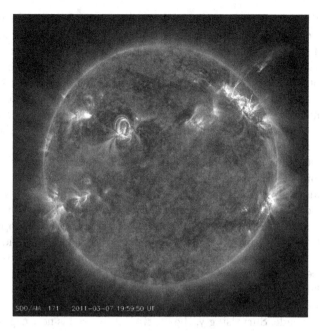

Fig. 3.1. Solar corona seen at 17.1 nm by the AIA instrument onboard SDO (Solar Dynamics Observatory) — NASA credits. The spatial resolution of the picture is about 700 km.

for example by SoHO/ESA (Chae *et al.*, 1998). These observations also give evidences that the line broadening is due to motions which are still not resolved neither in space, with scales smaller than the diameter of coronal loops, nor in time, with timescales shorter than the exposure time of the order of few seconds. These velocity measurements are very often interpreted as a signature of the MHD turbulence where small scales are produced naturally via a nonlinear cascade of energy. In light of the most recent observations, it seems fundamental to study, both theoretically and numerically, the impact of small-scale phenomena on the coronal heating. Note that recent Hinode/JAXA and SDO/NASA pictures seem to show a magnetic field controlled by plasma turbulence at all scales in which Alfvén waves are omnipresent (see e.g. Doschek *et al.*, 2007; Nishizuka *et al.*, 2008; Cargill and De Moortel, 2011). Thus, the turbulent activity of the corona is one of the key issues to understand the heating processes.

In the framework of turbulence, the energy supplied by the photospheric motions and transported by Alfvén waves through the corona is transferred towards smaller and smaller scales by nonlinear coupling between modes (the so-called energy cascade) until dissipative scales are reached from which the energy is converted into heating. The main coronal structures considered in such a scenario are the magnetic loops which cover the solar surface. Each loop is basically an anisotropic bipolar structure anchored in the photosphere. It forms a tube of magnetic fields in which the dense and hot matter is confined. Because a strong guiding magnetic field ($\mathbf{B_0}$) is present, the nonlinear cascade that occurs is strongly anisotropic with small scales mainly developed in the $\mathbf{B_0}$ transverse planes. Most of the models published deals with isotropic MHD turbulence (see e.g. Hendrix and Van Hoven, 1996) and it is only very recently that anisotropy has been included in the turbulent heating models (Buchlin and Velli, 2007).

The latest observations show that waves and turbulence are among the main ingredients of the solar coronal activity. The weak MHD turbulence is now invoked and has a possible regime for some coronal loops since a very small ratio is found between the fluctuating magnetic field and the axial component (Rappazzo *et al.*, 2007, 2008). Inspired by the observations and by recent direct numerical simulations of 3D MHD turbulence (Bigot *et al.*, 2008b), an analytical model of coronal structures has been proposed (Bigot *et al.*, 2008c) where the heating is seen as the end product of a wave turbulent cascade. Surprisingly, the heating rate found is non-negligible and may explain the observational predictions.

The coronal heating problem also concerns the regions where the fast solar wind is produced, i.e. the coronal holes (Hollweg and Isenberg, 2002; Cranmer *et al.*, 2007). Observations seem to show that the heating affects preferentially the ions in the direction perpendicular to the mean magnetic field. The electrons are much cooler than the ions, with temperatures generally less than or close to 10^6 K (see e.g. David *et al.*, 1998). Additionally, the heavy ions become hotter than the protons within a solar radius of the coronal base. Ion cyclotron waves could be the agent which heats the coronal ions and accelerates the fast wind. Naturally, the question of the origin of these high frequency waves arises. Among different scenarios, turbulence appears to be a natural and efficient mechanism to produce ion cyclotron waves. In this case, the Alfvén waves launched at low altitude with frequencies in the MHD range, would develop a (weak) turbulent cascade (Verdini *et al.*, 2012) to finally degenerate and produce ion cyclotron waves at much higher frequencies. In that context, the wave turbulence regime was considered in the weakly compressible MHD case at low-β plasmas (where β is the ratio between the thermal and magnetic pressure) in order to analyze the nonlinear three-wave interaction transfer to high frequency waves (Chandran, 2005). The wave turbulence calculation shows — in the absence of slow magnetosonic waves — that MHD turbulence provides a convincing explanation for the anisotropic ion heating.

3.3. Turbulence and Anisotropy

This section is first devoted to the comparison between wave and strong turbulence. In particular, we will see how the theoretical questions addressed at the end of the 20th century have led to the emergence of a large number of papers on wave turbulence in magnetized plasmas and to many efforts to characterize the fundamental role of anisotropy.

3.3.1. *Navier–Stokes turbulence*

Navier–Stokes turbulence is basically a strong turbulence problem in which it is impossible to perform a (nontrivial and) consistent linearization of the equations against a stationary homogeneous background. We remind that wave turbulence demands the existence of linear (dispersive) propagative waves as well as a large separation of linear and nonlinear (eddy-turnover) time scales (see e.g. Benney and Newell, 1969). In his third 1941 turbulence paper, Kolmogorov (1941) found that an exact and nontrivial relation may be derived from Navier–Stokes equations for the third-order longitudinal structure function (Kolmogorov, 1941): it is the so-called four-fifth's law. Because of the rarity of such results, the Kolmogorov's four-fifth's law is considered as one of the most important results in turbulence (Frisch, 1995). Basically, this law makes the following link in the 3D physical space between a two-point measurement, separated by a distance $\mathbf{r}$, and the distance itself[2]

$$-\frac{4}{5}\varepsilon^v r = \langle (v_L' - v_L)^3 \rangle, \tag{3.1}$$

where $\langle \rangle$ denotes an ensemble average, the longitudinal direction L is the one along the vector separation $\mathbf{r}$, v is the velocity and ε^v is the mean (kinetic) energy injection or dissipation rate per unit mass. To obtain this exact relation, the assumptions of homogeneity and isotropy are made (Batchelor, 1953). The former assumption is satisfied as long as we are at the heart of the fluid (far from the boundaries) and the latter is also satisfied if no external agent (like, for example, rotation or stratification) is present. Additionally, the long time limit is considered for which a stationary state is reached with a finite ε^v and the infinite Reynolds number limit (i.e. the viscosity $\nu \to 0$) is taken for which the mean energy dissipation rate per unit mass tends to a finite positive limit. Therefore, the exact prediction is valid in the asymptotic limit of a large inertial range. The Kolmogorov law is well supported by the experimental data (see e.g. Frisch, 1995).

---

[2]The Kolmogorov four-fifth's law can be written in an equivalent way as a four-third's law (Antonia *et al.*, 1997; Rasmussen, 1999), namely: $-(4/3)\varepsilon^v r = \langle (v_L' - v_L) \sum_i (v_i' - v_i)^2 \rangle$.

The four-fifth's law is a fundamental result used to develop heuristic spectral scaling laws like the famous — but not exact — 5/3-Kolmogorov energy spectrum. This point makes a fundamental difference with wave turbulence where the power-law spectra found are exact solutions of the asymptotically exact wave turbulence equations. Nevertheless, the term "Kolmogorov theory" is often associated to the $-5/3$ spectrum since there exists a theory behind in the physical space.

3.3.2. *Incompressible MHD turbulence*

3.3.2.1. *Strong turbulence*

The wave turbulence regime exists in the incompressible MHD. The main reason is that Alfvén waves are linear solutions when a stationary homogeneous background magnetic field $\mathbf{B_0}$ is applied. This statement seems to be obvious but we will see that the problem is subtle and the existence of an Alfvén wave turbulence theory was the subject of many discussions basically because those waves are only pseudo-dispersive (i.e. the frequency ω is proportional to a wavenumber).

The question of the existence of an exact relation for third-order structure functions is naturally addressed for strong (without $\mathbf{B_0}$) MHD turbulence. The answer was given by Politano and Pouquet only in 1998 (see also, Chandrasekhar, 1951) for the incompressible MHD turbulence. The presence of the magnetic field and its coupling with the velocity field renders the problem more difficult and, in practice, we are dealing with a couple of equations. In this case, the possible formulation in 3D is the four-third's law

$$-\frac{4}{3}\varepsilon^{\pm}r = \left\langle \left(z_L'^{\mp} - z_L^{\mp}\right) \sum_i \left(z_i'^{\pm} - z_i^{\pm}\right)^2 \right\rangle, \qquad (3.2)$$

where the direction L is still the one along the vector separation $\mathbf{r}$, $\mathbf{z}^{\pm} = \mathbf{v} \pm \mathbf{b}$ is the Elsässer fields (with $\mathbf{b}$ normalized to a velocity field) and $\varepsilon^{\pm}$ is the mean energy dissipation rate per unit mass associated to the Elsässer energies. To obtain these exact relations, the assumptions of homogeneity and isotropy are still made, and the long time limit for which a stationary state is reached with a finite $\varepsilon^{\pm}$ is also considered. The infinite kinetic and magnetic Reynolds number limit ($\nu \to 0$ and the magnetic diffusivity $\eta \to 0$) for which the mean energy dissipation rates per unit mass have a finite positive limit is eventually taken. Therefore, the exact prediction is again valid, at first order, in a wide inertial range. This prediction is currently often used in the literature to analyze space plasma data and determine the local heating rate $\varepsilon^{\pm}$ in the solar wind (see e.g. Sorriso-Valvo *et al.*, 2007; MacBride *et al.*, 2008).

Fig. 3.2.   Alfvén wave packets propagating along a large scale magnetic field $\mathbf{B_0}$.

3.3.2.2. *Iroshnikov–Kraichnan spectrum*

The isotropy assumption used to derive the four-third's law is stronger for magnetized than neutral fluids, since in most of the situations encountered in astrophysics, a large scale magnetic field $\mathbf{B_0}$ is present which leads to anisotropy (see Sec. 3.2). Although this law is a fundamental result that may be used to develop a heuristic spectral scaling law, the role of B_0 has to be clarified. Indeed, we have now two time-scales: the eddy-turnover time and the Alfvén time. The former is similar to the eddy-turnover time in Navier–Stokes turbulence and may be associated to the distortion of wave packets (the basic entity in MHD), whereas the latter may be seen as the duration of interaction between two counter-propagating Alfvén wave packets (see Fig. 3.2). During a collision, there is a deformation of the wave packets in such a way that the energy is transferred mainly at smaller scales. The multiplicity of collisions leads to the formation of a well-extended power-law energy spectrum whose index lies between $-5/3$ (Kolmogorov prediction) and $-3/2$ (Iroshnikov–Kraichnan prediction) according to the phenomenology used, i.e. with or without the Alfvén wave effect. Note the slightly different approaches followed by Iroshnikov and Kraichnan to derive the $-3/2$ spectrum. In the former case, the presence of a strong magnetic field is explicitly assumed, whereas it is not in the latter case where it is claimed that the small-scale fluctuations see the large-scales — in the sub-inertial range — as a spatially uniform magnetic field. However, they both assumed isotropy to finally propose the so-called Iroshnikov–Kraichnan spectrum for MHD turbulence. It is important to note that the exact isotropic relation (3.2) in physical space corresponds dimensionally to a $-5/3$ energy spectrum since it is cubic in z (or in v and b) and linear in r. It is, therefore, less justified to use the term "theory" for the Iroshnikov–Kraichnan spectrum than for the Kolmogorov one for neutral fluids.

3.3.2.3. *Breakdown of isotropy*

The weakness of the Iroshnikov–Kraichnan phenomenology is the apparent contradiction between the (in)direct presence of a strong uniform magnetic field and the assumption of isotropy. An important difference between neutral and magnetized fluids is the impossibility in the latter case to remove a large-scale (magnetic) field by a Galilean transform. The role of a uniform magnetic field has been widely discussed in the literature and, in particular, during the last three decades (see e.g. Montgomery and Turner, 1981; Shebalin *et al.*, 1983; Matthaeus *et al.*, 1996; Ng and Bhattacharjee, 1996; Verma, 2004). At strong B_0 intensity, one of the most clearly established results is the bi-dimensionalization of MHD turbulent flows with a strong reduction of nonlinear transfers along B_0. The consequence is an energy concentration near the plane $k \cdot B_0 = 0$, a result illustrated later on by direct numerical simulations in two and three space dimensions (Shebalin *et al.*, 1983).

The effects of a strong uniform magnetic field may be handled through an analysis of resonant triadic interactions (Shebalin *et al.*, 1983) between the wavevectors (k, p, q) which satisfy the vectorial relation

$$k = p + q, \tag{3.3}$$

whereas the associated wave frequencies satisfy the scalar relation

$$\omega(k) = \omega(p) + \omega(q). \tag{3.4}$$

For incompressible MHD, the Alfvén frequency is

$$\omega(k) = \pm k \cdot B_0 = \pm k_{\parallel} B_0, \tag{3.5}$$

where $\parallel$ defines the direction along B_0 ($\perp$ will be the perpendicular direction to B_0 which is written in velocity unit). The solution of the three-wave resonance condition gives, for example, $q_{\parallel} = 0$, which implies a spectral transfer only in the perpendicular direction. For a strength of B_0 well above the rms level of the kinetic and magnetic fluctuations, the nonlinear interactions of Alfvén wave packets may dominate the dynamics of the MHD flow leading to the regime of wave turbulence (see Sec. 3.4), where the energy transfer, stemming from three-wave resonant interactions, can only increase the perpendicular component of the wavevectors, while the nonlinear transfers are completely inhibited along B_0. The end result is a strongly anisotropic flow.

3.3.2.4. *Emergence of anisotropic laws*

An important issue discussed in the literature is the relationship between perpendicular and parallel scales in the anisotropic MHD turbulence (see e.g. Higdon, 1984; Goldreich and Sridhar, 1995; Boldyrev, 2006; Sridhar, 2010). In order to take into account anisotropy, Goldreich and Sridhar (1995) proposed a heuristic model based on the conjecture of a critical balance between the Alfvén and the eddy-turnover times, which are respectively

$$\tau_A \sim \ell_\parallel / B_0 \qquad (3.6)$$

and

$$\tau_{\text{eddy}} \sim \ell_\perp / u_\ell, \qquad (3.7)$$

where $\ell_\parallel$ and $\ell_\perp$ are typical length scales parallel and perpendicular to $\mathbf{B_0}$. The conjecture says that in the inertial range we have, $\tau_A = \tau_{\text{eddy}}$. The latter relation leads trivially to, $u_\ell \sim B_0 \ell_\perp / \ell_\parallel$. Following the Kolmogorov arguments, one ends up trivially with the Kolmogorov energy spectrum

$$E(k_\perp, k_\parallel) \sim k_\perp^{-5/3} k_\parallel^{-1}, \qquad (3.8)$$

(where $\mathbf{k} \equiv (\mathbf{k}_\perp, k_\parallel)$, $k_\perp \equiv |\mathbf{k}_\perp|$, $\iint E(k_\perp, k_\parallel) dk_\perp dk_\parallel = \int E(k_\perp) dk_\perp$), and with the nontrivial anisotropic scaling law (with $u_\ell^2 / \tau_{\text{eddy}} = $ constant)

$$k_\parallel \sim k_\perp^{2/3}. \qquad (3.9)$$

This heuristic prediction means that anisotropy is stronger at smaller scales.

A generalization of this result was proposed by Galtier *et al.* (2005) in order to model the MHD flows both in the wave and strong turbulence regimes, as well as for the transition between them. In this heuristic model, the constant time-scale ratio $\chi = \tau_A / \tau_{\text{eddy}}$ is not assumed to be necessarily equal to unity. The relaxation of the constraint ($\chi = 1$) allows us to recover the anisotropic scaling law (3.9) which now includes B_0,

$$k_\parallel \sim k_\perp^{2/3} / B_0, \qquad (3.10)$$

and to find a universal prediction for the total energy spectrum

$$E(k_\perp, k_\parallel) \sim k_\perp^{-\alpha} k_\parallel^{-\beta}, \qquad (3.11)$$

with

$$3\alpha + 2\beta = 7. \qquad (3.12)$$

Note that a classical calculation with a transfer time $\tau = \tau_{\text{eddy}}^2 / \tau_A = \tau_{\text{eddy}} / \chi$ leads trivially to the spectrum (3.8): the choice of τ fixes irreversibly the spectrum. It is only when the ansatz (3.11) is introduced that the nontrivial relation (3.12) emerges; this ansatz is believed to be weak since power-laws are the natural solutions of the problem. According to direct numerical simulations (see e.g. Cho and Vishniac, 2000; Maron and Goldreich, 2001; Ng *et al.*, 2003; Shaikh and Zank, 2007; Bigot *et al.*, 2008b), the anisotropic scaling law between parallel and perpendicular scales (3.9) seems to be a robust result and an approximately constant ratio χ, generally smaller than one, is found between the Alfvén and the eddy-turnover times. This sub-critical value of χ implies therefore a dynamics mainly driven by Alfvén waves interactions. Note that the presence of B_0 in relation (3.10) shows the convergence towards wave turbulence ($B_0 \to +\infty$, with respect to the fluctuations) for which the parallel transfer is totally frozen.

The question of the spectral indices is still a challenging problem in anisotropic turbulence (Sagaut and Cambon, 2008). The main conclusion of Bigot *et al.* (2008b) is that the difficulty to make the measurements is generally underestimated in a sense that the scaling prediction in $k_\perp$ may change significantly when $E(k_\perp, k_\parallel)$ is plotted at a given $k_\parallel$ instead of $E(k_\perp)$: indeed, the slow mode $E(k_\perp, k_\parallel = 0)$ may play a singular role in the dynamics with a scaling in $k_\perp$ different from the one given by the 3D modes $E(k_\perp, k_\parallel > 0)$. This comment holds primarily for direct numerical simulations where technically it is currently possible to make this distinction, observations being still far from this possibility. This point will be further discussed in the last section. Note finally that all these spectral predictions suffer from rigorous justifications and the word "theory" that we find very often in the literature is not justified at all. A breakthrough could be achieved if one could develop the equivalent of an exact four-fifth's law for anisotropic MHD turbulence. Then, a dimensional derivation from this law could lead to an anisotropic spectral prediction and in some sense, for the first time, the possibility of having a theoretical link between strong and wave turbulence. Recently, an attempt has been made in that direction by using the idea of a critical balance for third-order moments (Galtier, 2012). Moreover, it was shown that the introduction of the dynamic alignment conjecture into the exact relation for third-order moments (Politano and Pouquet, 1998) may give a $k_\perp^{-3/2}$ spectrum (Boldyrev *et al.*, 2009).

3.3.3. *Towards an Alfvén wave turbulence theory*

In view of the importance of anisotropy in natural magnetized plasma (see Sec. 3.2), Sridhar and Goldreich (1994) suggested that a plasma evolving in a medium permeated by a strong uniform magnetic field and in the regime of wave turbulence is characterized by four-wave nonlinear interactions. The essence of wave turbulence is the statistical study of large ensembles of weakly interacting waves via a systematic asymptotic expansion in the powers of small nonlinearity. This technique leads finally to the exact derivation of wave kinetic equations for the invariants of the system like the energy spectrum (see Sec. 3.4). In MHD, it is the presence of a strong uniform magnetic field $\mathbf{B_0}$ that allows to introduce a small parameter in the system, namely the ratio between the fluctuating fields and B_0. The result found by Sridhar and Goldreich in 1994 implies that the asymptotic development has no solution at the main order and that we need to go to the next order to describe the Alfvén wave turbulence. Several articles, using a phenomenology (see e.g. Montgomery and Matthaeus, 1995; Verma, 2004) or a rigorous treatment (Ng and Bhattacharjee, 1996), were published to contest this conclusion and sustain the nontrivial character of the three-wave interactions. In response, a detailed theory was finally given in 2000 (Galtier *et al.*, 2000; Nazarenko *et al.*, 2001; Galtier *et al.*, 2002) whose main prediction may be derived heuristically in few lines as follows. According to Fig. 3.2, the main process which happens in Alfvén wave turbulence is the stochastic collisions of wavepackets. To find the transfer time and then the energy spectrum, first, we shall evaluate the modification of a wavepacket produced by one collision. We have (for simplicity we only consider the balanced case for which $z_\ell^\pm \sim z_\ell \sim u_\ell \sim b_\ell$)

$$z_\ell(t + \tau_A) \sim z_\ell(t) + \tau_A \frac{\partial z_\ell}{\partial t} \sim z_\ell(t) + \tau_A \frac{z_\ell^2}{\ell_\perp}, \qquad (3.13)$$

where τ_A is the duration of one collision; in other words, after one collision the distortion of a wavepacket is $\Delta_1 z_\ell \sim \tau_A z_\ell^2 / \ell_\perp$. This distortion is going to increase with time in such a way that after N stochastic collisions, the cumulative effect may be evaluated like a random walk

$$\sum_{i=1}^{N} \Delta_i z_\ell \sim \tau_A \frac{z_\ell^2}{\ell_\perp} \sqrt{\frac{t}{\tau_A}}. \qquad (3.14)$$

The transfer time τ_{tr} that we are looking for is the one for which the cumulative distortion is of the order of one, i.e. of the order of the wavepacket itself

$$z_\ell \sim \tau_A \frac{z_\ell^2}{\ell_\perp} \sqrt{\frac{\tau_{\text{tr}}}{\tau_A}}, \qquad (3.15)$$

then we obtain

$$\tau_{\text{tr}} \sim \frac{1}{\tau_A} \frac{\ell_\perp^2}{z_\ell^2} \sim \frac{\tau_{\text{eddy}}^2}{\tau_A}. \qquad (3.16)$$

A classical calculation, with $\varepsilon \sim z_\ell^2 / \tau_{\text{tr}}$, leads finally to the energy spectrum

$$E(k_\perp, k_\parallel) \sim \sqrt{\varepsilon B_0}\, k_\perp^{-2} k_\parallel^{-1/2}, \qquad (3.17)$$

where $k_\parallel$ has to be seen as a parameter since no transfer along the parallel direction is expected (see Sec. 3.3.2.3). Note that this demonstration is traditionally used for deriving the Iroshnikov–Kraichnan spectrum, but since in this case isotropy is assumed, a $E(k) \sim \sqrt{\varepsilon B_0}\, k^{-3/2}$ is predicted.

First signatures that may be attributed to wave turbulence were found by Perez and Boldyrev (2008) in numerical simulations of a reduced form of the MHD equations (Galtier and Chandran, 2006). The detection of the wave turbulence regime from direct numerical simulations of the MHD equations is still a difficult task but recent results have been obtained in which temporal, structural and spectral signatures are reported (Bigot *et al.*, 2008a, 2008b; Bigot and Galtier, 2011). Currently, efforts are made to analyze the effects of other inviscid invariants, like the cross-helicity, on the scaling laws of wave turbulence (Lithwick and Goldreich, 2003; Chandran, 2008). It is worth noting that these works on the imbalanced wave turbulence have been followed by (sometimes controversial) research investigations in the strong turbulence regime which is of great relevance for the solar wind (see e.g. Chandran *et al.*, 2009; Beresnyak and Lazarian, 2010; Podesta and Bhattacharjee, 2010).

3.3.4. *Wave turbulence in compressible MHD*

Most of the investigations devoted to wave turbulence refers to isotropic media where the well-known conformal transform proposed by Zakharov and Filonenko (1966) may be applied to find the so-called Kolmogorov–Zakharov spectra (see Sec. 3.4.4). (Surprisingly, a similar transform was used in the meantime by Kraichnan (1967) to investigate the problem of 2D turbulence.) The introduction of anisotropy in plasmas was studied to a smaller extent: the first example is given by the magnetized ion-sound waves (Kuznetsov, 1972). The compressible MHD case was analyzed later by Kuznetsov (2001) for a situation where the plasma (thermal) pressure is small compared with the magnetic pressure (small β limit). In this case, the main nonlinear interaction involving the MHD waves is the scattering of a fast magneto-acoustic and Alfvén waves on slow magneto-acoustic waves. In this process, the fast and Alfvén waves act as high-frequency waves with respect to the slow waves. (To simplify the analysis, other three-wave interaction processes that do not involve slow waves are neglected.)

A variant of the wave turbulence analysis in compressible MHD was proposed by Chandran (2005) in the limit of a small β in which the slow waves are neglected and a constant density is imposed. The other (mathematical) difference is that the Hamiltonian formalism was used in the former analysis whereas a Eulerian description was employed in the latter case. Because the compressible regime is much more difficult to analyze than the incompressible one, simplifications have been made and to date, no general theory has been proposed for the compressible MHD wave turbulence.

3.3.5. *Wave turbulence in Hall and electron MHD*

Modeling the physics of a turbulent plasma beyond the MHD approximation, namely on spatial scales shorter than the ion inertial length d_i (but larger than the electron inertial length d_e) and time scales shorter than the ion cyclotron period $1/\omega_{ci}$, is a highly challenging problem even in the fluid approximation. (For kinetic models, we can mention the gyrokinetic approximation useful for weakly collisional plasmas in which time scales are supposed to be much larger than $1/\omega_{ci}$ (see e.g. Schekochihin *et al.*, 2009), whereas a fluid model including some kinetic effects with, in particular, a Landau fluid closure has been developed (see e.g. Hunana *et al.*, 2011).) In that context, the electron MHD approximation (Kingsep *et al.*, 1990) is often used: in such a limit, one assumes that ions do not have the time to follow electrons and provide a static homogeneous background on which electrons move. The electron MHD approximation is particularly relevant in the context of collisionless magnetic reconnection where a diffusion region is developed with multiscale structures corresponding to ion and electron characteristic lengths (Biskamp, 1997; Bhattacharjee, 2004; Yamada, 2007).

An important issue in the electron MHD turbulence is about the impact of whistler waves on the dynamics. Biskamp *et al.* (1999) argued that although whistler wave propagation effects are non-negligible in electron MHD turbulence, the local spectral energy transfer process is independent of the linear wave dispersion relation and the energy spectrum may be predicted by a Kolmogorov type argument. Direct numerical simulations were used to illustrate the theory but no mean magnetic field was introduced (Biskamp *et al.*, 1996; Ng *et al.*, 2003). Dastgeer *et al.* (2000) investigated the turbulence regime in the presence of a moderate background magnetic field B_0 and provided convincing numerical evidence that the turbulence is anisotropic. It was argued that although whistler waves may appear to play a negligible role in determining the spectral index, they are important in setting up an anisotropic turbulent cascade. The whistler wave turbulence regime was then investigated theoretically by Galtier and Bhattacharjee (2003, 2005) in the limit $B_0 \to +\infty$ (with respect to the fluctuations). It was shown that similar to the MHD case, anisotropy is a central feature of such a turbulence. Attempts to find an anisotropic law for the electron MHD was made by Cho and Vishniac (2004) (see also, Galtier *et al.*, 2005) and a scaling relation in $k_\parallel \sim k_\perp^{1/3}$ was found both heuristically and numerically.

Hall MHD is an extension of the standard MHD where the ion inertia is retained in Ohm's law. It provides a multiscale description of magnetized plasmas from which both standard and electron MHD approximations may be recovered. Hall MHD is often used to understand, for example, the magnetic field evolution in neutron star crusts (Goldreich and Reisenegger, 1992), the turbulent dynamo (Mininni *et al.*, 2003), the formation of filaments (Laveder *et al.*, 2002), the multiscale solar wind turbulence (Ghosh *et al.*, 1996; Krishan and Mahajan, 2004; Galtier, 2006a, 2006b), or the dynamics of the magnetosheath (Belmont and Rezeau, 2001). Anisotropy in Hall MHD is clearly less understood than in MHD or electron MHD mainly because the numerical treatment is more limited since a wide range of scales is necessary to detect any multiscale effects. From a theoretical point of view, it is only recently that a wave turbulence theory has been achieved for the incompressible case (Galtier, 2006b; Sahraoui *et al.*, 2007). For such a turbulence, a global tendency towards anisotropy was found (with, however, a weaker effect at intermediate scales) with nonlinear transfers preferentially in the direction perpendicular to the external magnetic field $\mathbf{B_0}$. The energy spectrum is characterized by two inertial ranges, separated by a knee, which are exact solutions of the wave kinetic equations. The position of the knee corresponds to the scale where the Hall term becomes sub/dominant. To date, the compressible Hall MHD is still an open problem in the regime of wave turbulence. (A first step was made by Sahraoui *et al.* (2003) who found the Hamiltonian system.)

3.4. Wave Turbulence Formalism

3.4.1. *Wave amplitude equation*

Wave turbulence is the study of the long time statistical behavior of a sea of weakly nonlinear dispersive waves. It is described by the wave kinetic equations. In this section, we present the wave turbulence formalism which leads to these nonlinear equations. We shall use the inviscid model equation

$$\frac{\partial \mathbf{u}(\mathbf{x}, t)}{\partial t} = \mathcal{L}(\mathbf{u}) + \varepsilon \mathcal{N}(\mathbf{u}, \mathbf{u}), \qquad (3.18)$$

where $\mathbf{u}$ is a stationary random vector, $\mathcal{L}$ is a linear operator which insures that waves are solutions of the linear problem, and $\mathcal{N}$ is a quadratic nonlinear operator (like for MHD-type fluids). The factor ε is a small parameter $(0 < \varepsilon \ll 1)$ that will be used for the weakly nonlinear expansion. For all the applications considered here, the smallness of the nonlinearities is the result of the presence of a strong uniform magnetic field $\mathbf{B_0}$; the operator $\mathcal{L}$ is thus proportional to B_0.

We introduce the 3D direct and inverse Fourier transforms

$$\mathbf{u}(\mathbf{x}, t) = \int_{\mathbf{R}^3} \mathbf{A}(\mathbf{k}, t) \exp(i\mathbf{k} \cdot \mathbf{x}) d\mathbf{k}, \qquad (3.19)$$

$$\mathbf{A}(\mathbf{k}, t) = \frac{1}{(2\pi)^3} \int_{\mathbf{R}^3} \mathbf{u}(\mathbf{x}, t) \exp(-i\mathbf{k} \cdot \mathbf{x}) d\mathbf{x}. \qquad (3.20)$$

Therefore, a Fourier transform of Eq. (3.18) gives for the j-component

$$\left(\frac{\partial}{\partial t} + i\omega(\mathbf{k})\right) A_j(\mathbf{k}, t) = \varepsilon \int_{\mathbf{R}^6} \mathcal{H}_{jmn}^{\mathbf{kpq}} A_m(\mathbf{p}, t) A_n(\mathbf{q}, t) \delta(\mathbf{k} - \mathbf{p} - \mathbf{q}) d\mathbf{p} d\mathbf{q},$$

(3.21)

where $\omega(\mathbf{k}) = \omega_k$ is given by the appropriate dispersion relation (within general $\omega(-\mathbf{k}) = -\omega(\mathbf{k})$) and $\mathcal{H}$ is a symmetric function in its vector arguments which basically depends on the quadratic nonlinear operator $\mathcal{N}$. Note the use of the Einstein's notation. We introduce

$$\mathbf{A}(\mathbf{k}, t) = \mathbf{a}(\mathbf{k}, t) e^{-i\omega_k t},$$

(3.22)

and obtain in the interaction representation

$$\frac{\partial a_j(\mathbf{k})}{\partial t} = \varepsilon \int_{\mathbf{R}^6} \mathcal{H}_{jmn}^{\mathbf{kpq}} a_m(\mathbf{p}) a_n(\mathbf{q}) e^{i\Omega_{k,pq} t} \delta_{k,pq} d\mathbf{p} d\mathbf{q},$$

(3.23)

where the Dirac delta function $\delta_{k,pq} = \delta(\mathbf{k} - \mathbf{p} - \mathbf{q})$ and $\Omega_{k,pq} = \omega_k - \omega_p - \omega_q$; the time dependence in fields, $\mathbf{a}$, is omitted for simplicity. Relation (3.23) is the wave amplitude equation whose dependence in ε means that the weak nonlinearities will modify the wave amplitude slowly, in time. By nature, the problems considered here (in MHD, electron and Hall MHD) involve mainly three-wave interaction processes as it is expected by the form of the wave amplitude equation. The exponentially oscillating term is essential for the asymptotic closure since we are interested in the long time statistical behavior for which the nonlinear transfer time is much greater than the wave period. In such a limit, most of the nonlinear terms will be destroyed by the random phase mixing and only a few of them — called the resonance terms — will survive. Before going to the statistical formalism, we note the following general properties that will be used

$$\mathcal{H}_{jmn}^{\mathbf{kpq}} = \left(\mathcal{H}_{jmn}^{-\mathbf{k}-\mathbf{p}-\mathbf{q}}\right)^*,$$

(3.24)

$$\mathcal{H}_{jmn}^{\mathbf{kpq}} \text{ is symmetric in } (\mathbf{p}, \mathbf{q}) \text{ and } (m, n),$$

(3.25)

$$\mathcal{H}_{jmn}^{\mathbf{0pq}} = 0,$$

(3.26)

where, *, stands for the complex conjugate.

3.4.2. *Statistics and asymptotics*

We turn now to the statistical description, introduce the ensemble average $\langle \ldots \rangle$ and define the density tensor for homogeneous turbulence

$$q_{jj'}(\mathbf{k}') \delta(\mathbf{k} + \mathbf{k}') = \langle a_j(\mathbf{k}) a_{j'}(\mathbf{k}') \rangle.$$

(3.27)

We also assume that on average $\langle \mathbf{u}(\mathbf{x},t) \rangle = 0$ which leads to the relation $\mathcal{H}^{0\mathbf{pq}}_{jmn} = 0$. From the nonlinear equation (3.23), we find

$$\frac{\partial q_{jj'}\delta(k+k')}{\partial t} = \left\langle a_{j'}(\mathbf{k}')\frac{\partial a_j(\mathbf{k})}{\partial t}\right\rangle + \left\langle a_j(\mathbf{k})\frac{\partial a_{j'}(\mathbf{k}')}{\partial t}\right\rangle$$

$$= \varepsilon \int_{\mathbf{R}^6} \mathcal{H}^{\mathbf{kpq}}_{jmn}\langle a_m(\mathbf{p})a_n(\mathbf{q})a_{j'}(\mathbf{k}')\rangle e^{i\Omega_{k,pq}t}\delta_{k,pq}d\mathbf{p}d\mathbf{q}$$

$$+ \varepsilon \int_{\mathbf{R}^6} \mathcal{H}^{\mathbf{k'pq}}_{j'mn}\langle a_m(\mathbf{p})a_n(\mathbf{q})a_j(\mathbf{k})\rangle e^{i\Omega_{k',pq}t}\delta_{k',pq}d\mathbf{p}d\mathbf{q}.$$

$$(3.28)$$

A hierarchy of equations will clearly appear which gives for the third-order moment equation

$$\frac{\partial\langle a_j(\mathbf{k})a_{j'}(\mathbf{k}')a_{j''}(\mathbf{k}'')\rangle}{\partial t}$$

$$= \varepsilon \int_{\mathbf{R}^6} \mathcal{H}^{\mathbf{kpq}}_{jmn}\langle a_m(\mathbf{p})a_n(\mathbf{q})a_{j'}(\mathbf{k}')a_{j''}(\mathbf{k}'')\rangle e^{i\Omega_{k,pq}t}\delta_{k,pq}d\mathbf{p}d\mathbf{q}$$

$$+ \varepsilon \int_{\mathbf{R}^6}\{(\mathbf{k},j) \leftrightarrow (\mathbf{k}',j')\}d\mathbf{p}d\mathbf{q} + \varepsilon\int_{\mathbf{R}^6}\{(\mathbf{k}'',j'') \leftrightarrow (\mathbf{k}',j')\}d\mathbf{p}d\mathbf{q},$$

$$(3.29)$$

where in the right-hand side, the second line means an interchange in the notations between two pairs with the first line as a reference, and the third line means also an interchange in the notations between two pairs with the second line as a reference. At this stage, we may write the fourth-order moment in terms of a sum of the fourth-order cumulant plus products of second-order ones, but a natural closure arises for times asymptotically large (see e.g. Newell *et al.*, 2001; Nazarenko, 2011; Newell and Rumpf, 2011). In this case, several terms do not contribute at large times like, in particular, the fourth-order cumulant which is not a resonant term. In other words, the nonlinear regeneration of third-order moments depends essentially on products of second-order moments. The time scale separation imposes a condition of applicability of wave turbulence which has to be checked *in fine* (see e.g. Nazarenko, 2007). After integration in time, we are left with

$$\langle a_j(\mathbf{k})a_{j'}(\mathbf{k}')a_{j''}(\mathbf{k}'')\rangle$$

$$= \varepsilon \int_{\mathbf{R}^6} \mathcal{H}^{\mathbf{kpq}}_{jmn}(\langle a_m(\mathbf{p})a_n(\mathbf{q})\rangle\langle a_{j'}(\mathbf{k}')a_{j''}(\mathbf{k}'')\rangle$$

$$+ \langle a_m(\mathbf{p})a_{j'}(\mathbf{k}')\rangle\langle a_n(\mathbf{q})a_{j''}(\mathbf{k}'')\rangle$$

$$+ \langle a_m(\mathbf{p})a_{j''}(\mathbf{k''})\rangle \langle a_n(\mathbf{q})a_{j'}(\mathbf{k'})\rangle) \Delta(\Omega_{k,pq})\delta_{k,pq}d\mathbf{p}d\mathbf{q}$$

$$+ \varepsilon \int_{\mathbf{R}^6} \{(\mathbf{k},j) \leftrightarrow (\mathbf{k'},j')\}d\mathbf{p}d\mathbf{q}$$

$$+ \varepsilon \int_{\mathbf{R}^6} \{(\mathbf{k''},j") \leftrightarrow (\mathbf{k'},j')\}d\mathbf{p}d\mathbf{q}, \tag{3.30}$$

where

$$\Delta(\Omega_{k,pq}) = \int_0^{t \gg 1/\omega} e^{i\Omega_{k,pq}t'}dt' = \frac{e^{i\Omega_{k,pq}t} - 1}{i\Omega_{k,pq}}. \tag{3.31}$$

The same convention as in (3.29) is used. After integration in wave vectors $\mathbf{p}$ and $\mathbf{q}$ and simplification, we get

$$\langle a_j(\mathbf{k})a_{j'}(\mathbf{k'})a_{j''}(\mathbf{k''})\rangle$$

$$= \varepsilon\Delta(\Omega_{kk'k''})\delta_{kk'k''}\left(\mathcal{H}_{jmn}^{\mathbf{k}-\mathbf{k'}-\mathbf{k''}}q_{mj'}(\mathbf{k'})q_{nj''}(\mathbf{k''})\right.$$

$$+ \mathcal{H}_{jmn}^{\mathbf{k}-\mathbf{k''}-\mathbf{k'}}q_{mj''}(\mathbf{k''})q_{nj'}(\mathbf{k'}) + \mathcal{H}_{j'mn}^{\mathbf{k'}-\mathbf{k}-\mathbf{k''}}q_{mj}(\mathbf{k})q_{nj''}(\mathbf{k''})$$

$$+ \mathcal{H}_{j'mn}^{\mathbf{k'}-\mathbf{k''}-\mathbf{k}}q_{mj''}(\mathbf{k''})q_{nj}(\mathbf{k}) + \mathcal{H}_{j''mn}^{\mathbf{k''}-\mathbf{k}-\mathbf{k'}}q_{mj}(\mathbf{k})q_{nj'}(\mathbf{k'})$$

$$\left.+ \mathcal{H}_{j''mn}^{\mathbf{k''}-\mathbf{k'}-\mathbf{k}}q_{mj'}(\mathbf{k'})q_{nj}(\mathbf{k})\right). \tag{3.32}$$

The symmetries (3.25) lead to

$$\langle a_j(\mathbf{k})a_{j'}(\mathbf{k'})a_{j''}(\mathbf{k''})\rangle$$

$$= 2\varepsilon\Delta(\Omega_{kk'k''})\delta_{kk'k''}\left(\mathcal{H}_{jmn}^{\mathbf{k}-\mathbf{k'}-\mathbf{k''}}q_{mj'}(\mathbf{k'})q_{nj''}(\mathbf{k''})\right.$$

$$\left.+ \mathcal{H}_{j'mn}^{\mathbf{k'}-\mathbf{k}-\mathbf{k''}}q_{mj}(\mathbf{k})q_{nj''}(\mathbf{k''}) + \mathcal{H}_{j''mn}^{\mathbf{k''}-\mathbf{k}-\mathbf{k'}}q_{mj}(\mathbf{k})q_{nj'}(\mathbf{k'})\right). \tag{3.33}$$

The latter expression may be introduced into (3.28). We take the long time limit (which introduces irreversibility) and find

$$\Delta(x) \to \pi\delta(x) + i\mathcal{P}(1/x), \tag{3.34}$$

with $\mathcal{P}$ the principal value of the integral. We finally obtain the asymptotically exact wave kinetic equations

$$\frac{\partial q_{jj'}(\mathbf{k})}{\partial t} = 4\pi\varepsilon^2 \int_{\mathbf{R}^6} \delta_{k,pq}\delta(\Omega_{k,pq})\mathcal{H}_{jmn}^{\mathbf{kpq}}\left(\mathcal{H}_{mrs}^{\mathbf{p}-\mathbf{q}-\mathbf{k}}q_{rn}(\mathbf{q})q_{j's}(\mathbf{k})\right.$$

$$\left.+ \mathcal{H}_{nrs}^{\mathbf{q}-\mathbf{pk}}q_{rm}(\mathbf{p})q_{j's}(\mathbf{k}) + \mathcal{H}_{j'rs}^{-\mathbf{k}-\mathbf{p}-\mathbf{q}}q_{rm}(\mathbf{p})q_{sn}(\mathbf{q})\right)d\mathbf{p}d\mathbf{q}. \tag{3.35}$$

These general 3D wave kinetic equations are valid, in principle, for any situation where three-wave interaction processes are dominant; only the form of $\mathcal{H}$ has to be adapted to the problem. Equation for the (total) energy is obtained by taking the trace of the tensor density, $q_{jj}(\mathbf{k})$, whereas other inviscid invariants are found by including nondiagonal terms.

3.4.3. *Wave kinetic equations*

Equation (3.35) is the wave kinetic equation for the spectral tensor components. We see that the nonlinear transfer is based on a resonance mechanism since we need to satisfy the relations

$$\omega_k = \omega_p + \omega_q, \tag{3.36}$$

$$\mathbf{k} = \mathbf{p} + \mathbf{q}. \tag{3.37}$$

The solutions define the resonant manifolds which may have different forms according to the flow. For example, in the limit of the weakly compressible MHD, when the sound speed is much greater than the Alfvén speed, it is possible to find (for the shape of the resonant manifolds in the 3D k-space) spheres or tilted planes for Fast-Fast-Alfvén and Fast-Fast-Slow wave interactions, and rays (a degenerescence of the resonant manifolds) for Fast-Fast-Fast wave interactions (Galtier *et al.*, 2001). We also find planes perpendicular to the uniform magnetic field $\mathbf{B_0}$ for Slow-Slow-Slow, Slow-Slow-Alfvén, Slow-Alfvén-Alfvén or Alfvén-Alfvén-Alfvén wave interactions; it is a similar situation to the incompressible MHD turbulence (since, at first order, slow waves have the same frequency as Alfvén waves) for which the resonant manifolds foliate the Fourier space.

The representation of the resonant manifolds is always interesting since it gives an idea of how the spectral densities can be redistributed along (or transverse to) the mean magnetic field direction whose main effect is the nonlinear transfer reduction along its direction (Matthaeus *et al.*, 1996). The previous finding was confirmed by a detailed analysis of the wave compressible MHD turbulence (Kuznetsov, 2001; Chandran, 2005) in the small β limit where the wave kinetic equations were derived as well as their exact power-law solutions. The situation for electron and Hall MHD is more subtle and there is no simple picture for the resonant manifolds like in MHD. In this case, it is nevertheless important to check if the resonance condition allows simple particular solutions in order to justify the domination of three-wave interaction processes over higher order (four-wave) processes.

The form of the wave kinetic equations (3.35) is the most general one for a dispersive problem as whistler wave turbulence in electron

MHD. The incompressible MHD system constitutes a unique example of pseudo-dispersive waves for which wave turbulence applies. In this particular case, some symmetries are lost and the principal value terms remain present. For that reason, incompressible MHD may be seen as a singular limit of incompressible Hall MHD (Galtier, 2006b).

3.4.4. *Finite flux solutions*

The most spectacular result of the wave turbulence theory is its ability to provide exact finite flux solutions. These solutions are found after applying to the wave kinetic equations a conformal transform proposed first by Zakharov (1965) for isotropic turbulence: it is the so-called Kolmogorov–Zakharov spectra. Because anisotropy is almost always present in magnetized plasmas, a bi-homogeneous conformal transform is more appropriate (Kuznetsov, 1972). This operation can only be performed if, first, one assumes axisymmetry. With this assumption, the wave kinetic equations (3.35) can be written as

$$
\frac{\partial \tilde{q}_{jj'}(k_\perp, k_\parallel)}{\partial t} = 4\pi\varepsilon^2 \int_{\mathbf{R}^4} \delta_{k,pq}\delta(\Omega_{k,pq})\tilde{\mathcal{H}}_{jmn}^{\mathbf{kpq}}\left(\tilde{\mathcal{H}}_{mrs}^{\mathbf{p-q-k}}\tilde{q}_{rn}(q_\perp, q_\parallel)\tilde{q}_{j's}(k_\perp, k_\perp)\right.
$$

$$
+ \tilde{\mathcal{H}}_{nrs}^{\mathbf{q-pk}}\tilde{q}_{rm}(p_\perp, p_\parallel)\tilde{q}_{j's}(k_\perp, k_\parallel)
$$

$$
\left. + \tilde{\mathcal{H}}_{j'rs}^{\mathbf{-k-p-q}}\tilde{q}_{rm}(p_\perp, p_\parallel)\tilde{q}_{sn}(q_\perp, q_\parallel)\right)dp_\perp dp_\parallel dq_\perp dq_\parallel, \quad (3.38)
$$

where $\tilde{q}_{jj'}(k_\perp, k_\parallel) = q_{jj'}(\mathbf{k})/(2\pi k_\perp)$ and $\tilde{\mathcal{H}}$ is a geometric operator. Note that we have performed an integration over the polar angle. Except for the incompressible MHD for which the wave kinetic equations simplify thanks to the absence of nonlinear transfer along the parallel direction (along $\mathbf{B_0}$), in general, we have to deal with a dynamics in the perpendicular and parallel directions with a relatively higher transfer transverse to $\mathbf{B_0}$ than along it. In this case, we may write the wave kinetic equations in the limit $k_\perp \gg k_\parallel$. The system of integro-differential equations obtained is then sufficiently reduced to allow us to extract the exact power-law solutions. We perform the conformal transformation to the equations for the invariant spectral densities (total energy, magnetic helicity, etc.)

$$
\begin{aligned}
p_\perp &\to k_\perp^2/p_\perp, \\
q_\perp &\to k_\perp q_\perp/p_\perp, \\
p_\parallel &\to k_\parallel^2/p_\parallel, \\
q_\parallel &\to k_\parallel q_\parallel/p_\parallel,
\end{aligned} \qquad (3.39)
$$

and we search for stationary solutions in the power-law form $k_\perp^{-n} k_\parallel^{-m}$. Basically, two types of solutions are found: the fluxless solution, also called the thermodynamic equilibrium solution, which corresponds to the equipartition state for which the flux is zero, and the finite flux solution which is the most interesting one. During the last decades, many papers have been devoted to the finding of these finite flux solutions for isotropic as well as anisotropic wave turbulence (Zakharov *et al.*, 1992; Nazarenko, 2011).

Recently, and thanks to high numerical resolutions, a new challenge has appeared in wave turbulence: the study of the incompressible Alfvén wave turbulence (Galtier *et al.*, 2000) reveals that the development of the finite flux solution of the wave kinetic equations is preceded by a front characterized by a significantly steeper scaling law. An illustration is given in Fig. 3.3 for the balance case ($E = E^+ = E^-$): the temporal evolution of the energy spectrum reveals the formation of a $k_\perp^{-7/3}$ front before the establishment of the Kolmogorov–Zakharov solution in $k_\perp^{-2}$. This finding is in contradiction with the theory proposed by Falkovich and Shafarenko (1991) on the nonlinear front propagation where the authors claimed that the Kolmogorov–Zakharov spectrum should be formed right behind the propagating front. The same observation was also made for the inverse cascade in the nonlinear Schrödinger equation (Lacaze *et al.*, 2001) and a detailed analysis was given by Connaughton *et al.* (2003) in the simplified case of strongly local interactions for which the usual wave kinetic equations

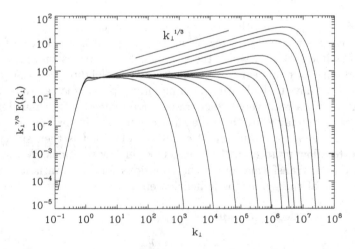

Fig. 3.3.   Time propagation (from left to right) of the compensated energy spectrum for incompressible Alfvén wave turbulence ($E = E^+ = E^-$): an anomalous scaling is observed before the formation of the Kolmogorov–Zakharov spectrum.

become simple PDEs (for MHD, see, Galtier and Buchlin, 2010). In spite of these attempts, the anomalous scaling is still an open problem without clear physical and proper mathematical answer.

3.5. Main Results and Predictions

3.5.1. *Alfvén wave turbulence*

We start to summarize the results of wave turbulence in the incompressible MHD case for which the wave kinetic equations are singular in the sense that the principal value terms remain except for the Elsässer energies $E^\pm$. As it was explained, the origin of this particularity is the pseudo-dispersive nature of the Alfvén waves which are a unique case where wave turbulence theory applies (see also Zakharov and Sagdeev, 1970). In the limit of strongly anisotropic turbulence for which $k_\perp \gg k_\parallel$, the wave kinetic equations (3.38) simplify. Then, the Elsässer energy spectra of the transverse fluctuations (shear-Alfvén waves) satisfy the asymptotic integro-differential equations

$$\frac{\partial E^\pm(k_\perp, k_\parallel)}{\partial t} = \frac{\pi \varepsilon^2}{B_0} \iint_\Delta \cos^2 \phi \sin \theta \frac{k_\perp}{q_\perp} E^\mp(q_\perp, 0)$$
$$\times \left[k_\perp E^\pm(p_\perp, k_\parallel) - p_\perp E^\pm(k_\perp, k_\parallel) \right] dp_\perp dq_\perp, \quad (3.40)$$

where Δ defines the domain of integration on which $\mathbf{k}_\perp = \mathbf{p}_\perp + \mathbf{q}_\perp$, ϕ is the angle between $\mathbf{k}_\perp$ and $\mathbf{p}_\perp$ and θ the angle between $\mathbf{k}_\perp$ and $\mathbf{q}_\perp$. A fundamental property of the Alfvén wave turbulence appears on equations (3.40): the evolution of the spectra $E^\pm$ is always mediated by an interaction with the slow mode $q_\parallel = 0$. This is a direct consequence of the resonance conditions discussed in Sec. 3.3.2.3. Then, the parallel wavenumber dependence does not affect the nonlinear dynamics: in other words, there is no transfer along $\mathbf{B_0}$ and $k_\parallel$ can be treated as an external parameter. Therefore, we introduce the notation $E^\pm(k_\perp, k_\parallel) = E^\pm(k_\perp) f_\pm(k_\parallel)$ where $f_\pm$ are arbitrary functions of $k_\parallel$ given by the initial conditions and with the assumption $f_\pm(0) = 1$, we obtain

$$\frac{\partial E^\pm(k_\perp)}{\partial t} = \frac{\pi \varepsilon^2}{B_0} \iint_\Delta \cos^2 \phi \sin \theta \frac{k_\perp}{q_\perp} E^\mp(q_\perp)$$
$$\times \left[k_\perp E^\pm(p_\perp) - p_\perp E^\pm(k_\perp) \right] dp_\perp dq_\perp, \quad (3.41)$$

which describes the transverse dynamics. The exact finite flux solutions of equations (3.41) for the stationary energy spectra are

$$E^\pm(k_\perp) \sim k_\perp^{n_\pm} \quad (3.42)$$

with

$$n_+ + n_- = -4 \,. \tag{3.43}$$

Additionally, the power-law indices must satisfy the condition of locality (i.e. the condition on the power-law solutions for which the wave kinetic equations remain finite)

$$-3 < n_\pm < -1 \,. \tag{3.44}$$

It is important to recall that wave turbulence is an asymptotic theory which must satisfy a condition of application. In our case, the transfer time must be significantly larger than the Alfvén time which means in terms of wavenumbers that $k_\parallel \gg \varepsilon^2 k_\perp$ (and that at small transverse scales turbulence becomes strong). At this point, a comment has to be made about the Kolmogorov–Zakharov solutions (3.43). These solutions imply the contribution of the wavevector $\mathbf{q}$ which is by nature a slow mode ($q_\parallel = 0$) whereas the other contributions $\mathbf{k}$ and $\mathbf{p}$ imply wave modes ($k_\parallel = p_\parallel > 0$). In Fig. 3.3, it is assumed that $E^+ = E^-$ and that the dynamics of the slow mode is given by the wave mode, hence, the stationary solution $n_+ = n_- = 2$. However, it may happen that the slow mode has its own dynamics which can belong to strong turbulence. Since relation (3.43) is exact, if the wave mode energy has a $-5/3$ scaling then the wave turbulence spectrum should be $-7/3$. Although this discussion was given in the original paper (Galtier *et al.*, 2000), it was not considered seriously until the most recent direct numerical simulations (Bigot *et al.*, 2008b; Bigot and Galtier, 2011) reveal that this situation may happen. We will come back to this point in the conclusion.

The locality of interactions (in Fourier space) is an important issue in MHD turbulence. According to some recent works in isotropic turbulence, nonlinear interactions seem to be more nonlocal in MHD than in a pure hydrodynamics in the sense that the transfer of energy from the velocity field to the magnetic field may be a highly nonlocal process in Fourier space (Alexakis *et al.*, 2005). The situation is different when we deal with anisotropic turbulence: in this case, interactions (between perpendicular wavevectors) are mainly local (Alexakis *et al.*, 2007). In wave turbulence, condition (3.44) has to be satisfied to validate the exact power-law solutions and avoid any divergence of integrals in the wave kinetic equations due to nonlocal contributions. In practice, numerical simulations of the wave kinetic equations have clearly shown that solutions (3.44) are attractive (Galtier *et al.*, 2000).

Recently, it was realized that the wave kinetic equations found in the anisotropic limit may be recovered without the wavenumber condition $k_\perp \gg k_\parallel$ if initially only the shear-Alfvén waves were considered (Galtier and

Chandran, 2006). Therefore, the finite flux solution may be extended to the entire wavenumber space which renders its detection easier. This idea was tested successfully against numerical simulations by Perez and Boldyrev (2008) who found spectral signatures of the Alfvén wave turbulence.

3.5.2. *Compressible MHD*

We turn now to the compressible regime for which two limits have been analyzed. The first case (case I) is the one for which the plasma (thermal) pressure is assumed to be small as compared to the magnetic pressure (small β limit) and where three-wave interaction processes that do not involve slow waves are neglected (Kuznetsov, 2001). In the second case (case II) for which we still have $\beta \ll 1$, slow waves are neglected and a constant density is imposed (Chandran, 2005). In both the situations, the general finite flux solutions are not obvious to express.

In case I, when only interactions between the Alfvén and slow waves are kept, a wave energy spectrum in $\sim k_\perp^{-2} k_\parallel^{-5/2}$ is found which corresponds to a (finite) constant energy flux solution. It is claimed that the addition of the interactions with the fast waves will lead to the same solution since the dynamics tends to produce strong anisotropic distributions of the waves concentrated in k-space within a narrow-angle cone in the $\mathbf{B_0}$ direction ($k_\perp \ll k_\parallel$). Under these conditions, the fast waves coincide with the Alfvén waves.

In case II, the general wave kinetic equations do not allow us to find exact power-law solutions. However, when only Fast-Fast-Fast interactions are kept, it is possible to find a finite flux solution for the fast wave 1D energy spectrum which scales as $\sim f(\theta) k^{-3/2}$, where $f(\theta)$ is an arbitrary function of the angle θ between the wave vector $\mathbf{k}$ and the uniform magnetic field $\mathbf{B_0}$. Numerical simulations of the general wave kinetic equations are made to find the behavior according to the angle θ. A solution close to $k_\perp^{-2}$ is found for the Alfvén wave 2D energy spectrum (for differently fixed $k_\parallel$) when $k_\perp \gg k_\parallel$. The k-spectra plotted at $\theta = 45°$ reveal a fast wave modal energy spectrum in $\sim k^{-3/2}$ and a steeper Alfvén wave spectrum, while for a small angle ($\theta = 7.1°$) both spectra follow the same scaling law steeper than $k^{-3/2}$ (Chandran, 2005).

3.5.3. *Whistler wave turbulence*

The electron MHD equations in the presence of a strong uniform magnetic field B_0 exhibit dispersive whistler waves. The wave turbulence regime was analyzed in the incompressible case by Galtier and Bhattacharjee (2003, 2005) who derived the wave kinetic equations by using a complex helicity decomposition. The strong anisotropic $(k_\perp \gg k_\parallel)$ finite flux solutions correspond to

$$E(k_\perp, k_\parallel) \sim k_\perp^{-5/2} k_\parallel^{-1/2} \tag{3.45}$$

for the magnetic energy spectrum, and

$$H(k_\perp, k_\parallel) \sim k_\perp^{-7/2} k_\parallel^{-1/2} \tag{3.46}$$

for the magnetic helicity spectrum. As for the other cases presented above, a direct cascade was found for the energy. In particular, it was shown that contrary to the MHD, the wave kinetic equations which involve three-wave interaction processes are characterized by a nonlinear transfer that decreases linearly with $k_\parallel$; for $k_\parallel = 0$, the transfer is exactly null. Thus, the 2D modes (or slow modes) decouple from the three-dimensional whistler waves. Such a decoupling is found in a variety of problems like rotating turbulence (Galtier, 2003; Sagaut and Cambon, 2008).

3.5.4. *Hall MHD*

The last example exposed in this chapter is the Hall MHD case which incorporates both the standard MHD and electron MHD limits. This system is much heavier to analyze in the regime of wave turbulence and it is only recently that a theory has been proposed (Galtier, 2006b). The general theory emphasizes the fact that the large scale limit of the standard MHD becomes singular with the apparition of a new type of terms, the principal value terms. Of course, the large scale and small scale limits tend to the appropriate theories (MHD and electron MHD wave turbulence), but in addition, it is possible to describe the connection between them at intermediate scales (scales of the order of the ion inertial length d_i). For example, moderate anisotropy is predicted at these intermediate scales whereas it is much stronger for other scales. It is also interesting to note that the small scale limit gives a system of equations richer than the pure electron MHD system (Galtier and Bhattacharjee, 2003) with the possibility to describe the sub-dominant kinetic energy dynamics. An exact power-law solution for the kinetic energy spectrum is found for ion cyclotron wave turbulence which scales as

$$E(k_\perp, k_\parallel) \sim k_\perp^{-5/2} k_\parallel^{-1/2} . \tag{3.47}$$

To date, the wave kinetic equations of Hall MHD have not been simulated numerically even in their simplified form (when helicity terms are discarded). It is an essential step to understand much better the dynamics at intermediate scales.

3.6. Conclusion and Perspectives

3.6.1. *Observations*

Waves and turbulence are two fundamental ingredients of many space plasmas. The most spectacular illustration of such characteristics is probably given by the observations of the Sun's atmosphere with the orbital solar observatory Hinode/JAXA and more recently by SDO/NASA launched in 2010. For the first time, detection of Alfvén waves is made through the small oscillations of many thin structures called threads (see Fig. 3.4). In the meantime, the highly dynamical nature of coronal loops is revealed by nonthermal velocities detected with spectrometers. These findings are considered as a remarkable step in our understanding of the solar coronal dynamics. Nowadays, it is believed that the Alfvén waves turbulence is a promising model to understand the heating of the solar corona (van Ballegooijen *et al.*, 2011).

The interplanetary medium is another example of magnetized plasma where waves and turbulence are detected. In this framework, the origin of the so-called "dissipative range", i.e. the extension of the turbulent inertial range beyond a fraction of hertz, is currently one of the main issues discussed in the community. Although a final answer is not given yet, wave turbulence is a promising regime to understand the inner solar wind dynamics in the sense that it gives exact results in regards to the possible multiscale behavior of magnetized plasmas as well as the intensity of the anisotropic transfer between modes.

The main feature of magnetized plasmas in the regime of wave turbulence is the omnipresence of anisotropy and the possibility to have different spectral scaling laws according to the space direction. To achieve a proper comparison between observational data and theoretical predictions, not only *in situ* measurements are necessary, but multipoints data also have

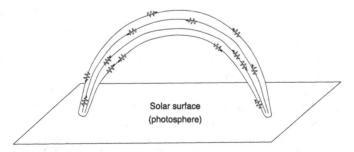

Fig. 3.4.   Schematic view of Alfvén wave turbulence on the Sun: coronal loops act as a resonant cavity for Alfvén wave packets which are transmitted from the deeper layers of the solar atmosphere.

to be accessible. It is at this price that the true nature of the magnetized plasmas will be revealed. The magnetosphere is the first medium where it is possible to perform such a comparison thanks to the Cluster/ESA mission (Sahraoui *et al.*, 2006). In the case of the Jupiter's magnetosphere, the large scale magnetic field is relatively strong. Actually, the first indirect evidence of the Alfvén wave turbulence was reported in Saur *et al.* (2002) by using five years set of Galileo/NASA spacecraft magnetic field data. (Since the work was based on a one point analysis a model was used to differentiate the perpendicular and parallel spectra.)

Perhaps, the first direct evidence of the Alfvén wave turbulence comes from solar (photospheric) magnetic field observations above active regions. Using data from SOHO/ESA, Abramenko (2005) observed that the magnetic energy spectra reduced to its radial component (along $\mathbf{B_0}$) have a transverse scaling significantly different from the Kolmogorov one with power-laws generally between -2 and -2.3. Since the magnetic field can only be detected in a thin layer (the temperature is too high in the chromosphere and in the corona to measure the magnetic field with the Zeeman effect), the magnetic spectra can be interpreted as a measure at a given $k_\parallel > 0$ (in practice, a small average is made in $k_\parallel$ since the data comes from slightly different altitudes). As we know from the Alfvén wave turbulence theory, parallel fluctuations follow the same dynamics as the perpendicular (shear-Alfvén) fluctuations and the addition of the power-law indices of the slow and wave mode spectra is -4. Then, these unexpected spectra may be seen as the direct manifestation of the Alfvén wave turbulence if the slow mode contribution scales like Kolmogorov (see the discussion on Sec. 3.5.1). If this interpretation is correct, it is probably the most important achievement of the Alfvén wave turbulence theory.

3.6.2. *Simulations*

Numerical simulation is currently the main tool to improve our knowledge on wave turbulence in magnetized plasmas since we are still limited by the observational (single point) data. Two types of simulations are available: the simulation of the wave kinetic equations and the direct numerical simulation of the original MHD-type equations. In the former case, it is a way to find, for instance, the spectral scaling laws when the wave kinetic equations are too complex to provide us the exact solutions after application of the usual conformal transform. An example is given by the compressible MHD: the numerical simulations revealed a variation of the power-law energy spectrum with the angle between $\mathbf{k}$ and $\mathbf{B_0}$ (Chandran, 2005). Simulations may also be useful to investigate wave turbulence when an external forcing is applied like in incompressible MHD (Galtier and Nazarenko, 2008).

For direct numerical simulation, the challenge is slightly different: indeed, in this case, the main goal is the measure of the transition between strong and wave turbulence, and thus between isotropic and anisotropic turbulence. The former regime has been extensively studied for more than three decades, whereas the latter is still a young subject. The main topic of such a simulation is also to find general properties that could help us to understand the single point measurements made in natural plasmas. We arrive here at the heart of current issues in wave turbulence. One of the most important points emphasized by recent direct numerical simulations in the incompressible MHD is the coexistence of wave and strong turbulence (Bigot *et al.*, 2008b). This characteristic should not be a surprise since basically wave turbulence is a perturbative theory which must satisfy conditions of applicability. In this case, the slow mode ($q_\parallel = 0$) may evolve differently from the wave modes ($k_\parallel, p_\parallel > 0$) since the former case may be characterized by strong turbulence and the latter by wave turbulence. This distinction is fundamental and not really taken into account in the community: Alfvén wave turbulence will not be fully revealed in space as well as simulated plasmas as long as strong and wave turbulence will not be separated. This point is illustrated with direct numerical simulations (Bigot *et al.*, 2008b; Bigot and Galtier, 2011) where steeper energy spectra may be found for the wave modes (between -2 and $-7/3$) compared to the slow mode (down to $-5/3$). The result seems to depend on the relative intensity of B_0: a strong magnetic background tends to increase the energy of the slow mode which, then, may evolve independently (the critical value of B_0 seems to be around 10 times the *r.m.s.* fluctuations). It is also found that the wave turbulence spectra are generally visible at a fixed $k_\parallel > 0$ and may be hidden in a $-5/3$ scaling if a summation over $k_\parallel$ is performed (this result could be due to the limited space resolution). The same investigation clearly shows that — as expected — an equipartition between the kinetic and magnetic energies is reached in the wave modes whereas the magnetic energy is dominant in the slow mode. Note that this Alfvén wave turbulence regime was also numerically found independently by Perez and Boldyrev (2008).

3.6.3. *Open questions*

In the light of such recent results, the following new questions may be addressed:

- Is the $-7/3$ spectrum more universal than -2 in the Alfvén wave turbulence?
- Is the slow mode ($q_\parallel = 0$) intermittent? Does it produce intermittency in the slave wave modes ($k_\parallel, p_\parallel > 0$)?

- Can we find direct signatures of the Alfvén wave turbulence in solar magnetic loops with three-dimensional magnetic field data?
- Is Alfvén wave turbulence the universal regime for stellar magnetic loops?
- Does the $-5/3$ energy spectrum observed in the solar wind correspond to a bias? What is the wave vector energy spectrum?
- Can we find an Alfvén wave turbulence regime in the youthful solar wind?
- Is the small energy ratio $E^u/E^b \sim 0.5$ found in the local solar wind (Bruno and Carbone, 2005) due essentially to the slow mode contribution in which case the wave modes may follow equipartition?
- Is the absence of intermittency and the -2.5 magnetic spectrum observed in the solar wind dispersive regime (Kiyani *et al.*, 2009) the signatures of Alfvén whistler wave turbulence?

Future missions like *Solar Orbiter* from the European Space Agency (scheduled for 2018) will certainly help us to answer these questions and make significant advances on plasma turbulence.

Bibliography

Abdurakhimov LV, Brazhnikov MY, Levchenko AA. 2008. Statistic of capillary waves on the surface of liquid hydrogen in a turbulent regime. *J. Low Temp. Phys.* 150: 431–434.

Abramenko VI. 2005. Relationship between magnetic power spectrum and flare productivity in solar active regions. *Astrophys. J.* 629: 1141–1149.

Alexakis A, Bigot B, Politano H, Galtier S. 2007. Anisotropic fluxes and nonlocal interactions in magnetohydrodynamic turbulence. *Phys. Rev. E* 76: 056313.

Alexakis A, Mininni PD, Pouquet A. 2005. Shell-to-shell energy transfer in magnetohydrodynamics. I. Steady state turbulence. *Phys. Rev. E* 72: 046301.

Alexandrova O, Carbone V, Veltri P, Sorriso-Valvo L. 2008. Small-scale energy cascade of the solar wind turbulence. *Astrophys. J.* 674: 1153–1157.

Antonia RA, Ould-Rouis M, Anselmet F, Zhu Y. 1997. Analogy between predictions of Kolmogorov and Yaglom. *J. Fluid Mech.* 332: 395–409.

Armstrong JW, Coles WA, Rickett BJ, Kojima M. 1990. Observations of field-aligned density fluctuations in the inner solar wind. *Astrophys. J.* 358: 685–692.

Aschwanden MJ, Poland AI, Rabin DM. 2001. The new solar corona. *Annu. Rev. Astron. Astrophys.* 39: 175–210.

Bale SD, Kellogg PJ, Mozer FS, Horbury TS, Reme H. 2005. Measurement of the electric field fluctuation spectrum of MHD turbulence. *Phys. Rev. Lett.* 94: 215002.

Barnes A. 1981. Interplanetary Alfvénic fluctuations: a stochastic model. *J. Geophys. Res.* 86: 7498–7506.

Batchelor GK. 1953. *The Theory of Homogeneous Turbulence.* Cambridge: Cambridge University Press.

Belcher JW, Davis L. 1971. Large-Amplitude Alfvén waves in the interplanetary medium. *J. Geophys. Res.* 76: 3534–3563.

Belmont G, Rezeau L. 2001. Magnetopause reconnection induced by magnetosheath Hall-MHD fluctuations. *J. Geophys. Res.* 106: 10751–10760.

Benney J, Saffman PG. 1966. Nonlinear interactions of random waves in a dispersive medium. *Proc. Roy. Soc. A* 289: 301–320.

Benney J, Newell AC. 1967. Sequential time closures for interacting random wavers. *J. Maths. Phys.* 46: 363–393.

Benney J, Newell AC. 1966. Random wave closures. *Stud. Appl. Math.* 48: 29–53.

Beresnyak A, Lazarian A. 2010. Scaling laws and diffuse locality of balanced and imbalanced magnetohydrodynamic turbulence. *Astrophys. J.* 722: L110–L113.

Bhattacharjee A. 2004. Impulsive magnetic reconnection in the earth's magnetotail and the solar corona. *Ann. Rev. Astron. Astrophys.* 42: 365–384.

Bieber JW, Wanner W, Matthaeus WH. 1996. Dominant two-dimensional solar wind turbulence with implications for cosmic ray transport. *J. Geophys. Res.* 101: 2511–2522.

Bigot B, Galtier S, Politano H. 2008a. Energy decay laws in strongly anisotropic MHD turbulence. *Phys. Rev. Lett.* 100: 074502.

Bigot B, Galtier S, Politano H. 2008b. Development of Anisotropy in Incompressible Magnetohydrodynamic Turbulence. *Phys. Rev. E* 78: 066301.

Bigot B, Galtier S, Politano H. 2008c. An anisotropic turbulent model for solar coronal heating. *Astron. Astrophys.* 490: 325–337.

Bigot B, Galtier S. 2011. Two-dimensional state in driven MHD turbulence. *Phys. Rev. E* 83: 026405.

Biskamp D. 1997. Collision and collisionless magnetic reconnection. *Phys. Plasmas* 4: 1964–1968.

Biskamp D, Schwarz E, Drake JF. 1996. 2D electron magnetohydrodynamic turbulence. *Phys. Rev. Lett.* 76: 1264–1267.

Biskamp D, Schwarz E, Zeiler A, Celani A, Drake JF. 1999. Electron magnetohydrodynamic turbulence. *Phys. Plasmas* 6: 751–758.

Boldyrev S. 2006. Spectrum of magnetohydrodynamic turbulence. *Phys. Rev. Lett.* 96: 115002.

Boldyrev S, Mason J, Cattaneo F. 2009. Dynamic alignment and exact scaling laws in magnetohydrodynamic turbulence. *Astrophys. J.* 699: L39–L42.

Brekke P, Kjeldseth-Moe O, Harrison RA. 1997. High-velocity flows in an active region loop system observed with the coronal diagnostic spectrometer on SOHO. *Sol. Phys.* 175: 511–521.

Bruno R, Carbone V. 2005. The solar wind as a turbulence laboratory. *Living Rev. Solar Phys.* 2: 1–186.

Buchlin É, Velli M. 2007. Shell models of RMHD turbulence and the heating of solar coronal loops. *Astrophys. J.* 662: 701–714.

Cargill P, De Moortel I. 2011. Waves galore. *Nature* 475: 463–464.

Chae J, Schühle U, Lemaire P. 1998. Sumer measurements of nonthermal motions: Constraints on coronal heating mechanism. *Astrophys. J.* 505: 957–973.

Chandran BDG. 2005. Weak compressible magnetohydrodynamic turbulence in the solar corona. *Phys. Rev. Lett.* 95: 265004.

Chandran BDG. 2008. Strong anisotropic MHD turbulence with cross helicity. *Astrophys. J.* 685: 646–658.

Chandran BDG, Quataert E, Howes GG, Hollweg JV, Dorland W. 2009. The turbulent heating rate in strong magnetohydrodynamic turbulence with nonzero cross helicity. *Astrophys. J.* 701: 652–657.

Chandrasekhar S. 1951. The invariant theory of isotropic turbulence in magneto-hydrodynamics. *Proc. Roy. Soc. Lond. A* 204: 435–449.

Cho J, Vishniac ET. 2004. The anisotropy of magnetohydrodynamic Alfvénic turbulence. *Astrophys. J.* 539: 273–282.

Cho J, Vishniac ET. 2004. The anisotropy of electron magnetohydrodynamic turbulence. *Astrophys. J.* 615: L41–L44.

Coleman PJ. 1968. Turbulence, viscosity and dissipation in the solar-wind plasma. *Astrophys. J.* 153: 371–388.

Connaughton C, Newell AC, Pomeau Y. 2003. Non-stationary spectra of local wave turbulence. *Physica D* 184: 64–85.

Coroniti FV, Kennel CF, Scarf FL. 1981. Whistler mode turbulence in the disturbed solar wind. *J. Geophys. Res.* 87: 6029–6044.

Cranmer SR, van Ballegooijen AA, Edgar RJ. 2007. Self-consistent coronal heating and solar wind acceleration from anisotropic magnetohydrodynamic turbulence. *Astrophys. J.* 171: 520–551.

Dastgeer S, Das A, Kaw P, Diamond PH. 2000. Whistlerisation and anisotropy in 2D electron magnetohydrodynamic turbulence. *Phys. Plasmas* 7: 571–579.

David C, Gabriel AH, Bely-Dubau F, Fludra A, Lemaire P, Wilhelm K. 1998. Measurement of the electron temperature gradient in a solar coronal hole. *Astron. Astrophys.* 336: L90–L94.

Denskat KU, Beinroth HJ, Neubauer FM. 1983. Interplanetary magnetic field power spectra with frequencies from $2.4\,10^{-5}$ Hz to 470 Hz from HELIOS-observations during solar minimum conditions. *J. Geophys.* 54: 60–67.

Doschek GA, Mariska JT, Warren HP, Brown CM, Culhane JL, Hara H, Waranabe T, Toung PR, Mason HE. 2007. Nonthermal velocities in solar active regions observed with the extreme-ultraviolet imaging spectrometer on Hinode. *Astrophys. J.* 667: L109–L112.

Dyachenko S, Newell AC, Pushkarev AN, Zakharov VE. 1992. Optical turbulence: weak turbulence, condensates and collapsing filaments in the nonlinear Schrödinger equation. *Physica D* 57: 96–160.

Elmegreen BG, Scalo J. 2004. Interstellar turbulence I: observations and processes. *Annu. Rev. Astron. Astrophys.* 42: 211–273.

Falcon É, Laroche C, Fauve S. 2007. Observation of gravity-capillary wave turbulence. *Phys. Rev. Lett.* 98: 094503.

Falkovich GE, Shafarenko AV. 1991. Nonstationary wave turbulence. *J. Nonlinear Sci.* 1: 457–480.

Forsyth RJ, Balogh A, Horbury TS, Erdös G, Smith EJ, Burton ME. 1996. The heliospheric magnetic field at solar minimum: Ulysses observations from pole to pole. *Astron. Astrophys.* 316: 287–295.

Frisch U. 1995. *Turbulence*. Cambridge: Cambridge University Press.

Galtier S, Nazarenko SV, Newell AC, Pouquet A. 2000. A weak turbulence theory for incompressible MHD. *J. Plasma Phys.* 63: 447–488.

Galtier S, Nazarenko SV, Newell AC. 2001 On wave turbulence in MHD. *Nonlin. Proc. Geophys.* 8, pp. 1–10.

Galtier S, Nazarenko SV, Newell AC, Pouquet A. 2002. Anisotropic turbulence of shear-Alfvén waves. *Astrophys. J.* 564: L49–L52.

Galtier S. 2003. A weak inertial wave turbulence theory. *Phys. Rev. E* 68: 015301.

Galtier S, Bhattacharjee A. 2003. Anisotropic weak whistler wave turbulence in electron magnetohydrodynamics. *Phys. Plasmas* 10: 3065–3076.

Galtier S, Bhattacharjee A. 2005. Anisotropic wave turbulence in electron MHD. *Plasma Phys. Control. Fusion* 47: B691–B701.

Galtier S, Pouquet A, Mangeney A. 2005. On the spectral scaling laws for incompressible anisotropic MHD turbulence. *Phys. Plasmas* 12: 092310.

Galtier S (2006a). Multi-scale turbulence in the solar wind. *J. Low Temp. Phys.* 145: 59–74.

Galtier S (2006b). Wave turbulence in incompressible magnetohydrodynamics. *J. Plasma Phys.* 72: 721–769.

Galtier S, Chandran BDG. 2006. Extended spectral scaling laws for shear-Alfvén wave turbulence. *Phys. Plasmas* 13: 114505.

Galtier S, Buchlin É. 2007. Multiscale Hall magnetohydrodynamic turbulence in the solar wind. *Astrophys. J.* 656: 560–566.

Galtier S, Nazarenko SV. 2008. Large-scale magnetic field sustainment by forced MHD wave turbulence. *J. Turbulence* 9: 1–10.

Galtier S, Buchlin E. 2010. Nonlinear diffusion equations for anisotropic MHD turbulence with cross-helicity. *Astrophys. J.* 722: 1977–1983.

Galtier S. 2012. Kolmogorov vectorial law for solar wind turbulence. *Astrophys. J.* 746: 184–187.

Ghosh S, Siregar E, Roberts DA, Goldstein ML. 1996. Simulation of high-frequency solar wind power spectra using Hall magnetohydrodynamic. *J. Geophys. Res.* 101: 2493–2504.

Goldreich P, Reisenegger A. 1992. Magnetic field decay in isolated neutron stars. *Astrophys. J.* 395: 250–258.

Goldreich P, Sridhar S. 1995. Toward a theory of interstellar turbulence. II. Strong alfvénic turbulence. *Astrophys. J.* 438: 763–775.

Goldstein ML, Roberts DA, Fitch CA. 1994. Properties of the fluctuating magnetic helicity in the inertial and dissipative ranges of solar wind turbulence. *J. Geophys. Res.* 99: 11519–11538.

Goldstein ML, Roberts DA. 1999. Magnetohydrodynamics turbulence in the solar wind. *Phys. Plasmas* 6: 4154–4160.

Hendrix DL, Van Hoven G. 1996. Magnetohydrodynamic turbulence and implications for solar coronal heating. *Astrophys. J.* 467: 887–893.

Higdon JC. 1984. Density fluctuations in the interstellar medium: Evidence for anisotropic magnetogasdynamic turbulence. I Model and astrophysical sites. *Astrophys. J.* 285: 109–123.

Hollweg JV, Isenberg PA. 2002. Generation of the fast solar wind: A review with emphasis on the resonant cyclotron interaction. *J. Geophys. Res.* 107: 1147-1–37.

Horbury TS, Forman M, Oughton S. 2008. Anisotropic scaling of magnetohydrodynamic turbulence. *Phys. Rev. Lett.* 101: 175005.

Howes GG, Cowley SC, Dordland W, Hammett GW, Quataert E, Schekochihin AA. 2008. A model of turbulence in magnetized plasmas: implications for the dissipation range in the solar wind. *J. Geophys. Res.* 113: A05103.

Hunana P, Laveder D, Passot T, Sulem P-L, Borgogno D. 2011. Reduction of compressibility and parallel transfer by Landau damping in turbulent magnetized plasmas. *Astrophys. J.* 743: 128.

Iroshnikov PS. 1964. Turbulence of a conducting fluid in a strong magnetic field. *Soviet Astron.* 7: 566–571.

Kingsep AS, Chukbar KV, Yankov VV. 1990. Electron magnetohydrodynamics. In *Reviews of Plasma Physics*, Vol. 16, pp. 243–291. New York: Consultant Bureau.

Kiyani KH, Chapman SC, Khotyaintsev YuV, Dunlop MW, Sahraoui F. 2009. Global scale-invariant dissipation in collisionless plasma turbulence. *Phys. Rev. Lett.* 103: 075006.

Klein L, Bruno, R, Bavassano B, Rosenbauer H. 1993. Anisotropy and minimum variance of magnetohydrodynamic fluctuations in the inner heliosphere. *J. Geophys. Res.* 98: 17461–17466.

Kolmakov GV, Pokrovsky VL. 1995. Stability of weak turbulence spectra in superfluid helium. *Physica D* 86: 456–469.

Kolmakov GV, Levchenko AA, Brahnikov MYu, Mezhov-Deglin LP, Silchenko AN, McClintock PVE. 2004. Quasiadiabatic decay of capillary turbulence on the charged surface of liquid hydrogen. *Phys. Rev. Lett.* 93: 074501.

Kolmogorov AN. 1941. The local structure of turbulence in incompressible viscous fluid for very high Reynolds numbers. *Dokl. Akad. Nauk. SSS* 30: 301–305.

Kraichnan RH. 1965. Inertial range spectrum in hydromagnetic turbulence. *Phys. Fluids* 8: 1385–1387.

Kraichnan RH. 1967. Inertial ranges in two-dimensional turbulence. *Phys. Fluids* 10: 1417–1423.

Krishan V, Mahajan SM. 2004. Magnetic fluctuations and Hall MHD turbulence in the solar wind. *J. Geophys. Res.* 109: A111051.

Krommes JA. 2002. Fundamental statistical descriptions of plasma turbulence in magnetic fields. *Phys. Reports* 360: 1–352.

Kuznetsov EA. 1972. Turbulence of ion sound in a plasma located in a magnetic field. *Sov. Phys. J. Exp. Theor. Phys.* 35: 310–314.

Kuznetsov EA. 2001. Weak magnetohydrodynamic turbulence of a magnetized plasma. *Sov. Phys. J. Exp. Theor. Phys.* 93: 1052–1064.

Lacaze R, Lallemand P, Pomeau Y, Rica S. 2001. Dynamical formation of a Bose-Einstein condensate. *Physica D* 152–153: 779–786.

Laveder D, Passot T, Sulem P-L. 2002. Transverse dynamics of dispersive Alfvén waves: I. direct numerical evidence of filamentation. *Phys. Plasmas* 9: 293–304.

Leamon RJ, Smith CW, Ness NF, Matthaeus WH. 1998. Observational constraints on the dynamics of the interplanetary magnetic field dissipation range. *J. Geophys. Res.* 103: 4775–4787.

Lithwick Y, Goldreich P. 2003. Imbalanced weak magnetohydrodynamic turbulence. *Astrophys. J.* 582: 1220–1240.

Lvov Y, Nazarenko SV, West R. 2003. Wave turbulence in Bose-Einstein condensates. *Physica D* 184: 333–351.

MacBride BT, Smith CW, Forman M. 2008. The turbulent cascade at 1 AU: Energy transfer and the third-order scaling for MHD. *Astrophys. J.* 679: 1644–1660.

Markovskii SA, Vasquez BJ, Smith CW. 2008. Statistical analysis of the high-frequency spectral break of the solar wind turbulence at 1 AU. *Astrophys. J.* 675: 1576–1583.

Maron J, Goldreich P. 2001. Simulations of incompressible magnetohydrodynamic turbulence. *Astrophys. J.* 554: 1175–1196.

Matthaeus WH, Ghosh S, Oughton S, Roberts DA. 1996. Anisotropic three-dimensional MHD turbulence. *J. Geophys. Res.* 101: 7619–7629.

Mininni PD, Gómez DO, Mahajan SM. 2003. Dynamo action in magnetohydrodynamics and Hall-magnetohydrodynamics. *Astrophys. J.* 587: 472–481.

Mininni PD, Alexakis A, Pouquet A. 2007. Energy transfer in Hall-MHD turbulence: Cascades, backscatter and dynamo action. *J. Plasma Physics* 73: 377–401.

Montgomery D, Turner L. 1981. Anisotropic magnetohydrodynamic turbulence in a strong external magnetic field. *Phys. Fluids* 24: 825–831.

Montgomery DC, Matthaeus WH. 1995. Anisotropic modal energy transfer in interstellar turbulence. *Astrophys. J.* 447: 706–707.

Morize C, Moisy F, Rabaud M. 2005. Decaying grid-generated turbulence in a rotating frame. *Phys. Fluids* 17: 095105.

Nazarenko SV, Newell AC, Galtier S. 2001. Non-local MHD turbulence. *Physica D* 152–153: 646–652.

Nazarenko SV. 2007. 2D enslaving of MHD turbulence. *New J. Phys.* 9: 307-1–14.

Nazarenko SV. 2011. *Wave Turbulence*, Lecture Notes in Physics, Berlin: Springer.

Newell AC, Nazarenko SV, Biven L. 2001. Wave turbulence and intermittency. *Physica D* 152–153: 520–550.

Newell AC, Rumpf B. 2011. Wave turbulence. *Annu. Rev. Fluid. Mech.* 43: 59–78.

Ng CS, Bhattacharjee A. 1996. Interaction of shear-Alfvén wave packets: Implication for weak magnetohydrodynamic turbulence in astrophysical plasmas. *Astrophys. J.* 465: 845–854.

Ng CS, Bhattacharjee A, Germaschewski K, Galtier S. 2003. Anisotropic fluid turbulence in the interstellar medium and solar wind. *Phys. Plasmas* 10: 1954–1962.

Nishizuka N, Shimizu M, Nakamura T, Otsuji K, Okamoto TJ, Katsukawa Y, Shibata K. 2008. Giant chromospheric anemone jet observed with Hinode and comparison with MHD simulations: Evidence of propagating Alfvén waves and magnetic reconnection. *Astrophys. J.* 683: L83–L86.

Perez J-C, Boldyrev S. 2008. On weak and strong magnetohydrodynamic turbulence. *Astrophys. J.* 672: L61–L64.

Podesta JJ, Bhattacharjee A. 2010. Theory of incompressible magnetohydrodynamic turbulence with scale-dependent alignment and cross-helicity. *Astrophys. J.* 718: 1151–1157.

Politano H, Pouquet A. 1998. von-Kárman-Howarth equation for MHD and its consequences on third-order longitudinal structure and correlation functions. *Phys. Rev. E* 57: R21–R24.

Pouquet A. 1993. Magnetohydrodynamic turbulence. In *Astrophysical Fluid Dynamics*, ed. J-P Zahn, J Zinn-Justin, pp. 139–227. Amsterdam: Elsevier science publishers.

Priest ER. 1992. *Solar Magnetohydrodynamics*. Dordrecht: D. Reidel Pub. Comp.

Rappazzo AF, Velli M, Einaudi G, Dahlburg RB. 2007. Coronal heating, weak MHD turbulence and scaling laws. *Astrophys. J.* 657: L47–L51.

Rappazzo AF, Velli M, Einaudi G, Dahlburg RB. 2008. Nonlinear dynamics of the Parker scenario for coronal heating. *Astrophys. J.* 677: 1348–1366.

Rasmussen HO. 1999. A new proof of Kolmogorov's 4/5s law. *Phys. Fluids* 11: 3495–3498.

Roberts DA, Goldstein ML, Klein LW, Matthaeus WH. 1987. Origin and evolution of fluctuations in the solar wind: Helios observations and Helios-Voyager comparison. *J. Geoph. Res.* 92: 12023–12035.

Rogers BN, Denton RE, Drake JF, Shay MA. 2001. Role of dispersive waves in collisionless magnetic reconnection. *Phys. Rev. Lett.* 87: 195004.

Sagaut P, Cambon C. 2008. *Homogeneous Turbulence Dynamics*. Cambridge: Cambridge University Press.

Sahraoui F, Belmont G, Rezeau L. 2003. Hamiltonian canonical formulation of Hall MHD: Toward an application to weak turbulence. *Phys. Plasmas* 10: 1325–1337.

Sahraoui F, Belmont G, Rezeau L, Cornilleau-Wehrlin N, Pinçon JL, Balogh A. 2006. Anisotropic turbulent spectra in the terrestrial magnetosheath as seen by the Cluster spacecraft. *Phys. Rev. Lett.* 96: 075002.

Sahraoui F, Galtier S, Belmont G. 2007. Incompressible Hall MHD waves. *J. Plasma Phys.* 73: 723–730.

Sahraoui F, Belmont G, Goldstein M. 2012. New insight into short-wavelength solar wind fluctuations from Vlasov theory. *Astrophys. J.* 748: 100.

Saur J, Politano H, Pouquet A, Matthaeus WH. 2002. Evidence for weak MHD turbulence in the middle magnetosphere of Jupiter. *Astron. Astrophys.* 386: 699–708.

Scalo J, Elmegreen BG. 2004. Interstellar turbulence II: Implications and effects. *Annu. Rev. Astron. Astrophys.* 42: 275–316.

Schekochihin AA, Cowley SC, Dorland W, Hammett GW, Howes GG, Quataert E, Tatsuno T. 2009. Astrophysical gyrokinetics: Kinetic and fluid turbulence cascades in magnetized weakly collisional plasmas. *Astrophys. J. Sup. Series* 182: 310–377.

Shaikh D, Zank G. 2007. Anisotropic cascades in interstellar medium turbulence. *Astrophys. J.* 656: L17–L20.

Shebalin JV, Matthaeus WH, Montgomery D. 1983. Anisotropy in MHD turbulence due to a mean magnetic field. *J. Plasma Phys.* 29: 525–547.

Smith CW, Hamilton K, Vasquez BJ, Leamon RJ. 2006. Dependence of the dissipation range spectrum of interplanetary magnetic fluctuations on the rate of energy cascade. *Astrophys. J.* 645: L85–L88.

Sorriso-Valvo L, Marino R, Carbone V, Noullez N, Lepreti F, Veltri P, Bruno R, Bavassano B, Pietropaolo E. 2007. Observation of inertial energy cascade in interplanetary space plasma. *Phys. Rev. Lett.* 99: 115001.

Sridhar S, Goldreich P. 1994. Toward a theory of interstellar turbulence 1. Weak alfvénic turbulence. *Astrophys. J.* 432: 612–621.

Sridhar S. 2010. Magnetohydrodynamic turbulence in a strongly magnetized plasma. *Astron. Nachr.* 331: 93–100.

Stawicki O, Gary PS, Li H. 2001. Solar wind magnetic fluctuations spectra: Dispersion versus damping. *J. Geoph. Res.* 106: 8273–8281.

van Ballegooijen AA, Asgari-Targhi M, Cranmer SR, Deluca EE. 2011. Heating of the solar chromosphere and corona by Alfvén wave turbulence. *Astrophys. J.* 736: 1–27.

Verdini A, Grappin R, Pinto R, Velli M. 2012. On the origin of the $1/f$ spectrum in the solar wind magnetic field. *Astrophys. J. Lett.* 750: L33.

Verma MK. 2004. Statistical theory of magnetohydrodynamic turbulence: Recent results. *Phys. Rep.* 401: 229–380.

Warren HP. 2008. Multithread hydrodynamic modeling of a solar flare *Astrophys. J.* 637: 522–530.

Yamada M. 2007. Progress in understanding magnetic reconnection in laboratory and space astrophysical plasmas. *Phys. Plasmas* 14: 058102-1–16.

Zakharov VE, Filonenko NN. 1966. The energy spectrum for stochastic oscillations of a fluid surface. *Doclady Akad. Nauk. SSSR* 170: 1292–1295 [*Sov. Phys. Docl.* 11: 881–884, 1967].

Zakharov VE. 1965. Weak turbulence in media with a decay spectrum. *J. Appl. Mech. Tech. Phys.* 6: 22–24.

Zakharov VE, Sagdeev RZ. 1970. Spectrum of a acoustic turbulence. *Sov. Phys. Dokl.* 15: 439–444.

Zakharov VE, L'vov V, Falkovich GE. 1992. *Kolmogorov Spectra of Turbulence I: Wave Turbulence.* Berlin, Germany: Springer-Verlag.

Chapter 4

Optical Wave Turbulence

S. K. Turitsyn*, S. A. Babin[‡,§], E. G. Turitsyna*, G. E. Falkovich[†],
E. V. Podivilov[‡,§] and D. V. Churkin[*,†]

*Aston University, UK

†Weizman Institute of Science, Israel

‡Institute of Automation and Electrometry, SB RAS, Novosibirsk, Russia

§Novosibirsk State University, Novosibirisk, Russia

We review recent progress in optical wave turbulence with a specific focus
on the fast growing field of fiber lasers. Weak irregular nonlinear interactions between a large number of resonator modes are responsible for
practically important characteristics of fiber lasers such as the spectral
broadening of radiation. Wave turbulence is a fundamental nonlinear
phenomenon which occurs in a variety of nonlinear wave-bearing physical
systems. The experimental impediments and the computationally inten-
sive nature of simulation of hydrodynamic or plasma wave turbulence
often make it rather challenging to collect a significant number of
statistical data. The study of turbulent wave behavior in optical devices
offers quite a unique opportunity to collect an enormous amount of
data on the statistical properties of wave turbulence using high-speed,
high precision optical measurements during a relatively short period of
time. We present recent theoretical, numerical and experimental results
in optical wave turbulence in fiber lasers ranging from weak to strong
turbulences for different signs of fiber dispersion. Furthermore, we report
on our studies of spectral wave condensate in fiber lasers that make
interdisciplinary links with a number of other research fields.

Contents

4.1. Optical Wave Turbulence: Introduction

Wave turbulence is a fundamental nonlinear phenomenon that occurs in a variety of nonlinear physical systems (see e.g. Zakharov *et al.* (1992), Nazarenko (2011) and references therein). Turbulence is a state of system with many degrees of freedom deviated far away from equilibrium. Even when an external excitation only effectively acts on one or a few waves, nonlinear interactions between waves lead to excitation of many waves and turbulence. Wave turbulence theory deals with the statistical behavior of a large number of interacting waves. Turbulence theory is one of the most challenging and truly fundamental theoretical problems with applications ranging from turbulent combustion in engines to turbulent effects in atmosphere and in outer space (Frisch, 1995; Falkovich, 2006; Falkovich and Sreenivasan, 2006). There are two primary types of physical systems with turbulent-like behavior. In the first case, mostly associated with using the term *turbulence*, is the so-called "fully developed wave turbulence" when scales, where waves are excited (energy is pumped into the system), are well separated from the scales where waves are dissipated. In this instance, the turbulent energy transfer between the spectral components in the inertial interval does not depend heavily on the details of excitation and dissipation (Zakharov *et al.*, 1992). In the second case, the scales at which the energy is pumped into the system and disappears cannot be fully separated and turbulence does not have a well-defined inertial interval. In this chapter, we present a rather interesting and important example of the nonlinear optical systems — fiber lasers — which exhibit wave turbulence behavior of the second type.

The term *optical turbulence* is used in literature in a variety of rather different research contexts. For instance, optical turbulence is often referred to as an atmospheric effect caused by fluctuations of the refractive index in the air that affects the propagation of light beams. This effect can significantly degrade infrared images or bit error rates in free-space lasers and ground-to-satellite communication systems and is one of the main causes that limit the spatial resolution attainable in ground-based astronomical long-range imaging systems. Optical turbulence has also been studied in experiments with a ring resonator synchronously driven by a train of pico-second light pulses (Mitschke *et al.*, 1996). Chaotic behavior of pulses resulted from nonlinear interactions and repetitive interference of a train. Furthermore, intra-pulse correlations have been used as a measure of the degree of complexity of generated temporal, spatio-temporal chaos and turbulence (Mitschke *et al.*, 1996). An interesting example of turbulent-like behavior of optical waves can be found in semiconductor lasers with delayed feedback. The latter is described by a generalized complex Swift–Hohenberg equation (Lega *et al.*, 1995) which is broadly used as a generic model in studies of filamentation in wide aperture lasers and pattern formation in the transverse section of semiconductor lasers. Finally, very important theoretical studies of optical turbulence have been undertaken in the general context of the chaotic behavior of system solutions of nonlinear partial differential equations. The turbulence associated with the so-called nonlinear Schrodinger equation (NLSE) is often called optical turbulence because of its relevance to the propagation of light beams in the transparent media with a nonlinear refractive index. In particular, in the frameworks of the 2D NLSE model (Dyachenko *et al.*, 1992; Newell *et al.*, 2001; Zakharov *et al.*, 2004; Dyachenko and Falkovich, 1996; Zakharov and Nazarenko, 2005), studies have focused on the so-called fully developed wave turbulence in which the number of active degrees of freedom is very large (corresponding in hydrodynamics to the large Reynolds number limit). Since NLSE has two integrals of motion, it allows for two cascades: a direct cascade of energy towards small scales and an inverse cascade of wave action (number of waves) towards large scales. An inverse cascade is arguably the most conceptually novel idea in turbulence since Kolmogorov. In a finite system, an inverse cascade leads to the creation of the spectral condensate i.e. the mode coherent across the whole system. In wave systems like NLSE, condensate corresponds to a monochromatic wave. One then distinguishes cases with a stable and unstable condensate (NLSE with repulsion or attraction, respectively). Using the focusing (attracting potential) 2D NLSE model, it has been shown that wave collapse might act as a mechanism for intermittency and a statistical behavior, which violates many of the traditional assumptions made by Kolmogorov (1941) (Lushnikov and Vadimirova, 2010). On the other hand, not much is known about the interaction of stable condensate and turbulence, the system that curiously has some quantum-like features since it combines coherence and fluctuations. Particularly, not much is known on how turbulence appears in a system where a coherent state is linearly stable (very much like fluid turbulence in pipe and channels, where laminar flows are linearly stable) This is still a topic of active research despite having over 100 year history. We shall see below that optical turbulence in fiber lasers sheds some light on such general problems of turbulence.

Optical manifestations of turbulence (Dyachenko *et al.*, 1992; Newell *et al.*, 2001; Zakharov *et al.*, 2004; Dyachenko and Falkovich, 1996; Zakharov and Nazarenko, 2005; Solli *et al.*, 2007; Falkovich *et al.*, 1996, 2001; Babin *et al.*, 2007a) are very closely linked to nonlinear optical phenomena such as self-phase and cross-phase modulation, self-focusing, four-wave-mixing (FWM), optical soliton radiation and interactions, nonlinear spectral broadening, supercontinuum (SC) generation and optical wave breaking (see e.g. Agrawal (1995); Skryabin *et al.* (2007), Dudley *et al.* (2006), Smirnov *et al.* (2006) and references therein). A variety of nonlinear optical effects and precision of optical experiments provide a great, but not yet fully explored potential for the study of turbulence properties through manifestations in optical problems and devices.

The main focus of this chapter is on our recent studies of a new type of optical wave turbulence that occurs in the context of Raman fiber lasers (RFLs) with cavities ranging from tens of meters to hundreds of kilometers. RFLs are stable and efficient light sources that have attracted a great deal of interest in recent years both because of their winning technical characteristics and as an important example of a nonlinear photonic device utilizing nonlinear science concepts. It has been demonstrated recently that multiple FWM interactions between very large numbers of cavity modes are primarily responsible for practical characteristics of RFLs, in particular, intensity fluctuations and spectral broadening of the generated radiation. These multiple interactions lead to strong de-phasing of modes and the randomization of dynamics. This process can be analyzed using the set of deterministic nonlinear ordinary differential equations describing the dynamics of modes, which are randomized due to the FWM nonlinear processes. Mathematically, the FWM interactions involve a huge number of rapidly oscillating terms with different amplitudes and phases making time evolution of any particular cavity mode extremely stochastic. Even weak nonlinear FWM interactions lead to a random energy exchange between waves and to dephasing because of a continuum of modes involved in nonlinear interactions. Optical wave dynamics in powerful fiber lasers is an interesting manifestation of a physical system where long-term average properties of the system are determined by a random behavior of a huge number of weakly interacting waves — *wave turbulence*. Therefore, theoretical analysis of the spectrum formation and fluctuations requires the use of statistical techniques and approaches instead of the dynamical formalism commonly used in fiber optics. In this chapter, we will overview recent developments in this field with a clear interdisciplinary link between fiber optics and the theory of turbulence.

4.2. Basics of Fiber Lasers

A powerful fiber laser presents both an interesting nonlinear physical system and a photonic device with an incredible range of practical applications (in telecommunications, medicine, metrology, spectroscopy, sensing, industrial cutting, welding and others). The RFL exploits the effect of stimulated Raman scattering (SRS). In the process of SRS, a pump photon gives up its energy to create a new photon at a longer wavelength, plus some residual energy, which is absorbed by the fiber material in the form of optical phonons. In contrast to bulk media where the light beam should be focused tightly into the substrate to observe the effect, the optical fibers that allow high-intensity light propagation over long distances, provide a much stronger SRS effect. More specifically, very low absorption of light ($\alpha \sim 0.2\,\mathrm{dB/km}$ at the wavelength $\lambda \sim 1.55\,\mu\mathrm{m}$ corresponding to the transparency window of silica glasses) and small fiber core diameter ($5\text{--}10\,\mu\mathrm{m}$) result in the propagation of high intensity light without significant attenuation, thus providing homogeneous intensity distribution in kilometer-scale fiber spans. At the same time, the SRS of the high-intensity wave treated as a pump, induces Raman amplification of the red-shifted Stokes wave with a coefficient $\sim 1\,\mathrm{dB/(kmW)}$ that becomes much higher than its attenuation already at relatively low power, $\sim 1\,\mathrm{W}$. In a conventional single mode fiber (SMF) with a germanium doped silica core, the Raman gain spectrum is rather broad with a maximum shift by $\sim 14\,\mathrm{THz}$ at the Stokes wavelength from the pump wavelength. The Raman gain medium may be placed in a cavity, which is typically formed by the insertion of two fiber Bragg gratings (FBGs) at each end of the fiber, playing the role of "mirrors" and trapping photons within the grating bandwidth, resonantly reflecting the forward and backward propagating Stokes waves. As a result, it is possible to achieve lasing in a fiber waveguide with the cavity length L to the order of several kilometers (Grubb *et al.*, 1995). Several frequency conversion cascades based on Stokes lines resonating in nested cavities shift the pump power deep into the longer wavelength region. Based on the Raman gain medium and cavity FBGs reflecting specific wavelengths, RFLs involving single as well as multiple order Stokes shifts can be designed to operate at almost any wavelength in the near-IR region (1.1–1.7 micron) pumped by high-power lasers at ~ 1.06 micron, e.g. by Yb-doped fiber laser providing all-fiber design. Furthermore, the number of conversion stages from short- to long-wavelength edge may be reduced using as Raman media P_2O_5-doped silica fibers with three-times larger Stokes shift (Dianov *et al.*, 1997). RFLs present excellent candidates for a variety of applications due to their unique attributes combining the wavelength tunability and the multi-wavelength operation (Mermelstein *et al.*, 2001) with the compactness and high-power continuous wave (CW) laser diode pumps.

Of a particular interest is the ability to fully cover the spectral region near the telecommunication windows of transparency, making RFLs very attractive pump sources for distributed Raman amplification. This is considered to be one of the most important enabling technologies in high-speed optical communication (Stolen and Ippen, 1973; Mollenauer *et al.*, 1986; Chernikov *et al.*, 1999; Vasilyev, 2003; Headley and Agrawal, 2004). Recently, an interesting realization of a first-order Raman ampli-fied communication system providing a quasi-lossless signal transmission based on the ultra-long Raman laser architecture has been proposed and implemented (see e.g. Ania-Castañón *et al.* (2006) and references therein). In this particular laser, the combined forward and backward propagating wave generated at the Stokes wavelength ($\sim$1455 nm) inside the RFL cavity (formed in the transmission SMF fiber itself) experiences reduced variations along the fiber span. Hence, the generated intra-cavity RFL power can be used as a homogeneous, stable secondary pump to provide a near constant (along the fiber) Raman gain for optical signal transmitted at 1550 nm. If the gain is nearly equal to optical loss, a quasi-lossless transmission is possible. The quasi-lossless fiber span can be implemented using symmetric second-order pumping schemes leading to the concept of ultra-long fiber lasers (Ania-Castañón *et al.*, 2006; Ellingham *et al.*, 2006; Babin *et al.*, 2007b; Turitsyn *et al.*, 2009), which promise new applications in trans-mission and secure communication (Scheuer and Yariv, 2006). Evidently, such a significant lengthening of the RFL cavity compared to that in usual lasers leads to a new interesting class of lasers with potentially different physical mechanisms underlying their operation. RFLs are also attractive continuous light sources for optical coherence tomography (Hsiung *et al.*, 2004), long distance remote sensing (Han *et al.*, 2005; Frazao *et al.*, 2009; Juarez *et al.*, 2005), efficient excitation of a mesospheric sodium laser guide star (Huang *et al.*, 2003) and other applications (see e.g. Kim (2008)). Despite existing and emerging practical applications of such lasers, some fundamental physical phenomena underlying their operation and nonlinear mechanisms determining properties of generated radiation are not yet fully understood.

We will present experimental and numerical analysis of spectral and temporal characteristics of both normal scale (several km) and ultra-long

Raman fiber lasers (URFL) with cavity length varied from 5 km to a record laser cavity length of 270 km. We will demonstrate that important characteristics of fiber lasers are determined by turbulent-like interactions of laser modes, whose number varies for considered cavity lengths between 1 and 100 million.

4.3. Key Mathematical Models

In this section, we introduce basic mathematical models focusing on the generic features of CW fiber lasers rather than on the particular properties of any specific systems. Without loss of generality, in what follows, we mostly consider the RFL schemes close to the one described by Ania-Castañón *et al.* (2006) which provides rather uniform spatial distribution of generated Stokes waves power along the cavity formed by FBGs. However, we would like to stress that most of our approaches and methods are also applied to a much broader and fundamental class of *CW lasers with a Fabry–Perot resonator*. In numerical simulations, we will use both the mathematical model, based on the standard evolution equation for the longitudinal modes (E_n) of the envelope that can be derived from the generalized Schrödinger equations for backward and forward Stokes waves with amplitudes A^+, A^- (Babin *et al.*, 2006; Babin *et al.*, 2007a) and direct modeling of the amplitude equations. The slow varying part (in the direction of propagation z and time t) of amplitude of the electromagnetic field E in the cavity of length L with reflecting boundary conditions can be represented as the sum of co- and counter-propagating waves:

$$E(z,t) = \frac{1}{\sqrt{2}} \{ A^+(z,t)\, e^{i(\omega_0 t - \beta_0 z)} + A^-(z,t)\, e^{i(\omega_0 t + \beta_0 z)} + c.c. \},$$

where $A^\pm$ corresponds to the forward and backward waves, respectively, propagating along the z-axis of the fiber, $\omega_0 = 2\pi c/\lambda_0$ is the frequency of the lasing (in the case of a RFL, typically defined by the central wavelength λ_0 of the reflectors-reflection maximum), the propagation constant $\beta(\omega)$ in the medium (e.g. fiber) of the spectral components of the modulated wave whose frequency slightly deviates from a carrier frequency ω_0 can be presented as: $\beta(\omega) = \beta_0 + \beta_1(\omega - \omega_0) + 0.5 \times \beta_2(\omega - \omega_0)^2 + \cdots$, c is the speed of light in a vacuum. We can define the total average intracavity Stokes wave power P (note that field intensity I in optical fiber is related to power P through the relation $P = I\, A_{\text{eff}}$, where A_{eff} is a characteristic effective area near the fiber core, where most of power is localized in direction transversal to the z-propagation) as:

$$P(z) = P^+(z) + P^-(z) = \langle |A^+|^2 \rangle + \langle |A^-|^2 \rangle. \tag{4.1}$$

The evolution of $A^{\pm}$ is governed by the equations (see e.g. Agrawal (1995) for more details):

$$\frac{1}{v_g}\frac{\partial A^+}{\partial t} + \frac{\partial A^+}{\partial z} = \frac{1}{2}\left(g(P) + i\beta_2\frac{\partial^2}{\partial t^2}\right)A^+ - i\gamma A^+(|A^+|^2 + 2|A^-|^2), \quad (4.2)$$

$$\frac{1}{v_g}\frac{\partial A^-}{\partial t} - \frac{\partial A^-}{\partial z} = \frac{1}{2}\left(g(P) + i\beta_2\frac{\partial^2}{\partial t^2}\right)A^- - i\gamma A^-(|A^-|^2 + 2|A^+|^2). \quad (4.3)$$

Here, $v_g = 1/\beta_1$ is the group velocity, and the terms in the right-hand side of these equations describe the following respectively: amplification/loss saturated with growth of power, dispersion and nonlinear interactions due to Kerr nonlinearity (self-phase modulation and cross-phase modulation); γ is the Kerr nonlinearity coefficient, β_2 is the fiber group velocity dispersion. Note that a more accurate model should include nonlinear terms responsible for the Raman effects and also evolution of optical fields at other wavelengths interacting with the considered waves through the Raman terms. However, here, the key effect resulting from the Raman power exchange between spectrally separated waves is included through the effective gain (e.g. gain minus distributed loss) coefficient $g(P)$. Also, the nonlinear coefficient γ includes the well-known contribution (Agrawal, 1995) from the Raman effect leading to small renormalization of the Kerr constant. The gain $g(P)$ is a positive function at small powers and is saturated with growing power approaching zero level and, thus, defining the power level of the generated radiation P.

These equations are considered in a resonator (laser cavity) of length L. Boundary conditions of Eqs. (4.2) and (4.3) correspond to reflectors at $z = 0$ and $z = L$ forming a Fabry–Perot resonator. In the spectral domain, the boundary conditions defined by the point reflectors (mirrors) at the edges of cavity read:

$$A^+(0, \omega) = r_0(\omega)A^-(0, \omega), \quad A^-(L, \omega) = r_L(\omega)A^+(L, \omega),$$

the power reflection coefficients are defined as: $R_{0,L}(\omega) = |r_{0,L}(\omega)|^2$.

Considering single round trip changes to be small, forward and backward propagating fields can be expanded in the resonator mode basis as:

$$A^+(z, t) = \frac{1}{\sqrt{2}}\sum_{-M/2}^{M/2} E_m(t)\exp[im(\Delta t - \pi z/L) + \tilde{\delta}_m z/(4L)],$$

$$(4.4)$$

$$A^-(z, t) = \frac{1}{\sqrt{2}}\sum_{-M/2}^{M/2} E_m(t)\exp[im(\pi z/L + \Delta t) - \tilde{\delta}_m z/(4L)],$$

here $\Delta = \pi c/(n_g L)$ is the frequency spectral mode separation, and an effective spectral-dependent distributed loss is introduced as $\tilde{\delta}_m = -\ln [R_0(m\Delta)R_L(m\Delta)]$. Note that the phase changes introduced by the reflectors can be important in some problems (Dalloz *et al.*, 2010). Whilst for the sake of clarity these are not considered here, however, this effect can be easily included in a similar manner. After straightforward manipulation, the standard slow evolution (over one round trip) equation for the longitudinal mth mode can be derived:

$$\frac{\tau_{rt}}{L}\frac{dE_m}{dt} = (G_m - i\beta_2\Omega_m^2)E_m - i\gamma \sum_{i,k} E_i E_k E_{i+k-m}^*. \qquad (4.5)$$

Here, the round-trip time is $\tau_{rt} = 2Ln_g/c$ (n_g is the refractive group index). The terms on the right-hand side describe, respectively, a combined effect of distributed gain and loss, and spectrally-dependent loss induced by point-action reflectors at the edges of the resonator, $G_m = g - \tilde{\delta}_m/(2L)$, which is a decreasing function of the total generated power, the dispersion, $\beta_2\Omega_m^2 = \beta_2(m\Delta)^2$, and nonlinear terms (accounting for self-phase modulation, cross-phase modulation and FWM) induced by the Kerr nonlinearity (γ).

Note that in a dissipative (with gain and loss) physical system such as a laser, not all linear resonator modes survive competition. First, only spectral modes with mode gain overcoming the loss are amplified. Next, the modes that are amplified more than others in field dynamics might dominate the building of the radiation due to saturation of gain in other modes. The number of modes generated in laser can be approximately estimated as B_L/Δ, where B_L is the laser linewidth that is related to effects limiting the spectral interval, over which the laser radiation is generated (e.g. FBGs — however, their bandwidth should be used only as an estimate of the linewidth B_L since it can depend on the total generated power). Taking that the spectral separation of modes is $\Delta \sim L^{-1}$, the number of generated modes can be very large in *long fiber lasers* (up to 100 million) making them an ideal test-bed for studies of physics far removed from the equilibrium of using one-dimensional wave turbulence.

4.4. Weak Optical Wave Turbulence in Fiber Lasers

Having a very large number of cavity modes in fiber lasers under the condition of a finite total power results in a very small amplitude of each mode. Nevertheless, experiments show that nonlinear effects are important in the considered laser systems. Thus, very weak interactions of small amplitude waves produce an observable overall effect, thus it is natural to start with a general kinetic approach already developed in a theory of weak wave turbulence (Zakharov *et al.*, 1992). Therefore, first of all, we present a detailed analytical self-consistent theory based on the wave kinetic equations. This theory describes the generation spectrum and output power of an RFL. As we will show both theoretically and experimentally, the quasi-degenerate FWM between different longitudinal modes is the main broadening mechanism in the RFL at high powers. We present examples of laser systems, in which the shape and power dependence of the intra-cavity Stokes wave spectrum are in excellent quantitative agreement with predictions of the theory. FWM-induced stochasticity of the amplitude and the phase of each of $\sim 10^6$ longitudinal modes generated in a RFL cavity is an example of light wave turbulence in a fiber.

4.4.1. *Theory of weak wave turbulence in the context of fiber laser*

In the case of the RFL, the gain/loss coefficients (4.2) and (4.3) take a specific form, e.g. for the Stokes wave an effective gain g depends on the averaged (over the fiber length) pump power $\bar{P}_p$ and the distributed loss as:

$$g = g_R \bar{P}_p - \alpha. \tag{4.6}$$

Here, $\alpha \, [\text{km}^{-1}]$ are distributed losses; $g_R [\text{W}^{-1}\text{km}^{-1}]$ is the Raman gain coefficient in the optical fiber, and the average pump power $\bar{P}_p$ can be approximately expressed through the generated Stokes power P (see Babin *et al.* (2003) for the details) as:

$$\bar{P}_p = P_p(0) \frac{1 - \exp\left(-\alpha_p L - \frac{\lambda}{\lambda_p} g_R L P\right)}{\left(\alpha_p + \frac{\lambda}{\lambda_p} g_R P\right) L}, \tag{4.7}$$

where $P_p(0)$ is the input (referring to the RFL) pump power, P is the generated Stokes wave power averaged over time and the fiber cavity; α_p is the pump wave optical losses coefficient, λ_p and λ are the pump and Stokes wavelengths, respectively. Because the RFL has a high-Q cavity, we can represent the Stokes wave as the sum of longitudinal cavity modes [similar to (4)]:

$$A^{\pm}(z,t) = \frac{1}{\sqrt{2}} \sum_m E_m(t) \exp(im \, \Delta t \mp i\kappa z m \pm \tilde{\delta}_m z/(4L)) \exp(-i\nu_m t), \tag{4.8}$$

where $\Delta = 2\pi/\tau_{\text{rt}}$ is the frequency spacing between adjacent Stokes wave longitudinal modes, $\tau_{\text{rt}} = 2Ln_g/c$ is the RFL cavity round-trip time for the Stokes wave and $\kappa = \pi/L$. The only difference to (4.4) here is an explicit introduction of $\nu_m = \beta_2 v_g (m\Delta)^2/2 + \gamma v_g P$ — a small frequency shift that takes into account the dispersion and the mean nonlinear phase shift. The factor $1/\sqrt{2}$ is chosen in order to normalize the total power in mth longitudinal mode to $P_m = |E_m|^2 = P_m^+ + P_m^-$.

It is convenient to rewrite the generalized nonlinear Schrödinger equation (4.5) for the amplitude E_m of mth longitudinal modes in a slightly different form, which is more suitable for further manipulations:

$$\tau_{rt} \frac{dE_m}{dt} - [g(P)L - \tilde{\delta}_m/2]E_m(t)$$

$$= -i\gamma L \sum_{l \neq 0} E_{m-l}(t) \sum_{k \neq 0} E_{m-k}(t) E^*_{m-k-l}(t) \exp(i\beta_2 kl\, \Delta^2 v_g t), \quad (4.9)$$

where we have taken into account the effective cavity mirror losses $\tilde{\delta}_m = -\ln[R_1(\Omega_m)R_2(\Omega_m)]$ for the mth longitudinal mode that is generated at the frequency detuned by $\Omega_m = m\Delta$ from the cavity mirror center frequency.

The right-hand side (RHS) of Eq. (4.9) for the specific RFL cavity length of 370 m considered in the experiment contains $\sim 10^{12}$ (even more in longer lasers!) different terms with random amplitudes and phases in each term that lead to a stochastic (turbulent) evolution in time (over many round trips) of the amplitude and phase of the longitudinal mode E_m. It is obvious that such an extremely complex nonlinear evolution can be described neither analytically nor numerically. For an adequate description of the Stokes wave spectrum, one has to use statistical analysis. The simplest form of such an analysis is based on the wave kinetic equations for the spectral power density rather than on the dynamic consideration based on deterministic equations for the multitude of longitudinal modes amplitudes E_n. Toward this end, we follow the methods of the weak wave turbulence introduced by Zakharov *et al.* (1992).

The approach is valid in the case of small nonlinearity and this smallness condition will be quantified below. To derive from Eq. (4.9) the simplified wave kinetic equation for the average time-independent Stokes wave spectrum $P(\Omega)$, let us multiply Eq. (4.9) by E_m^* and take the real part of the expression. As a result, one can obtain the equation describing small changes of the power of the mth longitudinal mode of the Stokes wave:

$$\tau_{rt} \frac{dP_m}{dt} - (2g(P)L - \tilde{\delta}_m)P_m(t)$$

$$= -\text{Re}\left[i2\gamma L \sum_{l \neq 0, k \neq 0} E_{m-l}(t) E_{m-k}(t) E^*_{m-k-l}(t) E^*_m(t) \exp(\beta_2 ikl\, \Delta^2 v_g t) \right].$$

$$(4.10)$$

Assuming that the phases of different longitudinal modes are random (uncorrelated), averaging the RHS of Eq. (4.10) in the main order leads to the mean zero value as all terms include a random phase difference. In order to take into account a phase correlation induced by the FWM processes, we consider the following standard approach (Zakharov *et al.*, 1992). Instead of reproducing with full details the standard steps described in the book by Zakharov *et al.* (1992), we only outline here the procedure and present the results. First, we find the correction to E_m by integrating the RHS of Eq. (4.9):

$$\delta E_m(t) = -i\gamma L \int_{-\infty}^{t} \sum_{l' \neq 0, k' \neq 0} E_{m-l'}(t') E_{m-k'}(t') E_{m-k'-l'}^*(t')$$

$$\times \exp(\beta_2 i k' l' \Delta^2 v_g t') \frac{dt'}{\tau_{\rm rt}}. \tag{4.11}$$

Second, we substitute this correction (with the index m changed to $m-l$) for E_{m-l} in the RHS of Eq. (4.10), then for E_{m-k} and so on. As a result, we obtain a rather cumbersome expression, and after averaging it over a huge number of spectral components ($\sim 10^4$ in the case of the reference experiment that will be described below) near the frequency $\Omega = m\Delta$, we derive a wave kinetic equation for the stationary Stokes wave spectrum $P(\Omega) = \langle P_m \rangle / \Delta$:

$$\tau_{\rm rt} \frac{dP(\Omega)}{dt} = [2gL - \delta(\Omega)]P(\Omega) + S_{\rm FWM}(\Omega), \tag{4.12}$$

where FWM-induced terms are

$$S_{\rm FWM}(\Omega) = -\delta_{\rm NL} P(\Omega)$$

$$+ (2\gamma L)^2 \int \frac{P(\Omega - \Omega_1) P(\Omega - \Omega_2) P(\Omega - \Omega_1 - \Omega_2)}{(3\tau_{\rm rt}/\tau) \cdot [1 + (2\tau L \beta_2 / 3\tau_{\rm rt})^2 \Omega_1^2 \Omega_2^2]} d\Omega_1 d\Omega_2 \tag{4.13}$$

and the nonlinear FWM-induced loss term $\delta_{\rm NL}$ has the following form:

$$\delta_{\rm NL} = (2\gamma L)^2 \int d\Omega_1 d\Omega_2$$

$$\times \frac{[P(\Omega - \Omega_1) + P(\Omega - \Omega_2)] P(\Omega - \Omega_1 - \Omega_2) - P(\Omega - \Omega_1) P(\Omega - \Omega_2)}{(3\tau_{\rm rt}/\tau) \cdot [1 + (2\tau L \beta_2 / 3\tau_{\rm rt})^2 \Omega_1^2 \Omega_2^2]}. \tag{4.14}$$

Here, we assumed an exponential decay of the correlation function $\langle E_l(t) E_l^*(t') \rangle = P_l e^{-|t-t'|/\tau}$ with the correlation time τ, and the Gaussian statistics for field $E_m(t)$.

The exponential decay of the correlation function results in the Lorentzian shape of peaks in the Stokes wave inter-mode beating radio-frequency (RF) spectrum:

$$F(\Omega) = \sum_{n \neq 0} \frac{DF_n}{\pi[D^2 + (\Omega - n\Delta)^2]}, \tag{4.15}$$

where $F_n = \sum_l P_l P_{l+n}$, and $D = 2/\tau$. So, the correlation time $\tau = 2/D$ can be found from the experimentally measured spectral width D of the peaks in the RF inter-mode beating spectrum of the Stokes wave.

The last term in the RHS of Eq. (4.13) describes an increase of intensity in the mode with frequency Ω due to the scattering of modes with frequencies $\Omega - \Omega_1$ and $\Omega - \Omega_2$ into modes with frequencies Ω and $\Omega - \Omega_1 - \Omega_2$, i.e. this term represents the FWM-induced *nonlinear gain*. The physical meaning of δ_{NL} is the round-trip *nonlinear attenuation* coefficient of the Stokes wave with frequency Ω induced by its scattering on other longitudinal modes with frequencies $\Omega - \Omega_1 - \Omega_2$. It is exactly this nonlinear attenuation that leads to the exponential decay of the correlation function. The nonlinear attenuation increases with power, and completely determines the correlation time at high Stokes power:

$$\frac{\tau_{\mathrm{rt}}}{\tau} \simeq \delta_{\mathrm{NL}}/2. \tag{4.16}$$

The latter condition makes equations (4.12)–(4.14) self-consistent. It is important to stress that the wave kinetic equation (4.12)–(4.16) is valid when the average effect of nonlinearity is much less than the average impact of the dispersion:

$$\gamma P \ll \beta_2 \overline{\Omega^2}, \tag{4.17}$$

where $\overline{\Omega^2}$ is a mean-square spectral half-width parameter:

$$\overline{\Omega^2} = \frac{\int \Omega^2 P(\Omega) d\Omega}{P}. \tag{4.18}$$

Condition (4.17) ensures a broad spectral distribution and a phase mismatch between remote spectral components (longitudinal modes). As a result, the FWM-induced phase synchronization is suppressed due to dispersion, and phases of far separated longitudinal modes can be considered as only weakly correlated.

By calculation of the integrals in Eqs. (4.13) and (4.14) in the limit of (4.17), one can find the spectral width and the spectral peak power of the Stokes wave assuming that the spectrum is bell-shaped, from the wave kinetic equation. For simplicity, but without loss of generality, we consider the case of the cavity reflectors (mirrors) as having Gaussian spectral function. In this case, the effective mirror losses have a parabolic form

$$\delta(\Omega) = \delta_0 + \delta_2 \Omega^2. \tag{4.19}$$

The integral over the frequency of $S_{\mathrm{FWM}}(\Omega)$ is equal to zero, which means that energy is conserved in FWM processes. It follows from Eq. (4.12) that

$$\delta_0 + \delta_2 \overline{\Omega^2} = 2gL = 2g_R L \bar{P}_p - 2\alpha L. \tag{4.20}$$

This equation has a simple physical meaning of equilibrium between the Stokes wave power gain and losses in a single round trip. In particular, Eq. (4.20) links the spectral half-width with the saturated gain g.

Further simplification of the kinetic equation-based analysis can be achieved by replacing frequencies $\Omega_{1,2}^2$ in all denominators of the integrands in the RHS of (4.13) and (4.14) by corresponding spectrally averaged characteristics. The validity of such a replacement should be justified in any specific problem. In particular, this approximation is expected to work well for narrow spectrum. Generally, this approach provides a very illustrative way to describe spectral power dependence. Equations (4.14)–(4.16) can then be simplified to the wave kinetic equation for the averaged stationary Stokes wave spectrum $P(\Omega)$:

$$[\delta(\Omega) + 2\alpha L + \delta_{\mathrm{NL}}]P(\Omega)$$

$$= 2g_R L \bar{P}_p P(\Omega) + \frac{\delta_{\mathrm{NL}}}{P^2} \int P(\Omega_1) P(\Omega_2) P(\Omega_1 + \Omega_2 - \Omega) d\Omega_1 \, d\Omega_2, \tag{4.21}$$

where δ_{NL} defines nonlinear FWM-induced losses, which can be written in the following form:

$$\delta_{\mathrm{NL}} = \sqrt{\frac{2}{3}} \frac{2\gamma P L}{\sqrt{1 + (4\beta_2 L \overline{\Omega^2}/3\delta_{\mathrm{NL}})^2}}. \tag{4.22}$$

Here, $P = \int P(\Omega) d\Omega$ is the total intra-cavity Stokes wave power. The round-trip dispersion-induced phase difference between the longitudinal modes is replaced here by the phase difference averaged over the spectrum, $\beta_2 \overline{\Omega^2} L$, where the mean-square spectral width $\overline{\Omega^2}$ is defined in (4.18).

In the case of a parabolic form of effective losses on cavity mirrors (4.19), Eq. (4.21) has a simple analytical solution for the intra-cavity Stokes wave spectrum:

$$P(\Omega) = \frac{2P}{\pi\Gamma \cosh(2\Omega/\Gamma)}, \qquad (4.23)$$

where the spectral width $\Gamma = 2\sqrt{2\delta_{\mathrm{NL}}/(\pi^2\delta_2)}$ and the spectral power density is maximum at $P(0) = 2P/(\pi\Gamma)$. This expression confirms the general prediction that the spectral width Γ increases as the square root of the Stokes wave power P. The maximum of the spectral power density $P(0)$ increases also as a square root of P. Finally, by substituting Eq. (4.23) into Eq. (4.21) and integrating over Ω, we can describe the dependence of the intra-cavity Stokes wave power P on the pump power $P_p(0)$:

$$\delta_0 + \frac{\delta_{\mathrm{NL}}(P)}{2} + 2\alpha L = 2g_R\, P_p(0) \times \frac{1 - \exp\left(-\alpha_p L - \frac{\lambda}{\lambda_p} g_R L P\right)}{\alpha_p + \frac{\lambda}{\lambda_p} g_R P}, \qquad (4.24)$$

where $\delta_{\mathrm{NL}} = 2\gamma P L \sqrt{2/3}/\left[\sqrt{1 + (2\beta_2 L/3\delta_2)^2}\right]$.

4.4.2. *Experiments*

To experimentally verify the predictions of the theory of weak wave turbulence, we studied the spectral broadening of a Stokes wave generated in a one-stage RFL based on a phosphosilicate fiber (Dianov *et al.*, 1997, 2000) as shown in Fig. 4.1. The following parameters characterize the experimental set-up: $\alpha_p = 2.5\,\mathrm{dB/km}$ (including lumped losses, i.e. losses on intra-cavity coupler, excess losses on FBGs, and splice losses), $\alpha = 2.5\,\mathrm{dB/km}$ (including lumped losses), the normal dispersion $\beta_2 = 13.3\,\mathrm{nm^{-2}km^{-1}}$, the nonlinear Kerr coefficient $\gamma \approx 1.5\,\mathrm{km^{-1}W^{-1}}$, the Raman gain coefficient $g_R = 1.3\,\mathrm{km^{-1}W^{-1}}$, the length $L = 370\,\mathrm{m}$. The phosphosilicate fiber has a distinct isolated P_2O_5-related Raman gain peak with a large Stokes shift, which is free from complications induced by overlapping different Raman gain peaks in germanosilicate fibers.

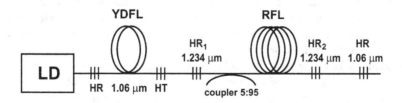

Fig. 4.1. Schematic diagram of one-stage RFL based on a phosphosilicate fiber.

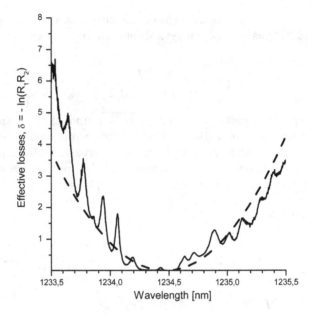

Fig. 4.2. Effective losses of cavity mirrors used in experiments (solid line) and parabolic approximation (dashed line).

The pump radiation was produced by a 16 meter long low-Q cavity ytterbium-doped fiber laser (YDFL) operating at the wavelength of 1060 nm. Pump radiation was converted to the first Stokes wave at 1234 nm by the SRS process in the high-Q RFL cavity of $L = 370$ m formed by P-doped fiber and two FBGs with high reflectivity.

Figure 4.2 shows that the spectral profile of losses induced by FBG reflectors $\delta = -\ln(R_1 R_2)$ is indeed well approximated by the parabolic shape. The fitted curvature of the effective FBGs losses is: $\delta_2 = 4\,\text{nm}^{-2}$. Using the intra-cavity coupler, we have measured, in details, the intra-cavity generated power and spectrum, and compare experimentally measured characteristics with the predictions of the weak wave turbulence theory.

First, let us consider the RF spectrum (Babin *et al.*, 2005). The intermode beating peaks in the RF spectrum are diffused, as can be seen from Fig. 4.3, the width of the inter-mode beating peaks increases with power and becomes comparable to the Stokes wave longitudinal modes spacing $\Delta \approx 0.3\,\text{MHz}$. This observation confirms that there are strong physical mechanisms that dephase the Stokes wave components during a round trip.

Equation (4.24) predicts the dependence of the generated power on the pumping power. Figure 4.4 shows remarkable agreement between the theoretically calculated and measured intra-cavity Stokes wave power P.

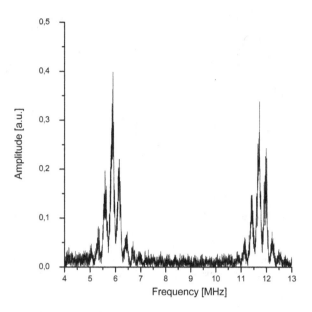

Fig. 4.3. Experimentally measured RF spectrum of the generated Stokes wave.

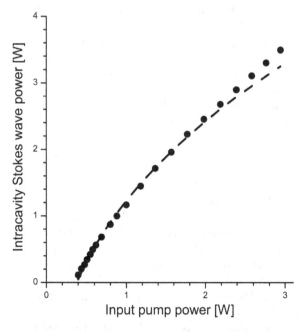

Fig. 4.4. Experimentally measured (dots) and theoretically calculated (dashed line) total intra-cavity Stokes wave power generated inside the RFL.

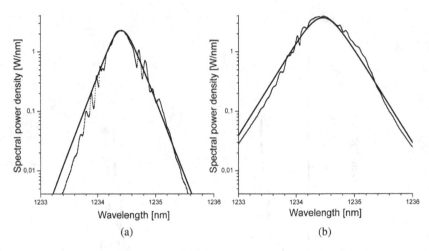

Fig. 4.5. Experimentally measured (dots) and theoretically calculated (solid line) total intra-cavity Stokes wave generation spectra at different pump power: (a) 2 W and (b) 3 W.

Using Eq. (4.23), one can easily calculate the Stokes wave spectral profiles $P(\Omega)$ at different intra-cavity Stokes wave power levels P. Figure 4.5 shows an excellent agreement between experimentally measured and theoretically calculated spectra in a broad power range.

We would like to stress that *no fitting parameters were used* in the comparison shown in Fig. 4.5. It is quite remarkable that a theory based on quite a few assumptions gives such a good analytical description for the observed RFL generation spectra. A specific feature of the spectra is their exponential tails. The oscillations in the experimentally measured profiles correspond to the ripples in the FBG losses profile. Note that the measured Stokes wave spectra are averaged over the standard resolution ~ 0.01 nm of a spectrum analyzer, that corresponds to $\sim 10^4$ longitudinal modes, while their total number is $\sim 10^6$. That justifies the averaging over $\sim 10^4$ longitudinal modes that was used in the derivation of the wave kinetic equation for the Stokes wave.

Figure 4.6 confirms the specific prediction of the wave turbulence theory: square-root dependence given by Eqs. (4.22) and (4.23) for the Stokes wave spectrum broadening with the growing generated power. The comparison between the experimental data and results of numerical calculation is shown both for the spectral width and a maximum of the spectral power density. Again, a very good agreement between the theory and experiment is demonstrated.

It is also important that one of the key RFL characteristic — the output RFL power $P_{0,L}^{\text{out}}(\Omega) = -\ln(R_{0,L})P^{\mp}(\Omega)$ — can also be easily found

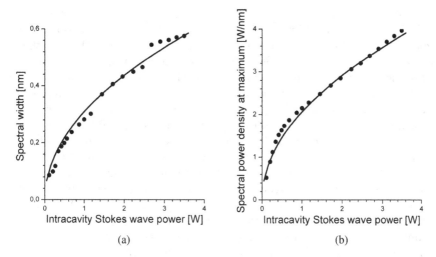

Fig. 4.6. Experimentally measured (dots) and theoretically calculated (solid line) spectral width (a) and spectral power density at maximum (b).

when applying the theory developed in previous sections. In the case of the identical Gaussian cavity mirrors, the running Stokes wave power is $P^{\pm}(\Omega) \simeq P(\Omega)/2$ which results in

$$P_{0,L}^{\text{out}}(\Omega) = \frac{\delta_0 + \delta_2 \Omega^2}{2} \frac{I}{\pi \Gamma \cosh(2\Omega/\Gamma)}. \tag{4.25}$$

A more detailed analysis of the output power in the case of highly transmitting output FBG is performed in Babin *et al.* (2008a).

4.4.3. *Statistical properties and optical rogue wave generation via wave turbulence in RFLs*

Extreme value statistics in nonlinear fiber optics have attracted great attention since the experimental demonstration of optical rogue waves in the supercontinuum (SC) generation (Solli *et al.*, 2007). The statistics of extreme events have already been studied in a number of optical systems including pulsed and CW pumped SC sources (Dudley *et al.*, 2008), silicon (Borlaug *et al.*, 2009) and fiber (Hammani *et al.*, 2008) Raman amplifiers, optical filamentation (Kasparian *et al.*, 2009), optical data transmission (Vergeles and Turitsyn, 2011) and others.

There is a variety of nonlinear optical systems with rather different physical mechanisms attributed to the emergence of the rare intense events. In the case of SC generation, the extreme events are defined by the complex interplay of the modulation instability (MI), third-order dispersion and collisions of Raman-shifted solitons (Dudley *et al.*, 2008; Mussot *et al.*, 2009). The analysis of coherent nonlinear structures e.g. the breathing solutions of the nonlinear Schrödinger equation provides useful insight into the physics of rogue waves in optical system (Akhmediev *et al.*, 2009; Akhmediev, Soto-Crespo and Ankiewicz, 2009; Dudley *et al.*, 2009) and classical problems of rogue waves appearance on the water surface (Kuznetsov, 1997; Kawata and Inoue, 1978; Akhmediev and Korneev, 1986; Peregrine, 1984; Shrira and Georgajev, 2007; Kharif *et al.*, 2009; Zakharov and Gelash, 2011; Dyachenko and Zakharov, 2011; Chabchoub *et al.*, 2011).

The rare occurrence of large amplitude waves, certainly, is not restricted to systems with coherent structures. In Raman amplifiers (Hammani *et al.*, 2008) and Raman parametric amplifiers (Hammani *et al.*, 2009), the main mechanism is the exponential transfer of the intensity fluctuations from the pump wave to the signal via a Raman response, which can be significantly enhanced in the case of the weak walk-off parameter (Hammani *et al.*, 2008). Similarly, in the optical filamentation processes, the noise transfer from the pump plays a key role (Kasparian *et al.*, 2009). It has been also observed that sporadic rogue wave events emerge from turbulent fluctuations and manifest themselves as bursts of light during the propagation of the optical wave along the long fiber (Hammani *et al.*, 2010).

It is likely that extreme rare events could also occur in the partially coherent (PC) continuous wave (CW) fiber lasers as the radiation of such systems presents a large number of longitudinal modes. Here, we examine the emergence of intense rare events through optical wave turbulence in the high-Q cavity RFL. The spectral, temporal and statistical properties of the RFLs can be calculated numerically from the system of coupled NLS equations (Agrawal, 1995):

$$\pm \frac{\partial A_p^\pm}{\partial z} + \beta_{1p} \frac{\partial A_p^\pm}{\partial t} + \frac{i}{2} \beta_{2p} \frac{\partial^2 A_p^\pm}{\partial t^2} + \frac{\alpha_p}{2} A_p^\pm$$

$$= i\gamma_p(|A_p^\pm|^2 + 2|A_s^\pm|^2)A_p^\pm - \frac{g_p}{2}(|A_s^\pm|^2 + \langle|A_s^\mp|^2\rangle)A_p^\pm$$

$$\pm \frac{\partial A_s^\pm}{\partial z} + \beta_{1s} \frac{\partial A_s^\pm}{\partial t} + \frac{i}{2} \beta_{2s} \frac{\partial^2 A_s^\pm}{\partial t^2} + \frac{\alpha_s}{2} A_s^\pm$$

$$= i\gamma_s(|A_s^\pm|^2 + 2|A_p^\pm|^2)A_s^\pm + \frac{g_s}{2}(|A_p^\pm|^2 + \langle|A_p^\mp|^2\rangle)A_s^\pm, \qquad (4.26)$$

where A is the complex field envelope, z is a coordinate, t stands for time, $v_s = 1/\beta_{1s}$ and $v_p = 1/\beta_{1p}$ are the Stokes and pump waves group-velocities, respectively, $\beta_2, \alpha, \gamma, g$ are dispersion, linear attenuation, Kerr and Raman coefficients, $\pm$ denote counter-propagating waves, "s" and "p" are used for the Stokes and pump waves. Angular brackets here denote averaging over time that is applied to reflect the fact of very large walk-off (difference in group velocities) between waves moving in opposite directions effectively leading to averaging of the impact from the counter-propagating waves. Alternatively, one can use the exponential approximation for the pump wave instead of a separate equation for pumping waves (see e.g. Randoux *et al.* (2011) and Falkovsky (2004) for details).

Currently, it is not practical to solve the Cauchy problem numerically to model the radiation generation in a typical fiber laser due to the level of required computational resources. Therefore, in (4.26), the powers of counter-propagating waves are included through their average effect. This is justified by the observation that counter-propagating waves move fast with regard to each other and, as a result, the overall impact is averaged out as in (4.26). Equations (4.26) are integrated along z using the iterative approach, i.e. when integrating equations for $A_{s,p}^+$ we used $A_{s,p}^-$ obtained from the previous iteration, and vice versa. Note that, strictly speaking, model (4.26) is not a system of equations in the mathematical sense, i.e. we do not solve these equations simultaneously. We solve the equations for forward and backward propagating waves one-by-one, i.e. forward propagating wave on the nth round trip interacts with backward propagating wave on the $(n-1)$th round trip which is already computed. The boundary conditions at the edges of the resonator are standard reflection conditions imposed by the FBGs or other reflectors. Similar iterative procedures have been used in the numerical modeling of RFL (Randoux *et al.*, 2011), YDFLs (Turitsyn *et al.*, 2011) and Brillouin fiber lasers (Preda *et al.*, 2011). Note that in this laser system the exponential transfer of fluctuations from the pump wave to the generation wave (XPM between pump and Stokes waves) does not make any significant contribution to the dynamics well above the generation threshold (see Fig. 4.7). Effectively, the corresponding terms proportional to $\gamma A_s A_p^2$ could be omitted in the equations (4.26).

Computations show that the generation becomes stable after about 10^2–10^4 round trips depending on the power. Results do not depend on the grid size at 2^{12}–2^{16} grid points. We checked that, neither decreasing the time steps, nor increasing the numerical window, affect the results. The following coefficients have been used in numerical modeling: $\alpha_p = 0.5\,(\text{km})^{-1}$, $\alpha_s = 0.83\,(\text{km})^{-1}$ (α values also include lumped losses on splices, couplers and so on), $\beta_{2p} = 17.9\,\text{ps}^2/\text{km}$, $\beta_{2s} = 7.17\,\text{ps}^2/\text{km}$, $\gamma_s = 3\,(\text{km*W})^{-1}$, $g_s = 1.3\,(\text{km*W})^{-1}$, the cavity length $L = 370\,\text{m}$. The computed generated total intra-cavity Stokes wave power agrees rather well with the experimental data (Fig. 4.7) and analytical theory (compare with Fig. 4.4).

Properties of radiation can be manipulated by using different spectral shapes of reflectors (see e.g. Turitsyna *et al.* (2010) and also Sec. 4.5.3).

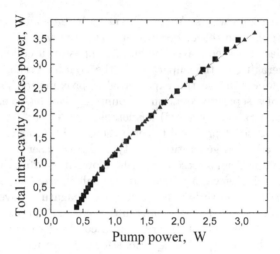

Fig. 4.7. Total generated intra-cavity power: experimental (squares), numerically calculated without XPM (dotted line) and with XPM (triangles).

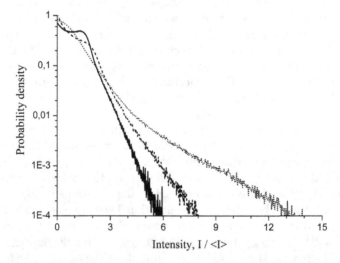

Fig. 4.8. Intensity PDF of the total intra-cavity (black line), reflected back to the cavity from the laser mirror (dashed line) and output (dotted line) laser radiation.

Considering the super-Gaussian FBGs of 0.5 nm width, we have found that the statistical properties of the generated radiation are different for the intra-cavity radiation incident on the laser mirror, the radiation reflected back to the cavity by the mirror and the output radiation as shown in Fig. 4.8. Randoux *et al.* (2011) observed that the amplitude PDF changes its shape during propagation from one laser mirror to another.

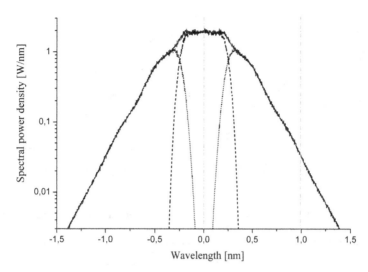

Fig. 4.9. Total intra-cavity (black line), reflected from the laser mirror (dashed line) and output (dotted line) spectrum generated in high-Q cavity RFL. Vertical dashed lines indicate the central positions of two spectral.

The origin of the difference in statistical properties before and after transmission/reflection at the laser mirror can be understood as follows. The generation spectrum well above the threshold is broader than the spectral profiles of the laser mirrors, as seen in Fig. 4.9. There are two principally different parts in the spectrum: the central part and the spectral wings. As the laser mirrors are highly reflective, the central part of the spectrum is built during many cavity trips. Turbulent-like FWM processes (Babin *et al.*, 2007a) lead to the generation of the spectral wings. Being detuned from the spectrum center by more than the laser reflector (mirror) bandwidth, the far spectral wings are out-coupled from the cavity at every round trip at each mirror. Therefore, effectively, they are generated at single pass only. Thus, the statistical properties of the intra-cavity radiation, in general, are defined mainly by the central part, rather than by the spectral wings. The power in the central part is several times higher than the power in the wings. On the contrary, the statistical properties of the out-going radiation are defined mainly by the spectral wings, as the central part is filtered out by the laser mirrors.

Since the intensity (power) PDF of the intra-cavity radiation is sufficiently nonexponential, it means that some correlations between different longitudinal modes do exist (Churkin *et al.*, 2010). Using spectral filters of different spectral widths centered at different spectral positions (filter 1 is

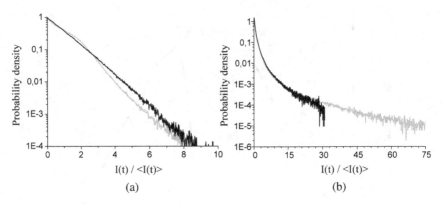

Fig. 4.10. Intensity PDF of the modes filtered out from the (a) spectrum center and (b) spectral wing (filter is detuned by 1 nm from the spectrum center). Filters spectral full widths are 0.1 nm (black) and 0.5 nm (gray).

centered at the maximum of the generated spectrum, filter 2 is detuned by 1 nm from the spectrum center), we can calculate the intensity PDF of a specific part of the generation spectrum. It is found that the intensity PDF of the modes generated on the flat top at the spectrum center is exponential (Fig. 4.10(a)). With the increase of the filter bandwidth, the PDF deviates more and more from the exponential. The statistics of the central part of the spectrum is close to the Gaussian one. However, the statistics of the distant spectral components are more influenced by nonlinear phenomena in the RFL. Deviations from the exponential form of the intensity PDF of spectral components at far spectral wings reveal the occurrence of rare events — waves with high amplitude occur with a probability higher than that given by the corresponding Gaussian distribution (Fig. 4.10(b)). The extreme value events are more pronounced in the output radiation of the high-Q cavity RFL. It is interesting to note that the self-filtering of extreme events by the laser mirror is similar to the designed spectral filtering approach used previously in Raman fiber amplifiers (RFAs) (Finot *et al.*, 2009).

Extreme value events can be detected from the analysis of the time series of the generated radiation. The temporal behavior of the total intra-cavity radiation is irregular with the occurrence of high amplitude fluctuations having an amplitude several times higher than the mean value as seen in Fig. 4.11(a). This kind of temporal intensity evolution is typical for stochastic signals, which consist of many independent modes with the Gaussian statistics. However, the situation is dramatically changed, considering that the output radiation that produces the rare fluctuations has the amplitude more than 100 times higher than the average output power in the given spectral region (Fig. 4.11(b)).

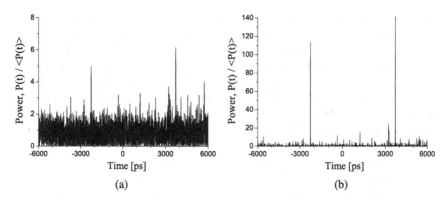

Fig. 4.11. (a) The time dynamics of the total intra-cavity radiation and (b) radiation generated within the 0.5 nm spectral width at the far spectral wing.

The typical temporal width of such an extreme rare event at the laser output is about 5–10 ps. This makes it difficult to detect them directly in real time, as such events occur irregularly.

Rogue waves of high amplitudes are generated in RFLs at far spectral wings, thus, such waves are generated in a one cavity pass only. However, the physical mechanism of rare events in RFLs is different from those observed in RFAs. In RFAs, the signal propagating along the fiber acquires the intensity fluctuations from the pump wave, which can evolve into extreme amplitude pulses under conditions of self-induced MI (Hammani *et al.*, 2008). In RFLs, the noise transfer from pump wave to the generation wave is negligible. The key role is played by the effect of turbulent FWM of a multitude of longitudinal modes in the generated Stokes waves (Babin *et al.*, 2007a). Occurrence of rogue waves is statistical, resulting from turbulent interactions leading to the generation of far spectral components of very high amplitudes. The turbulent mechanism of extreme waves generation in RFLs can be potentially described using the formalism of wave turbulence similar to that of propagating waves in a long fiber by Hammani *et al.* (2010). The similar mechanism of extreme wave generation can be important in other types of fiber lasers including lasers operating via random distributed feedback (Turitsyn *et al.*, 2010; Fotiadi *et al.*, 2011).

4.5. Optical Wave Turbulence in Ultra-Long Fiber Lasers

The number of generated modes increases with the lengthening of the fiber cavity that may influence their turbulent interaction and laser characteristics, as a result. Although the concept of ultra-long Raman fiber laser (URFL) architecture was initially proposed and implemented for development of a quasi-lossless signal transmission scheme, see e.g. Ania-Castañón *et al.* (2006) and references therein, the problem is more general: a substantial increase of the RFL cavity length leads to a new interesting class of lasers with potentially different physical mechanisms underlying their operation. In this context, various fundamental questions arise: What are the limits of the cavity length for laser operation? What are the particular spectral features of URFLs — cavity mode structure, output spectrum and corresponding temporal properties of the radiation generated inside the cavity?

We address these questions through the analysis of the spectral and temporal characteristics of URFLs with cavity lengths from 6.6 km to 84 km operating at 1455 nm, and up to 270 km for 1550-nm laser operating just in the fiber transparency maximum. We will show that important characteristics of such ultra-long fiber lasers can be directly explained by the occurrence of weak turbulent-like interactions of a huge number (∼100 millions in URFL) of longitudinal laser modes.

4.5.1. *Basics of ultra-long fiber lasers*

The basic design of the ultra-long laser cavity studied in this work is schematically depicted in Fig. 4.12. Without a loss of generality, as a first step, we consider a URFL designed to provide distributed signal amplification in the telecommunication window at 1550 nm similar to the scheme treated in Ania-Castañón *et al.* (2006). However, we would like to stress that the concept of URFL presented here is very general and is not limited to wavelengths considered in this paragraph.

The URFL system (see Fig. 4.12) consists of two equal-power depolarized primary Raman pumps with a nm-broad spectrum centered at 1365 nm which are launched from both ends of the standard single-mode fiber (SMF)

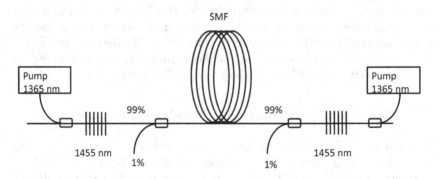

Fig. 4.12.   Schematic depiction of the URFL.

span. Two highly reflective FBGs (reflection coefficient is 98% at 1455 nm) were used at each end of the fiber span. This results in the formation of a high-Q cavity trapping the first Stokes counter-propagating waves between the two FBGs. When the power of the primary pumps is above the required threshold for the Raman gain to overcome the fiber attenuation, the laser starts to generate radiation at the Stokes wavelength of 1455 nm. Two 99:1 couplers were placed near the FBGs at the right (C1) and left (C2) ends of the span to monitor the generated intra-cavity Stokes wave power P and its optical and RF spectra.

The RF spectrum, monitored using a photodetector and an electrical spectrum analyzer, displays inter-mode beating peaks with a clear mode structure despite the large cavity length up to some critical power. The optical spectrum of the generated Stokes wave was monitored through an optical spectrum analyzer with a resolution of ∼0.01 nm. The Stokes wave power was measured by a power meter, whereas its temporal behavior was studied with a fast oscilloscope with a 50 ps resolution.

Figure 4.13(a) shows the total intra-cavity power P of the Stokes wave (1455 nm); measured via the 1% port at point C2, as a function of the total pump power at 1365 nm. The generated power exhibits typical laser behavior. Above the threshold required for the SRS to overcome fiber attenuation ($\alpha_{1455} \approx 0.25\,\mathrm{dB/km}$) and lumped losses of FBGs, couplers and connections ($\delta_0 \approx 1.4\,\mathrm{dB}$), the laser starts generation at 1455 nm. Experimental results are in good agreement with modeling (solid lines in Fig. 4.13(a)) of the cavity modes dynamics using ordinary differential equations similar to Eq. (26) describing FWM interactions (Babin *et al.*, 2008b).

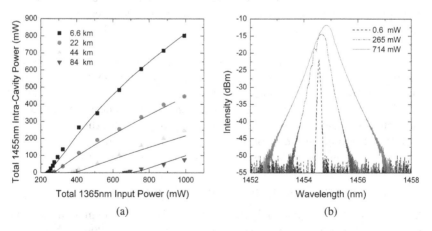

Fig. 4.13. (a) Total intra-cavity power at 1455 nm as a function of the total pump power at 1365 nm: experimental points and numerical simulation (solid curves). (b) 1455 nm spectrum measured at point C2, Stokes power = 0.6, 265 and 714 mW for $L = 6.6\,\mathrm{km}$.

Though the governing dynamical equations are deterministic, the solutions should be treated as a result of an irregular process: the FWM nonlinear process engages a huge number of rapidly oscillating terms with different amplitudes and phases. Above the threshold, the Stokes wave power grows nearly linearly with increasing pump power. As expected, the longer the cavity, the lower is the generated 1455 nm power, due to higher cavity losses caused by the attenuation of the pump ($\alpha_{1365} \approx 0.31\,\text{dB/km}$) and Stokes waves, which raise the threshold. The measured experimental threshold for the URFL (Babin *et al.*, 2008b) is in complete agreement with the simple theoretical RFL model, which was confirmed earlier by experiments with short fiber spans (Babin *et al.*, 2003). For longer spans, distributed losses remain the main factor and the threshold is nearly proportional to the length. Despite this, even for the longest cavity length of 84 km, the threshold pump power is rather small, $\sim$0.7 W.

The evolution of the intra-cavity Stokes wave optical spectra is shown in Fig. 4.13(b) for the 6.6 km case. The spectrum is rather narrow near the threshold and broadens with increasing pump power. Furthermore, the broadened spectrum acquires clear exponential tails for all cavity lengths studied, similar to a conventional RFL, see Fig. 4.5. To trace how the spectral structure is affected by the boundary condition, we used FBGs with ripples in the short-wave wing of their reflection profile. Figure 4.13(b) shows that the spectrum for Stokes power = 265 mW has the same low wavelength ripples as the FBGs, which are not seen for Stokes power = 714 mW: the generated spectra follow the FBGs reflection profile at low powers, but are not influenced by the boundary conditions at higher powers. A drift towards longer wavelengths with increasing power was observed, but the measurements of the gratings spectral response revealed a shift of the central wavelength of the FBGs — attributed to thermal expansion — as the cause.

The broadening mechanism and its dependence on the input power can be understood from the analysis of the interaction between the intra-cavity longitudinal modes based on the weak wave turbulence model of the RFL (Babin *et al.*, 2007a). The turbulent-like interaction of the modes also results in stochastic behavior of the integral power in the time domain exhibiting fluctuations in various time scales with an amplitude of noisy spikes reaching levels exceeding the average level by over 50%.

4.5.2. *Mode structure in ultra-long fiber lasers*

The existence of a longitudinal mode structure is one of the most important laser characteristics. It has been shown (Babin *et al.*, 2005) that for a Raman laser with $L = 0.37\,\text{km}$ cavity length, a corresponding RFL mode structure with spacing as small as $\Delta = c/2nL \sim 0.3\,\text{MHz}$ (n is refractive index, c is the speed of light in vacuum) is observable in its RF spectrum, see Sec. 4.4.2. At the same time, the observed peaks are broadened significantly with the increasing power of the intra-cavity generated Stokes wave indicating the impact of the nonlinear interactions of modes. An increase in the cavity length should proportionally reduce the mode spacing, which may be critical for resolving the mode structure. In this paragraph, we review the measurement techniques and the results of mode structure analysis in URFLs with cavity length varying from $L = 6.6$ km to $L = 84$ km presented in Babin *et al.* (2007b). To examine the mode structure of the URFL cavity, the signals from the 1% splitter ends (see Fig. 4.12) were analyzed using the fast photodiode and electrical spectrum analyzer with a resolution of <400 Hz, to obtain RF spectra in which inter-mode beating can be observed as being similar to conventional RFL (Babin *et al.*, 2005).

The experiment shows that the inter-mode beating peaks are resolvable for all cavity lengths studied, and that the spectral spacing Δ between them follows the classical formula $\Delta = c/2nL$. For the cavity length $L = 84$ km, the corresponding mode spacing is as low as $\Delta \approx 1.2$ kHz, while the width of the RF peaks is narrowed to hundreds of Hz. Furthermore, we have observed the broadening of the modal peaks for all the considered cavity lengths as the intra-cavity power is increased. Figure 4.14 shows the measured spectral

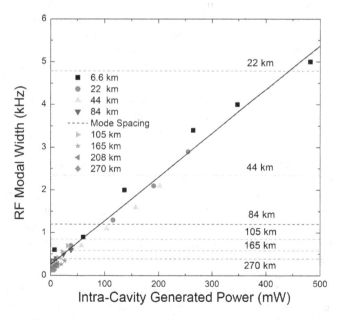

Fig. 4.14. Mode beating peak width (FWHM) for 6.6, 22, 44, 84, 105, 165, 208 and 270 km lasers (differently marked points) and their linear fit. Corresponding mode spacing values are shown by horizontal lines.

width (full width at half a maximum, i.e. spectral width at -3 dB level) of the RF peaks versus the total intra-cavity power. Here, several points (105 to 270 km cavity) have been added from the measurements of a different laser configuration that will be discussed in detail in Sec. 4.5.5. It is clearly seen that the spectral width of the RF peaks depends on the intra-cavity power in a nearly linear manner, and is almost independent of the fiber span length (Turitsyn *et al.*, 2009).

The width values extrapolated to zero power nearly corresponds to the nameplate resolution of the electric spectrum analyzer used. One quite intriguing observation from the presented results is that propagation in such a long span does not drastically impact the relative phases of the modes, resulting in narrow RF peaks at low powers. This illustrates the standing wave formation between the reflectors of the linear Fabry–Perot cavity with a length of ~ 100 km. The result is contrary to a natural expectation that various effects such as thermal noise and fiber span fluctuations would lead to mode dephasing in such long laser cavities.

It was expected that with the growing intra-cavity power, nonlinear optical effects would play an increasingly important role in determining laser characteristics. The role of nonlinear effects in the RF spectra of relatively short RFL with high-Q cavity has been treated by Babin *et al.* (2007a) in the framework of the weak turbulence theory, see Sec. 4.4.1. It has been suggested that with increasing intensity of laser radiation, the main contribution to dephasing of Stokes waves in the cavity comes from the effective "nonlinear attenuation" effect arising from the FWM-induced turbulent-like interaction of waves with different frequencies (corresponding to different cavity modes). This dephasing process is greatly enhanced in the ultra-long cavity because of the significantly larger number of interacting modes. In the case of the URFL, the mode number estimated as the ratio of the optical spectrum width and the mode spacing is really huge (10^7–10^8), which enhances stochastic effects.

We would like to stress that such a stochastic, turbulent-like behavior of the modes leads to a rather specific broadening of the RF spectra, proportional to the intensity generated. The width of the inter-mode beating peaks D appears almost independent of the cavity length and grows linearly with the increasing Stokes wave intensity P confirming the major role of nonlinear attenuation and the stochastic nature of the cavity mode interactions, see Eqs. (4.15) and (4.16).

At the same time, the mode spacing $\Delta = c/2Ln$ decreases with cavity length from $\Delta \sim 15.5$ kHz at $L = 6.6$ km to $\Delta \sim 1.2$ kHz at $L = 84$ km. The power limit for a resolvable mode structure may be defined as the value for which the modal width equals the mode spacing (i.e. $D \approx \Delta$). Higher power result in the generation of "modeless" stochastic spectra with an exponential-wing envelope.

It has been demonstrated by Turitsyn *et al.* (2009) that the mode structure of the ultra-long fiber laser can be resolved up to a record cavity length of 270 km. As we will show below, with further increase of of the laser cavity, the Rayleigh scattering starts to play a critical role in limiting resolvable mode operations of URFL. The very narrow RF peaks reveal that the relative phase fluctuations between the neighboring modes remains very small (at low power), even propagating through such a long span. These results prove the feasibility of a new class of lasers with ultra-long cavity. At the same time, significant turbulent dephasing occurs with an increase of the generated RFL power to the level of tens of mW inside the ultra-long cavity due to their interaction via multiple FWM nonlinear processes. As such, the model of optical spectra formation in URFLs based on the treatment of modes with random phases may be developed, which is done in the next paragraph.

4.5.3. *Nonlinear broadening of optical spectra*

Typical intra-cavity spectra generated by the URFL were shown in Fig. 4.13(b). Extending the laser cavity to $L > 100$ km greatly increases the number of modes ($\sim 10^8$), enhancing the effects of wave turbulence. This is confirmed by the strong broadening of the Stokes spectrum, which is significant even at mW level (see Fig. 4.15). The spectrum has well-defined exponential tails. The spectral width of the Stokes wave increases nonlinearly with power, as seen in Fig. 4.15, and does not vary

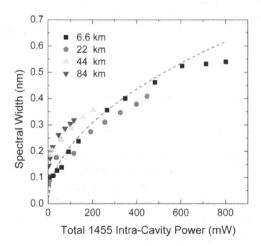

Fig. 4.15. Spectral widths at -3 dB of power distribution at 1455 nm as a function of the total 1455 nm power for different fiber cavity lengths $L = 6.6$, 22, 42 and 84 km. The dashed lines represent the fit by function $y = Ax^{1/2}$.

significantly with length for identical powers. A square-root fit appears to be a good approximation of the spectral width dependence on the power at $-3\,\mathrm{dB}$ level, as predicted analytically (Babin *et al.*, 2007). Note, however, that the analytical theory is not directly applicable in the case of URFLs.

The analysis of the impact of the FBG reflection profile demonstrates the washing out of the FBG-induced ripples after propagation through the fiber at high powers. Using an additional splitter at intermediate points, we observed that the spectrum acquired its characteristic exponential tails without ripples after <6 km and propagates further in the cavity without broadening. Therefore, we conclude that the system's memory of the FBG profile is lost in the nonlinear turbulent evolution of cavity modes. Experimental results are in good agreement with the modeling of modes dynamics using ordinary differential equations describing FWM interactions (Babin *et al.*, 2008b). It was also shown later by detailed numerical analysis (Soh *et al.*, 2010) that the nonlinear broadening (after kilometers of propagation) is stabilized by dispersion.

The new class of ultra-long lasers is characterized by a range of particular properties, such as the exponential wings of the generated optical spectra, which broaden nonlinearly (close to square-root) with increasing intensity. Their mode structure is resolvable in the RF spectra, in a limited power interval from the generation threshold up to a maximum power dependent on the cavity length. Above a given power value, the stochastic mode dephasing due to uncorrelated fluctuations of mode frequencies induced by FWM leads to broadening prevailing over mode separation and to the generation of a "modeless" or "quasi-continuous" spectrum. In extremely long cavities (>270 km) the spectrum is modeless even at the lasing threshold, because of the distributed feedback provided by the Rayleigh backscattering, see (Turitsyn *et al.*, 2009).

4.6. Developed Optical Wave Turbulence in Fiber Lasers

4.6.1. *The impact of fiber dispersion*

In treating a large number of interacting modes (up to 10^8 for long cavity fiber lasers) that share between them a finite generated power, one can assume the phases of different modes to be random and the interaction to be weak (weak-turbulence approach described in (Zakharov *et al.* (1992)). In quantitative terms, this can be expressed as the following condition on the effective nonlinearity/dispersion ratio

$$\xi = \gamma P / |\beta_2| \Omega_{\mathrm{rms}}^2 \ll 1. \tag{4.27}$$

Here $P = \sum_m |E_m|^2$ is total generated power,

$$\Omega_{\mathrm{rms}} = \sum_m m^2 |E_m|^2 \Big/ \sum_m |E_m|^2 * \Delta.$$

is spectral bandwidth, Δ — spectral separation between modes. In the weak wave turbulence approach, the results would be *insensitive to the sign of wave dispersion* β_2. However, it was demonstrated (Turitsyna *et al.*, 2009) that the change of the dispersion sign dramatically affects the spectral shape and statistics of the laser radiation. The observed wave turbulence cannot be treated in terms of the weak wave turbulence as the impact of nonlinear interactions is not small compared to other effects. Properties of the turbulent behavior of generated radiation present an example of the developed wave turbulence with strong nonlinear interactions between waves.

The main difference between the cases of normal ($\beta_2 > 0$) and anomalous ($\beta_2 < 0$) dispersion is an MI, which exists at $\beta_2 < 0$. For normal dispersion, the CW is stable with respect to infinitesimal perturbations. As we will show, the effect of dispersion on wave turbulence is rather nontrivial. The stable CW can be formed in the case of normal dispersion $\beta_2 > 0$, however, nonlinear interaction between several modes corresponding to real-world CW can lead to rather complicated dynamics leading effectively to a nonlinear instability of CW. Overall, this might add to a further spectrum broadening due to FWM. Spectrum broadening at $\beta_2 > 0$ is due to the competition between FWM and dispersion, so that one can expect that the width is determined by the balance of dispersion and nonlinearity, $|\beta_2|\overline{\Omega}_{\mathrm{rms}}^2 \cong \gamma P$, i.e. comparable to that determined by MI.

This is confirmed by direct numerical modeling of Eq. (4.5). Indeed, as seen in Fig. 4.16, ξ practically does not depend on $|\beta_2|L$ and the pump power. In the interval of pump powers between $500\,\mathrm{mW}$ and $100\,\mathrm{mW}$: $\xi = 1.7$ for $\beta_2 < 0$ and $\xi = 3.3$ for $\beta_2 > 0$. It is seen that the spectral width is such that the effective dispersion is almost twice larger for $\beta_2 > 0$ when it must balance nonlinearities, MI and FWM, acting together to widen the spectrum. Since $\xi > 1$, the system behavior cannot be explained by the weak-turbulence approximation in all these cases.

Figure 4.17 shows the average spectra of generated radiation after many round trips. While the integral characteristics such as the root-mean-square spectral width are comparable, the spectra for anomalous dispersion, $\beta_2 < 0$, are smoother and have a characteristic triangular shape with more narrow peaks compared to more concave and irregular spectra for normal dispersion, $\beta_2 > 0$. In the experiment, we have studied two RFLs with the cavities of $13.5\,\mathrm{km}$ built of two commercially available fibers — SMF, with anomalous dispersion $\beta_2 = -11.63\,\mathrm{ps}^2/\mathrm{km}$ (here $\beta_2 L \sim -157\,\mathrm{ps}^2$) and IDF

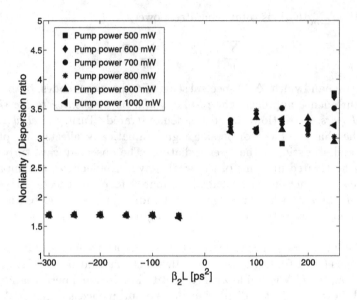

Fig. 4.16.   Nonlinearity/Dispersion ratio for different values of β_2.

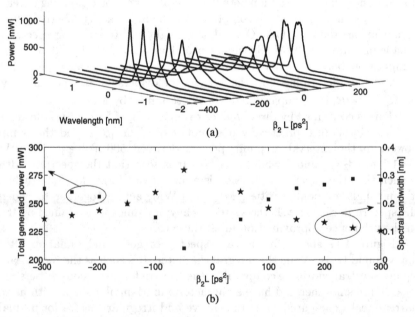

Fig. 4.17.   (a) Spectra for different $\beta_2 L$; (b) Spectral bandwidth and total generated power versus $\beta_2 L$. Total pump power 700 mW.

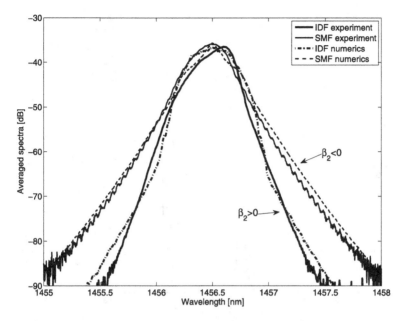

Fig. 4.18. Comparison of spectra of numerical simulation and experiment for SMF ($\beta2 < 0$) and IDF ($\beta_2 > 0$) fibers with total pump power 600 mW.

(inverse dispersion fiber) with normal dispersion $\beta_2 = 36.44\,\text{ps}^2/\text{km}$ (here $\beta_2 L \sim 492\,\text{ps}^2$).

We would like to stress that the theoretical model works rather well, for instance, Fig. 4.18 demonstrates a good agreement between numerical and experimental results for the generated spectra (here the total pump power was 600 mW) (Turitsyna *et al.*, 2011).

4.7. Spectral Condensate in Fiber Lasers

The evolution in the case of normal dispersion is peculiar but interesting. In the numerical simulations, we considered $L = 22$ km fiber laser and different values of pump powers (from 400–1000 mW) and dispersion ($\beta_2 L$ in the interval $[-300, 300]\,\text{ps}^2$). We have observed that in the case of anomalous dispersion ($\beta_2 < 0$), the generated spectra became quasi-steady after just a few round trips. In this case fluctuations in the generated power evolution are small compared to the average power level — of the order of few mW (dashed and dotted line in Fig. 4.19). For the normal dispersion ($\beta_2 > 0$), a very narrow condensate consisting of only few modes is formed initially.

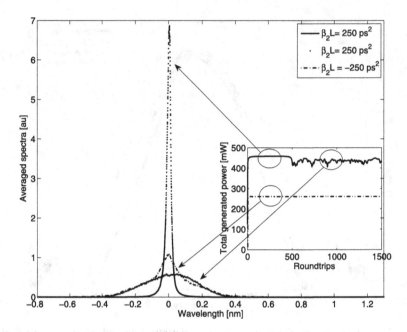

Fig. 4.19. Averaged spectra for normal and anomalous dispersion: solid line — normal dispersion condensate state, dotted line — normal dispersion destroyed condensate, dashed and dotted line — anomalous dispersion. Inset picture: the corresponding generated power evolution for normal and anomalous dispersions.

It persists for a time that depends on the number of modes and the absolute value of dispersion. During the condensate lifetime, the total intensity is constant with high accuracy (solid line in Fig. 4.19, inset picture). Then the condensate experiences nonlinear instability leading to its destruction. The condensate destruction is manifested by a sharp transition to a wider spectrum and a lower mean power. That new (statistically steady) state is accompanied by strong fluctuations which seem to be a sign of bi-stability (Turitsyna *et al.*, 2009).

Any transition like this one illustrates that the spectrum becomes broader and more top flat after about 500 round trips, with heavily oscillating tails (Fig. 4.20). It reflects the intensity curve evolution in Fig. 4.19 (inset picture). In the case of the opposite sign of dispersion (anomalous dispersion), we do not observe a few. In the case of anomalous dispersion, the spectrum is generated from the zero noise level, then after a few round trips (when the total power reaches a certain equilibrium level) it is stabilized and remains almost steady, with slightly oscillating tails (Fig. 4.21). Figure 4.22 characterizes fluctuations of the generated power by histograms,

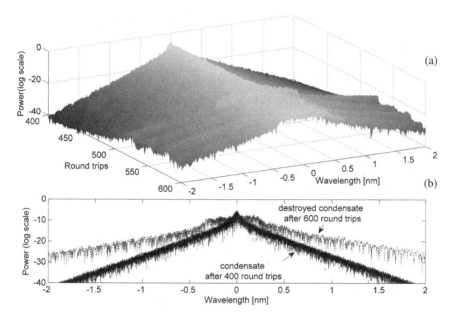

Fig. 4.20. (a) Spectra evolution for the case with normal dispersion, when the condensate destroys after 500 round trips; (b) the corresponding picture of the spectra for condensate (after 300 round trips) and after it is destroyed (after 600 round trips).

with the condensate distribution clearly being much narrower than both other histograms. Not surprisingly, the condensate is a less fluctuating state.

Note also that its histogram is very asymmetric with the highest value being most probable. It can be naturally interpreted as fluctuations being only "dips" in the condensate such as gray and dark solitons. An abrupt decay of the probability of intensities exceeding a certain value can be thought of as an interesting effect of nonlinear self-induced optical limiting. This prevents the laser power from random overshoot and might be of practical importance in power sensitive applications in laser systems providing for a stable mode condensate. On the contrary, when the condensate is destroyed, the histograms are much wider and more symmetric with substantial probability of fluctuations exceeding the mean value.

Now, we discuss the impact of a number of generated modes on the building of an equilibrium state. As explained in Sec. 4.3, one can estimate from this, that for FBG of 0.1 nm bandwidth (corresponds to $\delta_2 = 277\,\text{nm}^{-2}$) and 800 mW of total pump power, there will be ~32,280 modes with the spectral separation between modes $\Delta = \lambda^2/(2Ln) = 7.3 \times 10^{-7}\,\text{nm}$ (here $n = 1.45$ is a refractive index in silica). As we consider

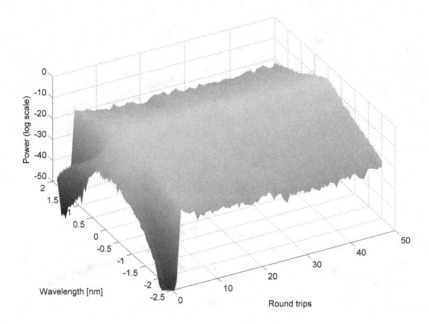

Fig. 4.21.   Spectra evolution for anomalous dispersion.

a spectral window eight times larger than the FBG bandwidth, it gives the total number of modes used in simulations $2^{20} = 1{,}048{,}576$. Figure 4.23 demonstrates the evolution of the generated spectra for fiber laser with normal dispersion cavity with this number of modes. As seen from Fig. 4.23, there is no clearly pronounced condensate state for this number of modes in the normal dispersion cavity.

Indeed, we have observed that the occurrence of a condensate state and properties of evolution do depend on the number of modes. For instance, Fig. 4.24 shows how evolution can be drastically changed by increasing the total number of interacting modes.

Transition to the developed turbulence from a linearly stable condensate is an interesting problem that will be discussed in full detail elsewhere. However, we present here a brief outline showing the most important nontrivial aspects of this nonlinear problem. We have investigated how resistant the condensate state may be to small structural perturbations. We extracted the field after the initial stage when the spectra became stable in the condensate state, and then imposed a small perturbation ($\sim$10%) to the 5th (from the center) mode (Fig. 4.25). The results of these observations, the following evolution of the spectra and the generated power of the perturbed condensate state are presented in Figs. 4.26 and 4.27.

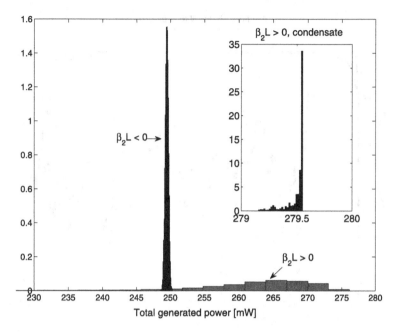

Fig. 4.22. Typical histogram of total generated power for normal (condensate and destroyed condensate state) and anomalous dispersion. Here $\beta_2 L = \pm 150 \, \text{ps}^2$.

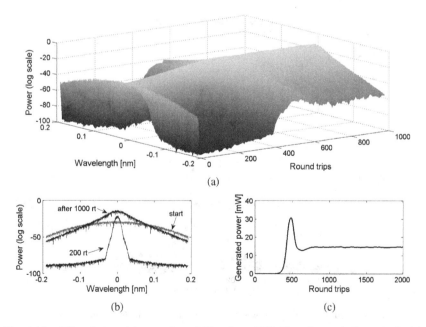

Fig. 4.23. Fiber laser with a cavity of 1 km long IDF fiber (normal dispersion). (a) spectra evolution for the first 1000 round trips; (b) spectra profiles on different stages — at the start, after 200 round trips and after 1000 round trips; (c) evolution of the generated power.

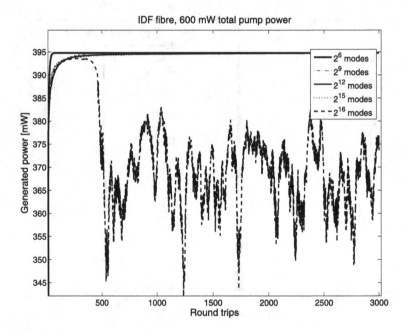

Fig. 4.24.   Generated power evolution for different number of modes for 600 mW total pump power, 13.5 km fiber cavity length.

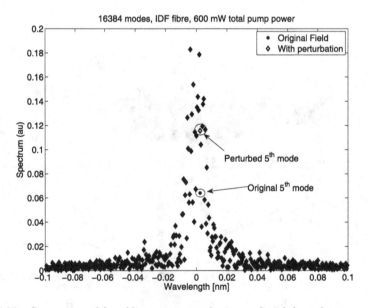

Fig. 4.25.   Spectrum used for adding some perturbation to the 5th from the center mode.

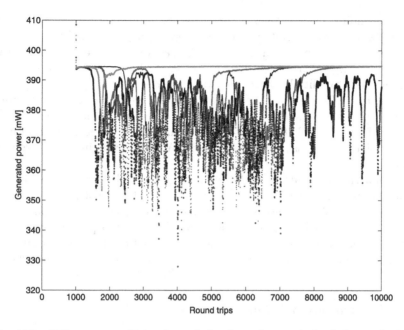

Fig. 4.26. Different types of intensity evolution depending on the level of perturbation.

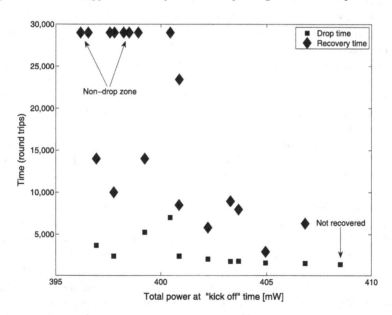

Fig. 4.27. Dependence of condensate's drop and recovery time on the total power at the time of perturbation.

Depending on the level of perturbation, we observed (as shown in Fig. 4.26) different evolutions in the generated intensity of the field: drop and recovery, nondrop and drop and not recovery. Figure 4.26 illustrates (corresponding quantitative data are given in Fig. 4.27) how the initial perturbation amplitudes affect the condensate.

Figure 4.27 quantifies the dynamics shown in Fig. 4.26 by presenting dependence of drop time and recovery time for condensate for IDF fiber, total power of 600 mW, 16,384 modes, with perturbed 5th mode. The generated power of the condensate state is 394.5 mW; and drop time measured at 393 mW.

Numerical simulations with a number of modes less than $2^{16} = 65,536$ demonstrated a rather long existence of the condensate state. In the case of a larger number of modes the condensate state can also be observed, however, it did not last long and experienced a sharp transition to a strongly fluctuating regime. Experimental observation of the condensate is a challenging problem and these studies will be presented in detail elsewhere.

4.8. Conclusions and Perspectives

We have presented an interesting interdisciplinary link between laser science and the theory of wave turbulence. Fiber lasers can have cavity lengths ranging from a few meters to several hundred kilometers. Generated fiber laser power is distributed between multitudes of very small amplitude cavity modes. However, the resonator is not linear, and nonlinear interactions between the modes do play an important role in the operation and performance of such lasers. Spectra of the generated radiation experience spectral broadening that is dependent on power. Thus, the nonlinear Kerr effect does affect the propagation of light in fiber lasers and leads to a nonlinear mixing of cavity modes. Any particular interaction between resonator modes can be considered as a very weak one — the properties of each wave are not changed substantially during a single interaction event. However, despite the small amplitudes of propagating waves, the overall effect can be important because of the huge number of modes involved in the interactions. The efficiency of interactions depends on resonance conditions between the phases of any four participating waves — the FWM process. Even weak nonlinear interactions through the FWM process lead to a random energy transfer between waves and to an enhancement of mode dephasing. This is because it involves a continuum of modes propagating with significant group velocity dispersion, resulting in exponential tales of the generated spectra. Note that other manifestations of the Kerr nonlinearity such as self-phase or cross-phase modulation effects might be important for overall optical field evolution, but they do not lead to the inter-mode phase difference as they change the mode phases synchronously. We stress that modes' dynamics are randomized not by noise, but due to the FWM nonlinear process that engages a huge number of deterministic interactions of rapidly oscillating modes with different amplitudes and phases making time evolution of any particular cavity mode extremely stochastic. Note that lasing is due to the Rayleigh backscattering results in a similar effect: the generation of a stochastic continuum of frequency components. As a result, the spectrum generated in a random distributed feedback laser is nearly the same as having a close bandwidth ($\sim$1 nm at half maximum) and similar exponential tails.

The situation with stochastic modes is somewhat similar to the classic physical problem in which long term average properties of the system are determined by a random behavior of a huge number of weakly interacting waves (which could be of quite different physical origins) — *weak wave turbulence* (Zakharov *et al.*, 1992). This link between the research areas of wave turbulence and laser physics also offers an opportunity to study the properties of wave turbulence in an optical device (laser) that allows very precise measurements. Note that recently such a similarity was successfully used in studies of optical rogue waves (Solli *et al.*, 2007). Our results demonstrated that the sign of cavity dispersion has a dramatic impact on optical wave turbulence that determines the spectral and temporal properties of generated radiation, directly related to the performance of fiber laser. For normal dispersion, we observe an intermediate state with an extremely narrow spectrum — spectral condensate, which experiences instability and a sharp transition to a strongly fluctuating regime. Power histograms show that normal dispersion increases the probability of a spontaneous generation of large amplitude pulses — optical rogue waves. For anomalous dispersion, we have observed triangular spectra and more coherent temporal behavior of the generated radiation.

Ultra-long lasers allow spatio-spectral transparency at long distances over a broad bandwidth even with only single-wavelength laser pumps, thanks to the extended signal gain bandwidth provided by the superposition of pump and cascading Stokes wave (Ellingham *et al.*, 2006; Ania-Castanon *et al.*, 2008). Recent results showing simultaneous transparency over 20 nm for standard fiber links of 20 km length (Ania-Castanon *et al.*, 2008) have opened the gate to a series of novel applications, from signal processing devices, in which different channels can be made to interact nonlinearly with each other in nondissipative conditions, to quantum communication systems, relying on continuous quantum variables, for which these trans-parent links might provide an ideal medium to perform experiments on self-phase modulation squeezing of quantum optical solitons (Drummond *et al.*, 1993). Quantum communications with transmission of a few photons are another option for which our transparent links seem to be suitable in principle, although this would, of course, require an extremely precise stabilization of the cavity characteristics, as well as of the energy of the Stokes wave trapped in the cavity. In addition, ultra-long laser cavities can be exploited in fundamentally new practical approaches for optical information transmission and secure communications. For instance, the concept of nonquantum secure key distribution based on establishing laser oscillations between the sender and receiver has been recently proposed in Scheuer (2006) Yariv and Zadok *et al.* (2008). Lasers with such a large number of cavity modes interacting through wave turbulent mechanisms have never been studied before. Therefore, we believe that our results may open an entirely new research field closely linked with different areas of physics such as nonlinear science, the theory of disordered systems, wave turbulence and others. In addition to such interesting connections with many areas of fundamental science, we also anticipate that new applications and technologies will keep emerging from the study of the physics of ultra-long fiber lasers.

Acknowledgments

We would like to acknowledge valuable discussions with E.A. Kuznetsov and the financial support of the Leverhulme Trust, the Royal Society, FP7 Marie Curie project IRSES, the European Research Council, and the research grants of the Russian Federation Government N11.G34.31.0035; N11.519.11.4001 and N11.519.11.4018.

Bibliography

Agrawal G. 1995. *Nonlinear Fibre Optics.* London: Academic Press.
Akhmediev N, Korneev V. 1986. Modulation instability and periodic solutions of the nonlinear Schrödinger equation. *Teoret. Mat. Fiz.* 69: 189–194.
Akhmediev N, Ankiewicz A, Soto-Crespo JM. 2009. Rogue waves and rational solutions of the nonlinear Schrödinger equation. *Phys. Rev. E* 80: 026601.
Akhmediev N, Soto-Crespo JM, Ankiewicz A. 2009. Extreme waves that appear from nowhere: On the nature of rogue waves. *Phys. Lett. A* 373: 2137–2145.
Ania-Castañón JD. 2004. Quasi-lossless transmission using second-order Raman amplification and fiber Bragg gratings. *Opt. Exp.* 12: 4372–4377.
Ania-Castañón JD, Ellingham TJ, Ibbotson R, Chen X, Zhang L, Turitsyn SK. 2006. Ultralong raman fiber lasers as virtually lossless optical media. *Phys. Rev. Lett.* 96: 023902.

Ania-Castanon JD, Karalekas V, Harper P, Turitsyn SK. 2008. Simultaneous spatial and spectral transparency in ultralong fiber lasers. *Phys. Rev. Lett.* 101: 123903.

Aschieri P, Garnier J, Michel C, Doya V, Picozzi A. 2011. Condensation and thermalization of classsical optical waves in a waveguide. *Phys. Rev. A* 83: 033838.

Babin SA, Churkin DV, Podivilov EV. 2003. Intensity interactions in cascades of a two-stage Raman fiber laser. *Opt. Comm.* 226: 329–335.

Babin SA, Churkin DV, Fotiadi AA, Kablukov SI, Medvedkov OI, Podivilov EV. 2005. Relative intensity noise in cascaded Raman fiberlasers. *IEEE Photon. Technol. Lett.* 17: 2553–2555.

Babin SA, Churkin DV, Ismagulov AE, Kablukov SI, Podivilov EV. 2006. Spectral broadening in Raman fiber lasers. *Opt. Lett.* 31: 3007–3009.

Babin SA, Churkin DV, Ismagulov AE, Kablukov SI, Podivilov EV. 2007a. FWM-induced turbulent spectral broadening in a long Raman fiber laser. *J. Opt. Soc. Am. B* 24: 1729–1738.

Babin SA, Karalekas V, Harper P, Podivilov EV, Mezentsev VK, Ania-Castañón JD, Turitsyn SK. 2007b. Experimental demonstration of mode structure in ultralong Raman fiber lasers. *Opt. Lett.* 32: 1135–1137.

Babin SA, Churkin DV, Ismagulov AE, Kablukov SI, Podivilov EV. 2008a. Turbulence-induced square-root broadening of the Raman fiber laser output spectrum. *Opt. Lett.* 33: 633–635.

Babin SA, Karalekas V, Podivilov EV, Mezentsev VK, Harper P, Ania-Castañón JD, Turitsyn SK. 2008b. Turbulent broadening of optical spectra in ultralong Raman fiber lasers. *Phys. Rev. A* 77: 033803.

Barviau B, Kibler B, Picozzi A. 2009. Wave-turbulence approach of supercontinuum generation: Influence of self-steepening and higher-order dispersion. *Phys. Rev. A* 79: 063840.

Bernard D, Boffetta G, Celani A, Falkovich G. 2006. Conformal invariance in two-dimensional turbulence. *Nat. Phys.* 2: 124–128.

Bernard D, Boffetta G, Celani A, Falkovich G. 2007. Inverse turbulent cascades and conformally invariant curves. *Phys. Rev. Lett.* 98: 024501.

Borlaug D, Fathpour S, Jalali B. 2009. Extreme value statistics in silicon photonics. *IEEE Photon. J.* 1: 33–39.

Bortolozzo U, Laurie J, Nazarenko S, Residori S. 2009. Optical wave turbulence and the condensation of light. *J. Opt. Soc. A* 26: 2280–2284.

Bouteiller JC. 2003. Spectral modeling of Raman fiber lasers. *IEEE Photon. Technol. Lett.* 15: 1698–1700.

Cao H, Xu JY, Zhang DZ, Chang S-H, Ho ST, Seelig EW, Liu X, Chang RPH. 2000. Spatial confinement of laser light in active random media. *Phys. Rev. Lett.* 84: 5584–5587.

Cao H. 2005. Review on latest developments in random lasers with coherent feedback. *J. Phys. A,* 38: 10497–10535.

Chabchoub A, Hoffmann N, Akhmediev, N. 2011. Rogue wave observation in a water wave tank. *Phys. Rev. Lett.* 106: 204502.

Chernikov SV, Zhu Y, Taylor JR, Gapontsev VP. 1997. Supercontinuumself-Q-switched ytterbium fiber laser. *Opt. Lett.* 22: 298–300.

Chernikov SV, Lewis AE, Taylor JR. 1999. 'Broadband Raman amplifiers in the spectral range of 1480–1620 nm. *Proc. Opt. Fibre Conf.*, WG6-1.

Churkin DV, Smirnov SV, Podivilov EV. 2010. Statistical properties of partially coherent cw fiber lasers. *Opt. Lett.* 35: 3288–3290.

Dalloz N, Randoux S, Suret P. 2010. Influence of dispersion of fiber Bragg grating mirrors on formation of optical power spectrum in Raman fiber lasers. *Opt. Lett.* 35: 2505–2507.

Dianov EM, Grekov MV, Bufetov IA, Vasiliev SA, Medvedkov OI, Plotnichenko VG, Koltashev V, Belov AV, Bubnov MM, Semjonov SL, Prokhorov AM. 1997. CW High power 1.24 μm and 1.48 μm Raman lasers based on low loss phosphosilicate fibre. *Electron. Lett.* 33: 1542–1544.

Dianov EM, Bufetov IA, Bubnov MM, Grekov MV, Vasiliev SA, Medvedkov OI. 2000. Three-cascaded 1407-nm Raman laser based onphosphorus-doped silica fiber. *Opt. Lett.* 25: 402–404.

Drummond PD, Shelby RM, Friberg SR, Yamamoto Y. 1993. Quantum solitons in optical fibers. *Nature* 365: 307–313.

Dudley JM, Genty G, Coen S. 2006. Supercontinuum generation in photonic crystal fiber. *Rev. Mod. Phys.* 78: 1135–1184.

Dudley JM, Genty MVG, Eggleton BJ. 2008. Harnessing and control of optical rogue waves in supercontinuum generation. *Opt. Exp.* 16: 3644–3651.

Dudley JM, Genty G, Dias F, Kibler B, Akhmediev N. 2009. Modulation instability, Akhmediev Breathers and continuous wave supercontinuum generation. *Opt. Exp.* 17: 21497–21508.

Dyachenko A, Newell AC, Pushkarev A, Zakharov VE. 1992. Optical turbulence: Weak turbulence, condensates and collapsing filaments in the nonlinear Schrödinger equation. *Physica D* 57: 96–160.

Dyachenko A, Falkovich G. 1996. Condensate turbulence in two dimensions. *Phys. Rev. E* 54: 5095–5099.

Dyachenko AI, Zakharov VE. 2011. Compact equation for gravity waves on deep water. *JETP Lett.* 93: 701–705.

Ellingham TJ, Ania-Castañón JD, Ibbotson R, Chen X, Zhang L, Turitsyn SK 2006. Quasi-lossless optical links for broad-band transmission and data processing. *IEEE Photon. Technol. Lett.* 18: 268–270.

Falkovich G, Kolokolov I, Lebedev V, Migdal A. 1996. Instantons and intermittency. *Phys. Rev. E* 54: 4896–4907.

Falkovich G, Kolokolov I, Lebedev V, Turitsyn S. 2001. Statistics of soliton-bearing systems with additive noise. *Phys. Rev. E* 63: 025601(R).

Falkovich GE. 2006. Introduction to turbulence theory. In *Lecture Notes on Turbulence and Coherent Structures in Fluids, Plasmas and Nonlinear Media*, Vol. 4, Chap. 1, Singapore: World Scientific, pp. 1–21.

Falkovich G, Sreenivasan KR. 2006. Lessons from hydrodynamic turbulence. *Phys. Today* 59: 43–49.

Falkovsky LA. 2004. Investigation of semiconductors with defects using Raman scattering. *Phys. Usp.* 47: 249272.

Fallert J, Dietz RJB, Sartor J, Schneider D, Klingshirn C, Kalt H. 2009. Coexistence of strongly and weakly localized random lasermodes. *Nat. Photon.* 3: 279–282.

Finot C, Hammani K, Fatome J, Dudley JM, Millot G. 2009. Selection of extreme events generated in Raman fibre amplifiers through spectral offset filtering. *IEEE J. Quantum Electron.* 46: 205–213.

Fotiadi AA, Kiyan RV. 1998. Cooperative stimulated Brillouinand Rayleigh backscattering process in optical fiber. *Opt. Lett.* 23: 1805–1807.

Fotiadi AA, Me'gret P, Blondel M. 2004. Dynamics of a self-Q-switched fiber laserwith a Rayleigh-stimulated Brillouin scattering ring mirror. *Opt. Lett.* 29: 1078–1080.

Fotiadi A, Preda E, Mégret P. 2011. Brillouin fibre laser with incoherent feedback. *Proc. CLEO: 2011 — Laser Applications to Photonic Applications*, OSA Technical Digest (CD), paper CTuI6. USA: Optical Society of America.

Frazao O, Correia C, Santos JL, Baptista JM. 2009. Raman fiber Bragg-gratinglaser sensor with cooperative Rayleigh scattering for strain-temperature measurement. *Meas. Sci. Technol.* 20: 045203.

Frisch U. 1995. *Turbulence: The Legacy of A. N. Kolmogorov.* Cambridge: Cambridge University Press.

Gottardo S, Sapienza R, García PD, Blanco A, Wiersma DS, López C. 2008. Resonance-driven random lasing. *Nat. Photon.* 2: 429–432.

Grubb SG, Strasser T, Cheung WY, Reed WA, Mizrachi V, Erdogan T, Lemaire EJ, Vengsarkar AM, Digiovanni DJ. 1995. High power 1.48 gm cascaded Raman laser in germanosilicate fibers. *Proc. Optical Amp. their Appl.* pp. 197–199.

Hammani K, Finot C, Dudley JM, Millot G. 2008. Optical rogue-wave-like extreme value fluctuations in fiber Raman amplifiers. *Opt. Exp.* 16: 16467–16474.

Hammani K, Finot C, Millot G. 2009. Emergence of extreme events in fiber-based parametric processes driven by a partially incoherent pump wave. *Opt. Lett.* 34: 1138–1140.

Hammani K, Kibler B, Finot C, Picozzi A. 2010. Emergence of rogue waves from optical turbulence. *Phys. Lett. A* 374: 3585–3589.

Hammani K, Picozzi A, Finot C. 2011. Extreme statistics in Raman fiber amplifiers: From analytical description to experiments. *Opt. Commun.* 284: 2594–2603.

Han Y-G, Tran TVA, Kim S-H, Lee SB. 2005. Development of a multiwavelength Raman fiber laser based on phase-shifted fiber Bragg gratings for long-distance remote-sensing applications. *Opt. Lett.* 30: 1114–1116.

Han Y-G, Moon DS, Chung Y, Lee SB. 2005. Flexibly tunable multiwavelength Raman fiber laser based on symmetrical bending method. *Opt. Exp.* 13: 6330–6335.

Haus HA, Nakazawa M. 1987. Theory of the fiber Raman soliton laser. *J. Opt. Soc. Am. B* 4: 652–660.

Headley C, Agrawal G. 2004. *Raman Amplification in Fibre Optical Communication Systems.* New York: Academic Press.

Herrmann J, Wilhelmi B. 1998. Mirrorless laser action by randomly distributed feedback in amplifying disordered media with scattering centers. *Appl. Phys. B* 66: 305–312.

Hsiung P-L, Chen Y, Ko T, Fujimoto J, de Matos C, Popov S, Taylor J, Gapontsev V. 2004. Optical coherence tomography using a continuous-wave, high-power, Raman continuum light source. *Opt. Exp.* 12: 5287–5295.

Huang SH, Feng Y, Shirakawa A, Ueda K. 2003. Generation of 10.5 W, 1178 nm laser based on phosphosilicate Raman fiber laser. *Jpn. J. Appl. Phys.* 42: L1439–L1441.

Imam H. 2008. Metrology: Broad as a lamp, bright as a laser. *Nat. Photon.* 2: 26–28.

Janssen PAEM. 2003. Nonlinear four-wave interactions and freak waves. *J. Phys. Oceanogr.* 33: 863–884.

Juarez JC, Maier EW, Kyoo NC, Taylor HF. 2005. Distributed fiber-opticintrusion sensor systems. *J. Lightwave Technol.* 23: 2081–2087.

Karpov V, Papernyi SB, Ivanov V, Clements WRL. 2004. Cascaded pump delivery for remotely pumpederbium doped fiber amplifiers. *Proc. SUBOPTIC Conf.*, 8.8.

Kasparian J, Béjot P, Wolf J-P, Dudley JM. 2009. Optical rogue wave statistics in laser filamentation. *Opt. Exp.* 17: 12070–12075.

Kawata T, Inoue H. 1978. Exact solutions of the derivative nonlinear Schrödinger equation under the nonvanishing conditions. *J. Phys. Soc. Jpn.* 44: 1968–1976.

Kharif C, Pelinovsky E, Slunyaev A. 2009. *Rogue Waves in the Ocean.* Berlin: Springer Verlag.

Kim NS. 2008. Review on the high-power pulse fibre laser technology and their industrial microelectronics applications. *Rev. Laser Eng. Supplemental Volume*, pp. 1115–1118.

Kivshar Yu. S, Agrawal GP. 2003. *Optical Solitons: From Fibres to Photonic Crystals.* New York: Academic Press.

Kobtsev S, Smirnov S. 2005. Modelling of high-power supercontinuum generation in highly nonlinear, dispersion shifted fibers at CW pump. *Opt. Exp.* 13: 6912.

Kolmogorov AN. 1941. Local structure of turbulence in an incompressible fluid at very high Reynolds numbers. *Doklady AN SSSR* 30: 299–303.

Kuznetsov EA. 1977. Solitons in a parametrically unstable plasma. *Sov. Phys. Dokl.* 22: 507–508.

Lawandy NM, Balachandran RM, Gomes ASL, Sauvain E. 1994. Laser actionin strongly scattering media. *Nature* 368: 436–438.

Lega J, Moloney JV, Newell AC. 1994. Swift-Hohenberg equation for lasers. *Phys. Rev. Lett.* 73: 2978–2981.

Lega J, Moloney JV, Newell AC. 1995. Universal description of laser dynamics near threshold. *Physica D* 83: 478–498.

Letokhov VS. 1968. Generation of light by a scattering medium with negativeresonance absorption. *Sov. Phys. JETP* 26: 835–840.

Lushnikov PM, Vladimirova N. 2010. Non-Gaussian statistics of multiple filamentation. *Opt. Lett.* 35: 1965–1967.

Ma Y-C. 1978. The complete solution of the long-wave-short-wave resonance equations. *Stud. Appl. Math.* 59: 201–221.

Markushev VM, Zolin VF, Briskina Ch. M. 1986. Powder laser. *Zh. Prikl. Spektrosk.* 45: 847–850.

Mermelstein MD, Headley C, Bouteiller J-C, Steinvurzel P, Horn C, Feder K, Eggleton BJ. 2001. Configurable three-wavelength Raman fiber laser for Raman amplification and dynamic gain flattening. *IEEE Photon. Technol. Lett.* 13: 1286–1288.

Michel C, Garnier J, Suret P, Randoux S, Picozzi A. 2011. Kinetic description of random optical waves and anomalous thermalization of a nearly integrable wave System. *Lett. Math Phys.* 96: 415–447.

Mitschke F, Steinmeyer G, Schwache A. 1996. Generation of one-dimensional optical turbulence. *Physica D* 96: 251–258.

Mollenauer LF, Gordon JP, Islam MN. 1986. Soliton propagation in long fibers with periodically compensated loss. *IEEE J. Quan. Electr.* 22: 157–173.

Mollenauer LF, Gordon JP. 2006. *Solitons in Optical Fibres: Fundamentals and Application.* Burlington, MA: Academic Press.

Mussot A, Kudlinski A, Kolobov M, Louvergneaux E, Douay M, Taki M. 2009. Observation of extreme temporal events in CW-pumped supercontinuum. *Opt. Exp.* 17: 17010–17015.

Nazarenko S. 2011. *Wave Turbulence*, Lecture Notes in Physics. Berlin: Springer-Verlag.

Newell AC, Nazarenko S, Biven L. 2001. Wave turbulence and intermittency. *Physica D* 152: 520–550.

Noginov MA. 2005. *Solid-State Random Lasers.* Berlin: Springer-Verlag.

Papernyi SB, Karpov VJ, Clements WRL. 2002. Third-order cascaded Raman amplification. *Proc. Optical Fibre Conf.*, FB4.

Preda CE, Ravet G, Fotiadi AA, Mégret P. 2011. Iterative method for Brillouin fiber ring resonator. *Proc. CLEO/Europe and EQEC 2011 Conf. Digest*, paper CJ_P27. USA: Optical Society of America.

Randoux S, Dalloz N, Suret P. 2011. Intracavity changes in the field statistics of Raman fiber lasers. *Opt. Lett.* 36: 790–792.

Ravet G, Fotiadi AA, Blondel M, Megret P. 2004. Passive Q-switching inall-fiber Raman laser with distributed Rayleigh feedback. *Electron. Lett.* 40: 528–529.

Rayleigh L, Strutt JW. 1899. On the transmission of light through anatmosphere containing small particles in suspension and on the origin of the blue sky. *Philos. Mag.* 47: 375–384.

Scheuer J, Yariv A. 2006. Giant fibre lasers (GFL): A new paradigm for secure key distribution. *Phys. Rev. Lett.* 97: 140502.

Shrira VI, Geogjaev VV. 2010. What makes the Peregrine soliton so special as a prototype of weak waves? *J. Eng. Math.* 67: 11–22.

Skryabin DV, Luan F, Knight JC, Russell PS. 2003. Soliton self-frequency shift cancellation in photonic crystal fibers. *Science* 301: 1705–1708.

Soh DBS, Koplow JP, Moore SW *et al.* 2010. The effect of dispersion on spectral broadening of incoherent continuous-wave light in optical fibers. *Opt. Exp.* 18: 22393–22405.

Solli DR, Ropers C, Koonath P, Jalali B. 2007. Optical rogue waves. *Nature* 450: 1054–1057.

Smirnov SV, Ania-Castanon JD, Ellingham TJ, Kobtsev SM, Kukarin S, Turitsyn SK. 2006. Optical spectral broadening and supercontinuum generation in telecom applications. *Opt. Fibre Techn.* 12: 122–147.

Stolen RH, Ippen EP, Tynes AR. 1972. Raman oscillation in glass optical waveguide. *Appl. Phys. Lett.* 20: 62–64.

Stolen R, Ippen E. 1973. Raman gain in glass optical waveguides. *Appl. Phys. Lett.* 22: 276–278.

Suret P, Randoux S. 2004. Influence of spectral broadening on steady characteristics of Raman fiber lasers: From experiments to questions about validity of usual models. *Opt. Commun.* 237: 201–212.

Türeci HE, Ge L, Rotter S, Stone AD. 2008. Strong interactions in multimode random lasers. *Science* 320: 643–646.

Turitsyn SK, Ania-Castanon JD, Babin SA, Karalekas V, Harper P, Churkin D, Kablukov SI, El-Taher AE, Podivilov EV, Mezentsev VK. 2009. 270-km ultralong Raman fibre laser. *Phys. Rev. Lett.* 103: 133901.

Turitsyn SK, Babin SA, El-Taher AE, Harper P, Churkin D, Kablukov SI, Ania-Castanon JD, Karalekas V, Podivilov EV. 2010. Random distributed feedback fiber laser. *Nat. Photon.* 4: 231–235.

Turitsyn SK, Bednyakova AE, Fedoruk MP, Latkin AI, Fotiadi AA, Kurkov AS, Sholokhov E. 2011. Modeling of CW Yb-doped fiber lasers with highly nonlinear cavity dynamics. *Opt. Exp.* 19: 1227–1230.

Turitsyna EG, Falkovich G, Mezentsev SK, Turitsyn SK. 2009. Optical turbulence and spectral condensate in long-fiber lasers. *Phys. Rev. A* 80, 031804(R).

Turitsyna EG, Turitsyn SK, Mezentsev VK. 2010. Numerical investigation of the impact of reflectors on spectral performance of Ramanfiber laser. *Opt. Exp.* 18: 4469–4477.

Turitsyna EG, Falkovich G, El-Taher A, Harper P, Shu X, Turitsyn SK. 2011. Optical turbulence and spectral condensate in fiber lasers. *Proc. IQEC/CLEO Pacific Rim 2011*.

Trulsen K, Dysthe KB. 1997. Freak waves — a three-dimensional wave simulation, *Proc. 21st Symp. Naval Hydrodynamics*, pp. 550–560.

Vasilyev M. 2003. Raman-assisted transmission: Toward ideal distributed amplification. *Proc. Optical Fibre Conf.*, p. 303.

Vergeles S, Turitsyn SK. 2011. Optical rogue waves in telecommunication data streams. *Phys. Rev. A* 83: 061801(R).

Wang Y, Po H. 2003. Impact of cavity losses on cw Raman fiber lasers. *Opt. Eng.* 42: 2872–2879.

Wiersma DS, Cavalieri S. 2001. A temperature tunable random laser. *Nature* 414: 708–709.

Wiersma DS. 2008. The physics and applications of random lasers. *Nat. Phys.* 4: 359–367.

Wiersma DS. 2009. Laser physics: Random lasers explained? *Nat. Photon.* 3: 246–248.

Zadok A, Scheuer J, Sendowski J, Yariv A. 2008. Secure key generation using an ultra-long fiber laser: transient analysis and experiment, *Opt. Exp.* 16: 16680–16690.

Zakharov VE, L'vov VS, Falkovich G. 1992. *Kolmogorov Spectra of Turbulence.* Berlin: Springer-Verlag.

Zakharov V, Dias F, Pushkarev A. 2004. One-dimensional wave turbulence. *Phys. Rep.* 398: 1–65.

Zakharov V, Nazarenko S. 2005. Dynamics of the Bose-Einstein condensation. *Physica D* 201: 203–211.

Zakharov VE, Gelash AA. 2011. Solitons on unstable condensate. arxiv.org: 1109.0620v2.

Chapter 5

Wave Turbulence in a Thin Elastic Plate: The Sound of the Kolmogorov Spectrum?

G. Düring* and N. Mordant[†]

*Center for Soft Matter Research, Department of Physics,
New York University, New York, NY 10003

[†]Laboratoire des Ecoulements Géophysiques et Industrial,
Université de Grenoble Alpes, Domaine Universitaire, BP53,
38041 Grenoble cedex France & Institut Universitaire de France

Thin metal plates have been used for ages to simulate the thunder noise in theatres. Turbulence of flexion waves in the plate radiates acoustic waves in the air. In this way it is possible to hear the Kolmogorov spectrum (as suggested by the title of the original article on the application of the Weak Turbulence theory to this issue)! In this chapter, we present the results of the application of the theory to flexion waves in a thin elastic plate. Then we compare the experimental results to the theoretical predictions. Advanced measurements of the plates deformation have ben implemented using a high-speed Fourier transform profilometry technique. This technique provides a 2D measurement of the deformation resolved in time. In this way a precise space and time analysis of wave turbulence could be performed for the first time in a real system. This analysis shows that the observed turbulence is in agreement with the phenomenology of the Weak Turbulence Theory but that there is a quantitative disagreement between observation and theoretical predictions. We discuss the possible reasons for such discrepancy: finite size effects or dissipation.

Contents

5.1. Weak Turbulence Theory for Thin Elastic Plates

5.1.1. *The Föppl–von Kármán equations for a thin elastic plate*

We aim at studying the statistical properties of thin elastic plates under external low frequency (compared with the plate modes) random forces. Several dissipative mechanisms can exist: dissipation through the boundary, thermoelasticity, viscoelasticity and sound radiation. Here, we assume that the forcing and the various dissipation mechanisms exist at very different length scales. This separation in the characteristic dissipation and injection scales provides an inertial region in which the dynamics is conservative, in direct analogy with hydrodynamic turbulence. The latter acts only as a transport mechanism for the conserved quantities between different scales.[1]

The dynamics of a conservative oscillating thin elastic plate can be formulated, as a first approximation, by the dynamical version of the Föppl–von Kármán equations (Föppl, 1907; von Kármán, 1910; Landau and Lifshitz, 1959) for the amplitude of the deformation $\zeta(x, y, t)$ and for the Airy stress function $\chi(x, y, t)$:

$$\rho \frac{\partial^2 \zeta}{\partial t^2} = -\frac{Eh^2}{12(1 - \sigma^2)} \Delta^2 \zeta + \{\zeta, \chi\}, \qquad (5.1)$$

$$\frac{1}{E} \Delta^2 \chi = -\frac{1}{2} \{\zeta, \zeta\}, \qquad (5.2)$$

where h is the thickness of the elastic sheet. The material has a mass density ρ, a Young modulus E and its Poisson ratio is σ. $\Delta = \partial_{xx} + \partial_{yy}$ is the usual Laplacian and the bracket $\{\cdot, \cdot\}$ is defined by $\{f, g\} \equiv f_{xx}g_{yy} + f_{yy}g_{xx} - 2f_{xy}g_{xy}$, which is an exact divergence, so Eq. (5.1) preserves the momentum of the center of mass, namely $\partial_{tt}(h\rho \int \zeta(x, y, t)\, dx dy) = 0$. The first term on the right-hand side of (5.1) represents the bending. The second term accounts for nonlinear stretching, a direct consequence of the Gaussian curvature.

---

[1]This assumption has been partially validated in experiments, however, its consequence on the dynamics is still unclear and matter of current research.

Equation (5.2) for the Airy stress function $\chi(x, y, t)$ may be seen as the compatibility equation for the in-plane stress tensor. In the derivation of equations (5.1) and (5.2), we have omitted the inertia of the in-plane modes of oscillations, or, in other words, we assume that the in-plane displacements are negligible and static equilibrium holds. The linear regime of Eq. (5.1) sustains small plane-wave perturbations which are dispersive with the usual ballistic behavior of bending waves (Rayleigh, 1945) $\omega_{\mathbf{k}} = \sqrt{\frac{Eh^2}{12(1-\sigma^2)\rho}} |\mathbf{k}|^2$.

The Föppl–von Kármán equations (5.1) and (5.2) may be derived from a variational principle. They correspond to the Euler–Lagrange equations obtained from the Lagrangian density

$$\mathcal{L} = h \left(\frac{\rho}{2} \dot{\zeta}^2 - \mathcal{F}[\zeta, \chi] \right), \tag{5.3}$$

where $\mathcal{F}$ is the free energy per unit volume of the plate

$$\mathcal{F}[\zeta, \chi] = \left[\frac{h^2 E}{24(1 - \sigma^2)} (\Delta \zeta)^2 - \frac{1}{2E} (\Delta \chi)^2 - \frac{1}{2} \chi \{\zeta, \zeta\} \right]. \tag{5.4}$$

The problem that we shall describe concerns the evolution of the wave amplitudes for different normal modes of oscillation; then, it is natural to use variables in Fourier space:

$$\zeta(\mathbf{r}, t) = \frac{1}{2\pi} \int \zeta_{\mathbf{k}}(t) e^{i\mathbf{k} \cdot \mathbf{r}} d\mathbf{k},$$

$$\chi(\mathbf{r}, t) = \frac{1}{2\pi} \int \chi_{\mathbf{k}}(t) e^{i\mathbf{k} \cdot \mathbf{r}} d\mathbf{k}.$$

Since $\zeta(\mathbf{r}, t)$ and $\chi(\mathbf{r}, t)$ are real fields, one has $\zeta_{\mathbf{k}} = \zeta_{-\mathbf{k}}^*$ and $\chi_{\mathbf{k}} = \chi_{-\mathbf{k}}^*$. The Airy stress function $\chi(x, y, t)$ is a passive field that only follows the deformation, which possesses its own dynamics. From Eq. (5.2), it is easy to see that $\chi_{\mathbf{k}}$ can be explicitly written in terms of $\zeta_{\mathbf{k}}$.

The Hamiltonian structure for this system can now be easily obtained under a Legendre transformation in terms of $\zeta_{\mathbf{k}}$ and its conjugate variable $p_{\mathbf{k}} = \rho h \dot{\zeta}_{\mathbf{k}}$. It is convenient to rewrite the Hamiltonian in terms of the complex canonical variables $(A_{\mathbf{k}}, A_{\mathbf{k}}^*)$, which are the classical analogs of creation-annihilation operators in quantum mechanics:

$$A_{\mathbf{k}} = \frac{1}{\sqrt{2}} \left(X_{\mathbf{k}}^{-1} \zeta_{\mathbf{k}} + i X_{\mathbf{k}} p_{\mathbf{k}} \right) \quad \text{where} \quad X_{\mathbf{k}} = \frac{1}{\sqrt{\omega_{\mathbf{k}} \rho h}}. \tag{5.5}$$

A straightforward calculation leads us to the Hamilton equations for the dynamics in terms of the new canonical variables

$$\frac{dA_{\mathbf{k}}^s}{dt} = -is\omega_{\mathbf{k}}A_{\mathbf{k}}^s$$

$$- \frac{is}{(2\pi)^2} \sum_{s_1 s_2 s_3} J_{-\mathbf{k}\mathbf{k}_1\mathbf{k}_2\mathbf{k}_3} A_{\mathbf{k}_1}^{s_1} A_{\mathbf{k}_2}^{s_2} A_{\mathbf{k}_3}^{s_3} \delta(\mathbf{k}_1 + \mathbf{k}_2 + \mathbf{k}_3 - \mathbf{k})d\mathbf{k}_{123}.$$

$$(5.6)$$

We shall use the notation: $A_{\mathbf{k}}^s$ with $s = +, -$, which means that $A_{\mathbf{k}}^+ = A_{\mathbf{k}}$ and $A_{\mathbf{k}}^- = A_{-\mathbf{k}}^*$. Then, $A_{-\mathbf{k}}^{-s} = A_{\mathbf{k}}^{s*}$ and the sum takes the values $s_i = -1$ and $+1$. The abbreviation $d\mathbf{k}_{123} = d\mathbf{k}_1 d\mathbf{k}_2 d\mathbf{k}_3$ and the notation $k = |\mathbf{k}|$ will be used. The nonlinear interaction coefficients read

$$J_{\mathbf{k}_1\mathbf{k}_2\mathbf{k}_3\mathbf{k}_4} = \frac{1}{3}X_{\mathbf{k}_1}X_{\mathbf{k}_2}X_{\mathbf{k}_3}X_{\mathbf{k}_4}(T_{\mathbf{k}_1\mathbf{k}_2;\mathbf{k}_3\mathbf{k}_4} + T_{\mathbf{k}_1\mathbf{k}_3;\mathbf{k}_2\mathbf{k}_4} + T_{\mathbf{k}_1\mathbf{k}_4;\mathbf{k}_3\mathbf{k}_2}),$$

where the elastic scattering amplitude is

$$T_{\mathbf{k}_1\mathbf{k}_2;\mathbf{k}_3\mathbf{k}_4} = \frac{E}{8}\left(\frac{1}{2|\mathbf{k}_1 + \mathbf{k}_2|^4} + \frac{1}{2|\mathbf{k}_3 + \mathbf{k}_4|^4}\right)(\mathbf{k}_1 \times \mathbf{k}_2)^2(\mathbf{k}_3 \times \mathbf{k}_4)^2.$$

We note that no approximation is made in deriving the nonlinear wave equation (5.6); it is completely equivalent to the Föppl–von Kármán equations. However, the universal form of this expression allows us to directly utilize the wave turbulence formalism.

5.1.2. *Kinetic equation and spectra*

Weak turbulence theory is valid for statistically homogeneous systems where the deformation amplitudes are small. Therefore, the local variations in the deformation ζ can be used as the small parameter. We denote the order of magnitude of the deformation by ϵ; rescaling $\zeta_{\mathbf{k}} \to \epsilon\zeta_{\mathbf{k}}$ so that $A_{\mathbf{k}} \to \epsilon A_{\mathbf{k}}$, one concludes that the rescaled Hamilton equations (5.6) at the lowest order correspond to a linear system of bending waves. The wave interactions are only introduced through the cubic nonlinearities. Note the absence of a quadratic term, which is coherent with the $z \leftrightarrow -z$ symmetry of the system.

If the wave interactions are weak, there is a natural separation between the fast linear time scale of the bending waves, and the slow nonlinear time evolution. The fast propagation of linear waves decorrelates the interacting modes, driving the system to a quasi-Gaussian distribution. In parallel, the small nonlinear interactions regenerate higher order correlations through cumulative effects, which are responsible for the slow amplitude evolution. This observation can be mathematically implemented using a multi-scale perturbation analysis, leading to a kinetic equation for the statistical evolution of the wave amplitudes (Hasselmann, 1962). These kinetic equations represent the core of wave turbulence theory, and can be derived in different but closely related ways (Benney and Saffman, 1966; Benney and Newell, 1969; Zakharov *et al.*, 1992; Choi *et al.*, 2005; Newell *et al.*, 2001; Newell, 1968).

The kinetic equation is expressed in term of the wavenumber $n_{\mathbf{k}}$, which is related to the second-order moment[2] $\langle A_{\mathbf{k}} A_{\mathbf{p}}^* \rangle = n_{\mathbf{k}} \delta(\mathbf{k} - \mathbf{p})$. The wave system thus satisfies a Boltzmann-type kinetic equation describing the exchange of energy from one mode to another through four wave resonances:

$$\frac{dn_{\mathbf{k}}}{dt} = 12\pi \int |J_{\mathbf{k}\mathbf{k}_1\mathbf{k}_2\mathbf{k}_3}|^2 \sum_{s_1 s_2 s_3} n_{\mathbf{k}_1} n_{\mathbf{k}_2} n_{\mathbf{k}_3} n_{\mathbf{k}} \left(\frac{1}{n_{\mathbf{k}}} + \frac{s_1}{n_{\mathbf{k}_1}} + \frac{s_2}{n_{\mathbf{k}_2}} + \frac{s_3}{n_{\mathbf{k}_3}} \right)$$
$$\times \, \delta(\omega_k + s_1\omega_{k_1} + s_2\omega_{k_2} + s_3\omega_{k_3}) \, \delta(\mathbf{k} + s_1\mathbf{k}_1 + s_2\mathbf{k}_2 + s_3\mathbf{k}_3) \, d^2\mathbf{k}_{123}. \tag{5.7}$$

As for the usual Boltzmann equation, the kinetic equation conserves "formally"[3] the total momentum per unit area $\mathbf{P} = h \int \mathbf{k} n_{\mathbf{k}}(t) \, d\mathbf{k}$ and the kinetic energy per unit area $\mathcal{E} = \int \omega_k n_{\mathbf{k}}(t) \, d\mathbf{k}$. Unlike some of the four wave interaction kinetic equations (i.e. gravity wave in fluids or the nonlinear Schrödinger equation) the "wave action" $\mathcal{N} = \int n_{\mathbf{k}}(t) d^2k$ is not conserved. This is due to the decaying type dispersion relation (i.e $\omega \sim k^\alpha$ with $\alpha > 1$) and that the original dynamical equation (5.6) is not phase invariant. In practice, the existence of a resonance process which satisfies $\omega_k + s_1\omega_{k_1} + s_2\omega_{k_2} + s_3\omega_{k_3} = 0$ with $s_1 s_2 s_3 = -1$ forbids the "wave action" conservation. This type of interaction will be referred to as $3 \leftrightarrow 1$ process. The other ones, which indeed preserve the wavenumber in the collision, will be named $2 \leftrightarrow 2$ interaction process. The existence of two interaction types is actually a rather general situation in systems with a dominant cubic nonlinearity (e.g. capillary waves in symmetric medias (Düring and Falcon, 2009)).

The nonlinear dynamics not only affects the wave amplitude, but also modifies the dispersion relation. Wave turbulence theory predicts a correction which depends on the wave amplitude (Newell *et al.*, 2001). The renormalized dispersion relation then reads

$$\omega_{\mathbf{k}}^{\text{Ren}} = \sqrt{\frac{Eh^2}{12(1 - \sigma^2)\rho} |\mathbf{k}|^2 + 6 \int J_{-\mathbf{k}\mathbf{k}-\mathbf{p}\mathbf{p}} n(\mathbf{p}) d\mathbf{p}}. \tag{5.8}$$

---

[2]As we are considering a statistically homogeneous system, one can consider spatial averages.

[3]Here, "formally" means that the proof requires convergence of any simple integral to the exchange of integration order by Fubinis theorem (Newell *et al.*, 2001).

An H-theorem can also be derived for this system: let $\mathcal{S}(t) = \int \ln(n_{\mathbf{k}}) \, d^2 k$ be the nonequilibrium entropy, then $d\mathcal{S}/dt \geq 0$. The kinetic equation (5.7) thus describes an irreversible evolution of the wave spectrum towards the Rayleigh–Jeans *equilibrium* distribution[4] $n_{\mathbf{k}}^{\text{eq}} = \frac{T}{\omega_k}$, where T is called, by analogy with thermodynamics, the temperature (with units of energy/length, i.e. a force) which is naturally related to the initial energy by $\mathcal{E}_0 = h \int \omega_k n_{\text{eq}} d^2 \mathbf{k} = hT \int d^2 \mathbf{k}$. The quantity $\int d^2 \mathbf{k}$ is the number of degrees of freedom per unit surface. This result is a natural consequence of the equipartition theorem. Trivially, for an infinite system, the number of degrees of freedom diverges (as well as the energy). This classical Rayleigh–Jeans catastrophe is always suppressed due to some physical cut-off which corresponds here to dissipation processes for wavelengths smaller than h. Numerical simulations, where a cut-off always exists, show how the Föppl–von Kármán equations (5.1) and (5.2), evolves toward the thermal equilibrium distribution (Fig. 5.1). In real experiments, the situation is different, the dissipation at small scales turns out to be very efficient in preventing the system from thermalization, hence, the Rayleigh–Jeans distribution is hardly observed.

Notwithstanding, there exists an additional set of stationary solutions, known as Kolmogorov–Zakharov (KZ) spectra (Zakharov, 1967; Zakharov *et al.*, 1992). The great success of the wave turbulence theory stems from the prediction of these (KZ) stationary out of equilibrium spectra. These turbulent-type solutions are of major importance in the nonequilibrium energy transfer between different modes. They can often be guessed via a dimensional analysis argument, but they are, indeed, exact solutions of the kinetic equation.

The eight terms of the collisional integral in the r.h.s. of Eq. (5.7) can be decomposed into two collisional terms depending on the wave interaction process $2 \leftrightarrow 2$ or $3 \leftrightarrow 1$. This can be formally expressed as $\frac{dn}{dt} = \text{Coll}_{2 \leftrightarrow 2} + \text{Coll}_{3 \leftrightarrow 1}$. In order to find a stationary spectra, power law solutions of the form $n_k \sim k^{-\alpha}$ are considered. Under a remarkable conformal transformation first introduced by Zakharov, one finds that the collisional terms scale as $\text{Coll}_{2 \leftrightarrow 2} = C_+(\alpha) k^{2-3\alpha}$ and $\text{Coll}_{3 \leftrightarrow 1} = C_-(\alpha) k^{2-3\alpha}$, where the coefficients $C_\pm(\alpha)$ are pure real functions depending only on α. The stationary solutions are then given by the zeros of $C_+(\alpha) + C_-(\alpha)$. Clearly the equilibrium spectrum $1/\omega_k$ is a solution, but in general, out of equilibrium spectra (KZ) also exist.

For the thin elastic plate system, both coefficients vanish with double degeneracy at $\alpha = 2$ indicating that the Kolmogorov spectrum: $n_k^{\text{KZ}} \sim \frac{1}{k^2}$

---

[4]For simplicity, consider $\mathbf{P} = 0$.

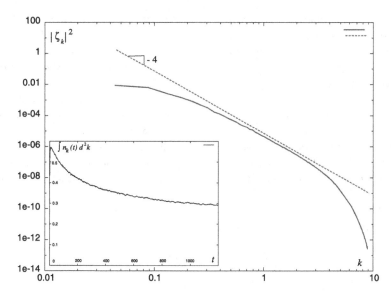

Fig. 5.1. Numerical simulation for a 512 square plate using 1024^2 modes with a mesh size $dx = 1/2$. Figure from Düring *et al.* (2006). The initial condition for the amplitude is $|\zeta_k| = 0.02\mathrm{e}^{-k^2}$ and the phase is randomly distributed. The figure shows the power spectrum of mean deflection $\langle|\zeta_k|^2\rangle$ versus wavenumber k after 1200 time units. The line plots the Rayleigh–Jeans power-law $7 \times 10^{-6}/k^4$ which gives $T \approx 2 \times 10^{-6}$ in agreement with the equipartition of the initial energy. The inset plots the evolution of the wave action with time.

coincides with the Rayleigh–Jeans solution. Hence, no power-law solution can support a finite energy flux. In fact, this degeneracy reveals the existence of a logarithmic correction, as was already observed for the nonlinear Schrödinger equation[5] (NLS) in 2D (Dyachenko *et al.*, 1992). This stationary solution with finite energy flux can be found perturbatively, however, a new characteristic scale k_* must be introduced. The physical meaning of this "cut-off" scale is not clear at all, but seems to stress the effect of the injection or dissipation regions, which have been neglected so far. Accordingly, in the limit $|\log(k/k_*)| \gg 1$ one obtains the spectrum (Düring *et al.*, 2006)

$$n_k^{\mathrm{KZ}} = C \frac{hP^{1/3}\rho^{2/3}}{(1-\sigma^2)^{2/3}} \frac{|\ln(k/k_*)|^{1/3}}{k^2}. \qquad (5.9)$$

---

[5]The logarithmic correction produces a divergent result for NLS. In our case, it is possible to show that all integrals are finite, indicating a finite energy flux.

Here, P is the energy flux involved in the energy cascade between the long-wave length scales and the short ones. Its value, as the one for k_*, should be found by matching the solution (5.9) to the injection and dissipation region. C is a pure real number that can be calculated from the functions $C_+(\alpha)$ and $C_-(\alpha)$.

For $\alpha = 0$ and $3\alpha - 4 = 0$, the collisional part $\text{Coll}_{2\leftrightarrow2}$ also vanishes. This solution corresponds to the wave action equipartition ($\alpha = 0$) with a second KZ spectrum $n_k \sim 1/k^{4/3}$ related to wave action inverse cascade. However, this spectrum does not lead to the vanishing of the second part of the collision term $\text{Coll}_{3\leftrightarrow1}$, in agreement with the nonconservation of the wave action mentioned above. Therefore, only the energy cascade exists.

The spectrum for the original deformation field can now be expressed using the canonical transformation (5.5), as

$$\langle |\zeta_{\mathbf{k}}|^2 \rangle = X_k^2\, n_k^{\text{KZ}} \sim \frac{|\ln(k/k_*)|^{1/3}}{k^4}.$$

The validity of this scenario has been confirmed numerically (Fig. 5.2) for the cases where injection and dissipation scales are well defined and far from each other.

Once we have the spectrum in hand, we can compute all the relevant quantities of the system. In particular, the renormalized frequency (5.8) in the limit $|\log(k/k_*)| \gg 1$ is (Düring, 2010)

$$\omega_{\mathbf{k}}^{\text{Ren}} = \sqrt{\frac{Eh^2}{12(1-\sigma^2)\rho}k^2 + \frac{9\pi}{2}\frac{C}{h^3}\left(\frac{(1-\sigma^2)P}{\rho}\right)^{1/3}|\ln(k/k_*)|^{1/3}}, \quad (5.10)$$

where the correction to the dispersion relation of bending waves has a weak dependence on k.

5.2. Experimental Setup

Experiments on wave turbulence on a thin elastic plate have been performed by two groups (Boudaoud et al., 2008; Mordant, 2008, 2010; Cobelli et al., 2009a). Both groups used a stainless steel plate 2 m by 1 m with a 0.4 or 0.5 mm thickness. The waves are excited by an electromagnetic shaker exerting a localized forcing at a given (low) frequency 20 or 30 Hz (see Fig. 5.3 for the setup of Mordant (2010)). Measurements are either single or dual point using a laser vibrometer (Boudaoud et al., 2008; Mordant, 2008) or a high speed space and time resolved profilometry technique (Cobelli et al., 2009a; Mordant, 2010). The principle of the latter technique is the

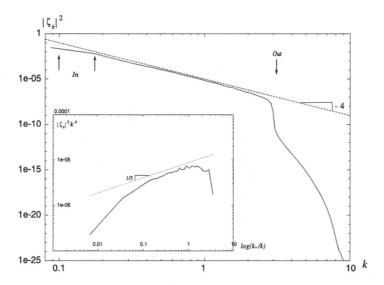

Fig. 5.2. Average power spectrum $\langle |\zeta_k|^2 \rangle$ for the energy cascade computed with a 512 square plate using 1024^2 modes and a mesh size $dx = 1/2$. Figure from Düring *et al.* (2006). The injection scale is $k_{\text{in}} \in (0.1, 0.25)$ while the dissipation is introduced through a linear damping for $k > k_{\text{out}}$ with $k_{\text{out}} = 3$. The line plots the power-law $1/k^4$. Inset plots $k^4 \langle |\zeta_k|^2 \rangle$ vs. $\log(k/k_*)$ in logarithmic scale with $k_* = k_{\text{out}}$.

following: a sine intensity pattern is projected on the surface of the plate by a videoprojector (Fig. 5.3). A high speed camera records the deformation of the pattern (due to the deformation of the plate) as seen from a different optical axis. The deformation of the pattern is coded as a phase modulation of the pattern recorded by the camera and can be extracted from the movies by a 2D phase demodulation (Cobelli *et al.*, 2009b). Snapshots of the deformation and the normal velocity of the plate are shown in Fig. 5.4. Fourier analysis of such movies can be performed both in space and time so as to study the structure of the waves, in particular, the dispersion relation.

5.3. The Fourier Spectrum of Turbulent Flexion Waves

The first measurements of this system were published by Boudaoud *et al.* (2008) using a laser vibrometer and confirmed shortly later by Mordant (2008). Both studies showed that the scaling of the observed spectrum was not in agreement with the theoretical prediction. The spectra show a short inertial regime with a power-law scaling with an exponent close to -0.5. A fast exponential decay is observed at high frequency with a

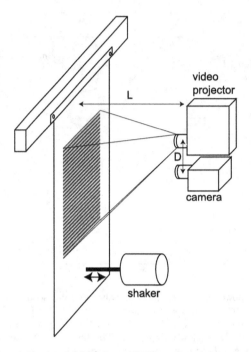

Fig. 5.3. Experimental setup of Mordant (2010): a flat stainless steel plate $1 \times 2\,\text{m}^2$, 0.4 mm thick is hanging under its own weight. The motion of the plate is excited by a electromagnetic shaker vibrating at 30 Hz. A sine intensity pattern is projected on the plate and its deformation is recorded by a Photron SA1 high speed camera.

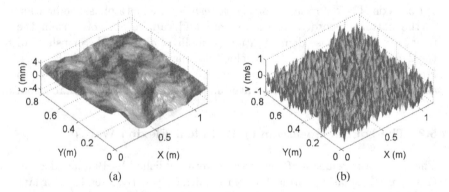

Fig. 5.4. Snapshots of deformation and velocity obtained by the time resolved profilometry technique. Figure from Mordant (2010).

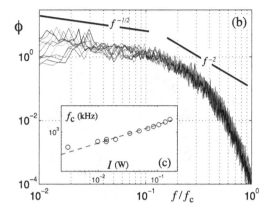

Fig. 5.5.   Frequency spectrum of the normal velocity of the plate measured at a single point with a laser vibrometer. The frequency has been rescaled by the cutoff frequency f_c of the fast decaying high frequency regime. The spectra have been rescaled by $P^{1/2}$, P being the average input power. Figure courtesy Boudaoud *et al.* (2008).

cut-off frequency f_c scaling with the injected power P as $P^{1/3}$. Spectra corresponding to various forcing strengths follow a master curve when the frequency is rescaled by f_c and the spectrum is normalized by $P^{1/2}$ (Fig. 5.5) yielding a scaling $E(\omega) \propto P^{2/3}\omega^{-0.5}$ in the inertial range quite different from the theoretical prediction. The scaling in P is neither $P^{1/2}$ nor $P^{1/3}$ expected for 3- or 4-wave resonance, respectively.

What is the origin of this disagreement between experiment and theory? A possibility suggested by Boudaoud *et al.* is that the motion could be dominated by singular structures such as ridges or D-cones. This hypothesis is discarded by the two-point measurement by Mordant (2008) and the space–time measurement using the high speed profilometry technique (Cobelli *et al.*, 2009a). These measurements show that the turbulent motion is made of the superposition of weakly nonlinear waves as expected in the weak turbulence theory. Figure 5.6 displays the space–time spectrum $E(\mathbf{k}, \omega)$ of the normal velocity. The energy of the turbulent field is localized on a single surface in the $(\mathbf{k}, \omega)$ Fourier space corresponding to the nonlinear dispersion relation. As will be shown in detail in the next section, the dispersion relation is weakly altered by nonlinearity as compared to the linear dispersion relation. This supports the fact that the amplitude of the waves is weak. In contrast to this situation, the same measurement operated on surface water waves shows that the energy is concentrated on several surfaces due to harmonics generation by singular structures (sharp crests) (Herbert *et al.*, 2010). An alternative explanation for the mismatch of the measured spectrum with the theory could be the dissipation or finite size effects. Measurement of the damping times at various frequencies

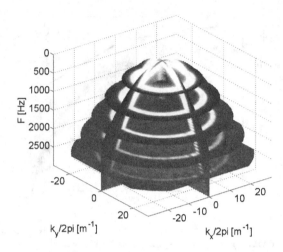

Fig. 5.6. Space–time spectrum $E(\mathbf{k}, \omega)$ of the normal velocity of the plate. The spectrum is made visible by planar cuts at $k_x = 0$, $k_y = 0$ and various cuts at constant frequency (500, 1000, 1500, 2000 and 2500 Hz). The logarithm of the energy is coded in gray scale. Figure from Mordant (2010).

show that dissipation is not strongly localized at small scales as expected by the theory (Boudaoud *et al.*, 2008). The finite size of the system affects the energy transfer and may alter the spectra due to the discreteness of the frequencies of the cavity modes (Kartashova, 2009). Such effects are discussed in Sec. 5.6.

Cuts of the spectrum at given values of the frequency are shown in Fig. 5.7. At low frequency, the energy spectrum is clearly anisotropic due to the forcing of the motion by the shaker. Note that although strongly localized in space, the forcing is effectively localized in scale at the wavelength corresponding to the forcing frequency (30 Hz). As the Kolmogorov–Zakharov energy cascade operates to higher and higher frequency, isotropy is restored by the wave interactions, the energy is spread in all directions and is isotropic at frequencies over 400 Hz. This behavior is expected in the framework of the Kolmogorov cascade of energy in which the scales of the inertial range are supposed to be widely separated from the forcing scales. Thus, the details of the forcing scheme are "forgotten" once the energy has cascaded to small enough scales.

5.4. The Dispersion Relation

Figure 5.8 displays the space–time spectrum $E(k, \omega)$ obtained from $E(\mathbf{k}, \omega)$ by summing over the directions of the wavevector $\mathbf{k}$. In this representation,

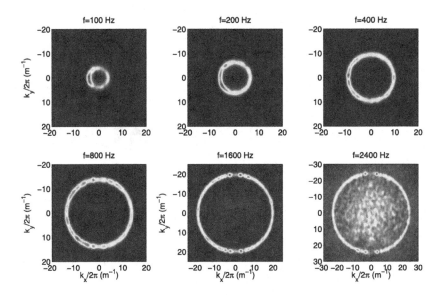

Fig. 5.7. Cuts of the space–time spectrum $E(\mathbf{k}, \omega)$ at given frequencies (100, 200, 400, 800, 1600 and 2400 Hz). Figure from Mordant (2010).

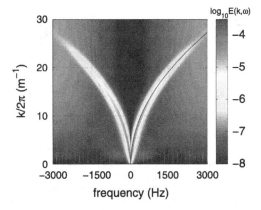

Fig. 5.8. Space–time spectrum $E(k, \omega)$ of the normal velocity of the plate obtained from $E(\mathbf{k}, \omega)$ after integration over the direction of $\mathbf{k}$. The solid black line corresponds to the linear dispersion relation $\omega = 0.641k^2$. Figure from Mordant (2010).

all the energy is seen to be concentrated on a line in the (k, ω) plane: the nonlinear dispersion relation. The linear dispersion relation (black line) lies in the vicinity of the nonlinear dispersion relation. The absence of harmonics and the closeness of the two dispersion relation curves is the sign that the amplitude of the wave is weak.

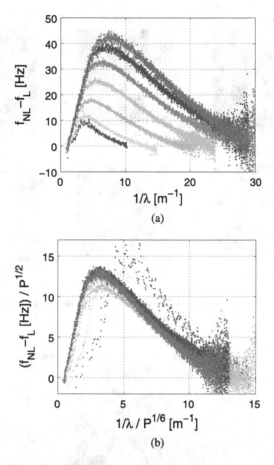

Fig. 5.9. Shift of the observed nonlinear dispersion relation with respect to the linear dispersion relation. λ is the wavelength. The nonlinear frequency is extracted from the spectrum $E(k,\omega)$ by the position of the maximum of the energy profile at each given value of the wavenumber. (a) Difference between the nonlinear frequency observed at a given wavenumber and the linear frequency observed at the same wavenumber at various forcing intensities ($P = 0.25, 1, 4, 9, 16, 25, 36$ in arbitrary units from bottom to top). (b) Same curves in normalized representation: the wavenumber is divided by $P^{1/6}$ and the frequency shift by $P^{1/2}$. Figure from Mordant (2010).

The position of the crest of the energy line is extracted from the spectrum $E(k,\omega)$ and compared to the linear dispersion relation. Figure 5.9(a) shows the shift between linear and nonlinear dispersion relations. A small separation of the two dispersion relations is observed at small wavenumbers. At high wavenumbers, the shift goes to zero. The small shift is due to nonlinear effects and is predicted by the theory as the frequency renormalization (see Eq. (5.10)) (Newell *et al.*, 2001; Düring, 2010). The shift is increasing with the strength of the forcing. The shifts corresponding to the various forcing intensities can be collapsed by rescaling the wavenumber by $P^{1/6}$ (corresponding to the scaling of the wavenumber at f_c) and dividing the shift by $P^{1/2}$ (Fig. 5.9(b)). This rescaling is the same as that of the spectrum $E(f)$. The theory indeed predicts that the frequency renormalization should display the same scaling in P as the spectrum (Düring, 2010) (Eq. (5.8)). In the experiment, the shift and the spectrum show the same scaling despite the fact that the observed scaling of the spectrum is not the one predicted by the theory.

In the framework of the weak turbulence theory, nonlinear effects are small so that the time scales of the energy transfer are long compared to the period of the waves. One effect of the energy exchanges is the degradation of the spectral width of the modes. This effect appears as a "thickness" of the dispersion relation. Such a thickness can be estimated from the data by fitting a gaussian shape on the energy concentration around the nonlinear dispersion. The fit can be performed either at a given frequency or a given wavenumber. The result of both fits is displayed in Fig. 5.10. An intrinsic width of the wavenumber exists due to the resolution of the Fourier transform over the size of the picture used to measure the deformation of the plate. This intrinsic width is proportional to the inverse of the picture

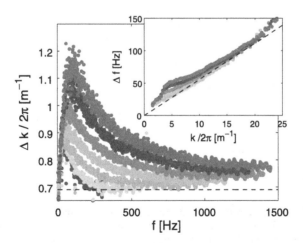

Fig. 5.10. "Width" of the dispersion relation. A Gaussian energy profile is fitted on the spectrum $E(k,\omega)$ at given frequency providing a wavenumber width Δk (main figure) or at constant wavenumber providing a frequency width Δf (insert). The width is shown for various values of the forcing intensity ($P = 0.25, 1, 4, 9, 16, 25, 36$ in arbitrary units from bottom to top). The dashed line corresponds to the intrinsic width due to the finite size of the measurement picture. Figure from Mordant (2010).

size. This translates into a frequency width through the dispersion relation $\Delta f \propto k\Delta k$. The nonlinear widening of the width of the dispersion relation is visible above this intrinsic width. The width is increasing with P (at given frequency) but its exact scaling is unclear so far. The width is a measure of the magnitude of the nonlinearity. It can be seen that the nonlinearity is increasing in frequency to a maximum in the inertial range and decays gradually to zero at the highest wavenumbers.

In most wave turbulence experiments, only single point measurement is performed. The only spectrum that can be estimated is a frequency one. Theoretical predictions are expressed in terms of the wave vector. A link between wave vector and frequency is used, assuming isotropy of the spectrum and using the dispersion relation of the wave to change variables. We show in Fig. 5.11 the comparison of the frequency spectrum measured directly from the profilometry data and estimated from the wavenumber spectrum through a change of variable using either the linear dispersion relation or the nonlinear one. The spectra obtained by the change of variable are indeed very close to the ones estimated directly (with a slightly better agreement when using the nonlinear dispersion relation). This check validates the usual artifice used to compare single point measurements to theory.

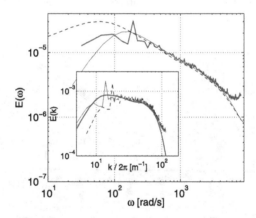

Fig. 5.11.   Comparison of the wavenumber spectrum $E(k)$ and the frequency spectrum $E(\omega)$ obtained either directly or from one another through a change of variable using the dispersion relation. Main figure: frequency spectrum (blue: direct measurement, red: from the wavenumber spectrum using the nonlinear dispersion relation, dashed black line: from the wavenumber spectrum using the linear dispersion relation). Insert: wavenumber spectra (thick solid line: direct measurement, thin solid line: from the wavenumber spectrum using the nonlinear dispersion relation, dashed line: from the wavenumber spectrum using the linear dispersion relation). Figure from Cobelli *et al.* (2009a).

5.5. Distributions and Correlations

In the weak turbulence theory, the statistics of the wave amplitudes are expected to be close to Gaussian. The distribution of the experimental wave magnitude is shown in Fig. 5.12. The distributions at all wavenumbers are seen to be close to Rayleigh distribution (exponential) confirming the Gaussian statistics. Boudaoud *et al.* studied the distribution of the single point velocity increments $v(\mathbf{r}, t+\tau) - v(\mathbf{r}, t)$ in the spirit of the classical tools of hydrodynamical turbulence studies (Frisch, 1996). They observed that the distribution of the velocity increments remains Gaussian whatever the scale τ of analysis (Boudaoud *et al.*, 2008). The structure function displays a normal scaling showing the absence of intermittency in this system.

Another key point of the theory is the absence of correlations of two waves of different wavenumbers:

$$\langle A(\mathbf{k}_1, t) A^*(\mathbf{k}_2, t) \rangle = n_{k_1} \delta(\mathbf{k}_1 - \mathbf{k}_2). \tag{5.11}$$

Figure 5.13 displays such correlations for four values of $\mathbf{k}_2$. It can be seen that no correlation is observed except for $\mathbf{k}_1 \approx \mathbf{k}_2$.

The Gaussian statistics and the absence of any two wave correlations are at the basis of the weak turbulence theory. In this framework, weak correlations should appear only between resonant waves. For thin elastic plates, such correlations are expected for four resonant waves. Because of the large number of waves and the expected weakness of the correlations, the measurement of such correlations is challenging and would require extremely large amount of data.

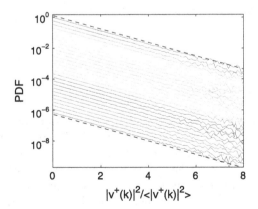

Fig. 5.12. Distribution of the wave amplitudes for various values of $k/2\pi$ between 2.2 and 25 m^{-1} ($|v^+(\mathbf{k})|^2 \propto k^2|A_\mathbf{k}|^2$). The wave amplitudes have been normalized to variance unity. Figure from Mordant (2010).

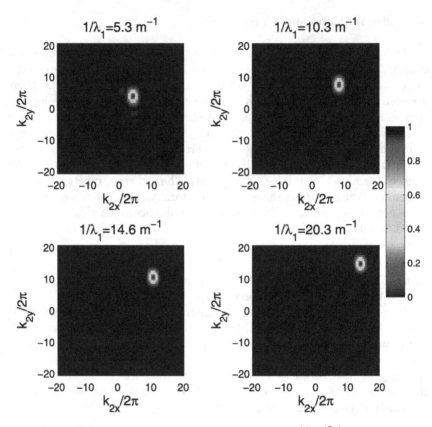

Fig. 5.13.   Correlations coefficients of wave magnitudes $\dfrac{\langle A_{\mathbf{k}_1} A_{\mathbf{k}_2}^* \rangle}{\sqrt{\langle |A_{\mathbf{k}_1}|^2 \rangle \langle |A_{\mathbf{k}_2}|^2 \rangle}}$. $\mathbf{k}_1$ is chosen on the diagonal of the axes and its magnitude is given in the title. Figure from Mordant (2010).

5.6. Finite Size Effects

The observed dispersion relation and the Gaussianity of the wave amplitudes are compatible with the weak nonlinearity required by the weak turbulence theory. No singular structure has been observed so far. The other major hypothesis of the theory that could be violated in the elastic plate experiments (as for almost any other laboratory system) can be the asymptotically large size. In the theory, the size of the system is supposed to be large so as to avoid the quantization of the frequencies of the cavity modes. In the theory, the energy transfers occur through multiple resonant waves. If the finite size effects are important, the discreteness of

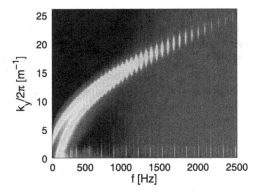

Fig. 5.14. Cut of the spectrum $E(\mathbf{k}, \omega)$ for $k_x = 0$. The existence of discrete frequencies of the modes is clearly visible. Figure from Mordant (2010).

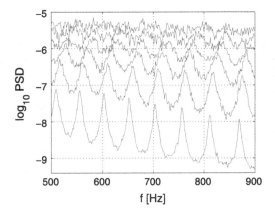

Fig. 5.15. Value of the power spectrum $E(\mathbf{k}, \omega)$ along the dispersion relation in the plane $k_x = 0$ for various values of P ($P = 0.25, 1, 4, 9, 16, 25, 36$ in arbitrary units from bottom to top). Figure from Mordant (2010).

the frequencies may prevent from finding resonant waves as the conservation equations relating the frequencies of the resonant waves may not have any solutions or a strongly reduced number of solutions. The energy transfer would then be strongly altered (Kartashova, 2009). Figure 5.14 shows a cut of the spectrum $E(\mathbf{k}, \omega)$ for $k_x = 0$. The line of high energy concentration (dispersion relation) is continuous at low frequencies and is made of spots at higher frequencies. The spots correspond to the discrete modes of the plate. Their frequency width is decreasing with the frequency (the wavenumber width is actually imposed by the size of the pictures used by the measurement technique as explained above). The frequency width of a mode is related to the inverse nonlinear time and the nonlinearity is clearly decaying with the frequency (at large values of the frequency).

Figure 5.15 displays the evolution of the crest line of energy in a given frequency interval for various values of the forcing strength. At very low forcing, the modes are clearly separated with a very small width. As the forcing is increased, the peaks are becoming wider. For a strong enough forcing, the peaks disappear and the spectrum is continuous. The discreteness of the peaks appears very clearly in the plane $k_x = 0$ because it corresponds to waves propagating along the smallest side of the plate (1 m) and thus the separation of the mode is the largest. In the $k_y = 0$ direction, the separation of the peaks should be twice as low and thus the transition to a continuous spectrum occurs much faster when the forcing is increased.

5.7. Conclusions

The thin elastic plate system is a very rich one. The wave amplitude evolution is given by a kinetic equation through a 4-wave resonance process. Wave turbulence theory shows the convergence toward statistical equilibrium for a free system and an energy cascade when forcing and dissipation are introduced. Both situations have been verified by numerical simulations. No wave action cascade is observed for this 4-wave system as predicted by the theory.

From the experimental point of view, use of the time resolved profilometry provides a wealth of new information that was available only in numerical simulations so far. The space–time Fourier spectrum is a tool which is extremely valuable to test the structure of the waves. We observed through several tests (nonlinear dispersion relation, distributions, 2-wave correlations) that the flexion wave turbulence is weakly nonlinear. Nevertheless, the scaling of the experimental spectra is not in agreement with the theoretical predictions. Two hypotheses can be invoked to explain such a fact. First, finite size effects have been observed that may alter the energy transfer. Second, the dissipation may not be strongly localized at high frequency as required by the theory so that the energy cascade could be "leaking" (Boudaoud *et al.*, 2008). A decisive test could be to study the energy flux among wavenumber and the possible existence of 3- or 4-wave correlations.

Bibliography

Benney D, Saffman P. 1966. Nonlinear interactions of random waves in a dispersive medium. *Proc. R. Soc. Lond.* A 289: 301–320.

Benney D, Newell A. 1969. Random wave closures. *Studies Appl. Math.* 48: 29.

Boudaoud A, Cadot O, Odille B, Touzé C. 2008. Observation of wave turbulence in vibrating plates. *Phys. Rev. Lett.* 100: 234504.

Choi Y, Lvov Y, Nazarenko S. 2005. Joint statistics of amplitudes and phases in wave turbulence. *Physica D* 201: 121–149.

Cobelli P, Petitjeans P, Maurel A, Pagneux V, Mordant N. 2009a. Space-time resolved wave turbulence in a vibrating plate. *Phys. Rev. Lett.* 103: 204301.

Cobelli PJ, Maurel A, Pagneux V, Petitjeans P. 2009b. Global measurement of water waves by Fourier transform profilometry. *Exp. Fluids* 46: 1037.

Düring G, Josserand C, Rica S. 2006. Weak turbulence for a vibrating plate: Can one hear a Kolmogorov spectrum? *Phys. Rev. Lett.* 97: 025503.

Düring G, Falcon C. 2009. Symmetry induced four-wave capillary wave turbulence. *Phys. Rev. Lett.* 103: 174503.

Düring G. 2010. Non-equilibrium dynamics of nonlinear wave systems: Turbulent regime, breakdown and wave condensation. Ph.D. thesis, Université Pierre et Marie Curie, Paris, France.

Dyachenko S, Newell A, Pushkarev A, Zakharov V. 1992. Optical turbulence — weak turbulence, condensates and collapsing filaments in the nonlinear Schrodinger-equation. *Physica D* 57: 96–160.

Föppl A. 1907. *Vorlesungen über technische Mechanik.*

Frisch U. 1996. *Turbulence: The Legacy of A. N. Kolmogorov.* Cambridge: Cambridge University Press.

Hasselmann K. 1962. On the non-linear energy transfer in a gravity-wave spectrum .1. General theory. *J. Fluid Mech.* 12: 481–500.

Herbert E, Mordant N, Falcon E. 2010. Observation of nonlinear dispersion relation and spatial statistics of wave turbulence on the surface of a fluid. *Phys. Rev. Lett.* 105: 144502.

Kartashova E. 2009. Discrete wave turbulence. *EPL* 87: 44001.

Landau L, Lifshitz E. 1959. *Theory of Elasticity.* New York: Pergamon.

Mordant N. 2008. Are there waves in elastic wave turbulence? *Phys. Rev. Lett.* 100: 234505.

Mordant N. 2010. Fourier analysis of wave turbulence in a thin elastic plate. *Eur. Phys. J. B* 76: 537–545.

Newell A. 1968. Closure problem in a system of random gravity waves. *Rev. Geophys.* 6: 1.

Newell AC, Nazarenko S, Biven L. 2001. Wave turbulence and intermittency. *Physica D* 152–153: 520.

Rayleigh J. 1945. *The Theory of Sound.* New York: Dover.

von Kármán T. 1910. *Encyk. d. Math. Wiss.* 4: 8.

Zakharov V, L'vov V, Falkovich G. 1992. *Kolmogorov Spectra of Turbulence I, Nonlinear Dynamics.* Berlin: Springer-Verlag.

Zakharov V. 1967. Weak-turbulence spectrum in a plasma without a magnetic field. *Sov. Phys. J.* 24: 455.

Chapter 6

Gravity Wave Turbulence in a Large Flume

R. Bedard, S. Lukaschuk* and S. Nazarenko[†]

*Department of Engineering,
Hull University, Hull, HU6 7RX, UK

†Mathematics Institute, Warwick University,
Coventry, CV4 7AL, UK

We overview past and new experimental results on gravity wave turbulence obtained at The Deep flume facility in Hull. The relatively large size of this flume, $12\,\mathrm{m} \times 6\,\mathrm{m} \times 1.5\,\mathrm{m}$, allows us to study the pure gravity waves without being concerned with modifications caused by the capillary or finite depth effects. We analyze wave spectra, probability density functions of wave heights and their increments, and structure functions to characterize both random weak waves and strong singular coherent structures. We see some evidence in favor of the Zakharov–Filonenko spectrum at larger wave intensities and we see signature of the finite size effects for weaker waves. Our new experiments are devoted to evolving wave turbulence, both the stages of forming the steady state and its decay, including experiments with a tilted wall.

"When casting pebbles into water, look at the ripples being formed thereby. Otherwise this activity will be an idle amusement."
— *Kozma Prutkov*

Contents

6.1. Introduction

Watching waves at sea, along with looking at flames in a campfire, have always drawn humans as if by magic, by its intricate moving patterns, chaotic and regular, simple and relaxing to watch and yet so complex to comprehend and describe. Even if they did not have any practical importance, the water waves would still be well worth studying, just as the most beautiful and familiar, fundamentally basic and, at the same time, enormously rich nonlinear system. However, unsurprisingly, the gravity water waves and their mutual nonlinear interactions are of huge importance for navigation and other activities at seas, e.g. oil drilling, for the transport of momentum, air and moisture through the water surface affecting the weather and climate evolutions (see, for example, Janssen (2004)). Field observations of the sea surface, laboratory experiments in wave flumes and numerical simulations are efficient and complementary tools for studying such random nonlinear waves and for testing existing theoretical models. Obvious advantage of the field observations is that they deal directly with the system we want to know about, rather than model it in a scaled-down laboratory experiment or in numerical simulation. In comparison with field measurements, laboratory experiments and numerical simulations allow more control over the physical conditions and over the quantities we measure, especially in the numerical simulations which allows us to access a much broader range of diagnostics than in experiments.

On the other hand, the laboratory experiments enable observations of much larger range of wave scales than it is possible in numerical simulations under the current level of resolution and, therefore, they allow to obtain cleaner power-law spectra and other scalings. Furthermore, laboratory experiments are much more realistic than numerics in reproducing the strongly nonlinear events because most numerical methods are based on weakly nonlinear truncations of the original fluid equations. Finally, they also have a natural dissipation mechanism as in open seas, wave breaking, in contrast to an artificial hyper-viscous dissipation which is usually used in numerics.

In this chapter, we summarize the results on the gravity wave turbulence in laboratory experiments obtained at The Deep facility at Hull starting at 2005 up until now. For the previous research publications on these experiments, see Denissenko *et al.* (2007), Lukaschuk *et al.* (2009) and Nazarenko *et al.* (2010). Our experiments were geared specifically at studying the wave turbulence regimes and testing the theory of weak wave turbulence (Nazarenko, 2011; Zakharov *et al.*, 1992), and as such they are complimentary to the experiments of Falcon *et al.* reported elsewhere in the present book. Their experiments were performed in a smaller wave tank, with the advantage of greater flexibility in trying different setups and in employing more versatile diagnostic techniques, e.g. mapping a two-dimensional (2D) surface. On the other hand, being large in size, $12\,\text{m} \times 6\,\text{m} \times 1.5\,\text{m}$, the Hull flume is unique for these type of experiments because it allows to study the pure gravity wave turbulence without being affected by the surface tension and the finite depth effects, and with minimized effects of finite horizontal size. This feature is important if one wants to obtain a reliable measurement of a wave spectrum over a sufficiently long scaling range so that a comparison with the idealized theory (assuming infinite depth and horizontal dimensions and no surface tension) could be legitimately made. Our results clearly indicate that, although some evidence in favor of the classical weak wave turbulence theory can be found under carefully selected conditions, the behavior of wave turbulence is typically much richer with abundance of physical processes that still await their theoretical explanations and descriptions. Following the overall spirit of the present book, our aim here will be to highlight such phenomena and to suggest questions to address future theory and experiment.

To pre-empt our main descriptions, we would like to formulate brief take-home messages, which we will later explain and justify.

- Ten-meter scale rectangular wave tanks are not sufficient for achieving the scalings predicted by the weak wave turbulence theory, particularly Zakharov–Filonenko (ZF) spectrum (Zakharov and Filonenko, 1967).

The finite-size effects are important and they are even more severe in smaller wave tanks. The spectrum and the structure functions appear to be nonuniversal and amplitude-dependent, although some (inconclusive) trend toward the ZF scaling could be argued to be seen in some special cases.

- At low amplitudes, the resonant wave interactions (the main driver of wave turbulence) seem to be suppressed due to the fact that the set of wave modes in bounded systems is discrete and one would need sufficiently large wave intensities to activate the quasi-resonances. Some interesting states of discrete and mesoscopic wave turbulence may occur at low amplitudes, when only isolated clusters of resonant modes are switched on and actively evolving.

- At larger amplitudes, random waves coexist with strongly nonlinear coherent structures — sharp crests, breaking waves and vertical splashes. These components can be effectively detected and characterized by measuring the structure functions of both the time and the physical space signals.

- The rate of formation of the statistically steady wave field appears to be consistent with the characteristic evolution time predicted by the standard four-wave kinetic equation of the weak turbulence theory. The likely cause for this is that the wave phases become random rather quickly due to the multiple reflections from the flume walls. This is an additional mechanism of phase randomization with respect to the nonlinearity, and it is absent in numerical simulations with periodic boundary conditions and in other more involved simulation techniques as in Shrira and Annenkov (2012). As a result, such simulations have often shown a faster initial time scale (dynamical time rather than the kinetic one).

- The time-scale of the decay phase, after turning the wave makers off, also appears to be consistent with the kinetic equation at it initial phase. The subsequent decay process is particularly interesting because, when waves gradually weaken, the wave turbulence crosses from the kinetic to the mesoscopic and possibly to the discrete regimes. For these regimes, we observe typical, for low-dimensional systems, recursive/oscillitary behavior with occasional bursty sandpile-like spills into the high-frequency tail (see Nazarenko, 2006; L'vov and Nazarenko, 2010; Nazarenko, 2010 for previously suggested scenarios in discrete and mesoscopic turbulence).

- Turbulence decays via the cascade mechanism for only a rather short time after the forcing is switched off. After this, decay of the total energy does not follow the power-law energy decrease predicted by the weak wave turbulence theory. Instead, it follows an exponential law, as if there was a linear dissipation mechanism draining energy of the largest scales rather than (or in addition to) the energy cascade mechanism. A possible candidate for such a linear mechanism is the wall friction.

- Turbulence decay appears to be faster when one of the flume walls is inclined. We tentatively attribute it to a possibility that the discrete-mode structure in such a trapezoidal flume is different from the one in the perfectly rectangular flume in such a way that the wave resonance (or quasi-resonance) conditions are satisfied for a larger set of modes, also the wall friction is slightly larger for trapezoidal flume than for rectangular ones of the same size.

Our review is organized as follows. In Sec. 6.2, we describe the relevant theories and predictions for the surface wave turbulence. In Sec. 6.3, we describe the experimental facility and the measurement techniques. In Sec. 6.4, we present the experimental results along with their discussion in the context of the theoretical predictions and possible interpretations. In Sec. 6.5, we present a summary of our findings and an outlook for future work.

6.2. Theoretical Background

We will start with a short description of background theory of the gravity wave turbulence. Further details could be found in Nazarenko (2011), Zakharov *et al.* (1992), Denissenko *et al.* (2007), Lukaschuk *et al.* (2009) and Nazarenko *et al.* (2010). In addition, we will present some new theoretical estimates relevant to the nonstationary evolution, such as the estimates for the deterministic and kinetic time scales which we will use later when discussing the spectrum formation and the decay stages, as well as the estimates for the energy decrease rate at the free decay stage.

6.2.1. *Spectra*

The power spectra are the most commonly discussed objects in the wave turbulence (and in the general turbulence) literature. The wave energy spectrum in the frequency domain is defined as

$$E_\omega = \int e^{i\omega t'} \langle \eta(\mathbf{x}, t) \eta(\mathbf{x}, t + t') \rangle \, dt', \qquad (6.1)$$

and the one-dimensional (1D) energy spectrum in the wavenumber domain, respectively, is

$$E_k = \int e^{ikz} \langle \eta(\mathbf{x}, t) \eta(\mathbf{x} + \mathbf{w}z, t) \rangle \, dz, \qquad (6.2)$$

where $\eta(\mathbf{x}, t)$ is the surface elevation at time t and location in the horizontal plane $\mathbf{x} = (x, y)$. The integration in (6.1) is taken over a time window, and in (6.2) over an interval of a straight line in the 2D plane (in our case a line illuminated by the laser sheet; see Sec. 6.3, Experimental setup) with $\mathbf{w}$ being a unit vector along this line. Angle brackets denote averaging over realizations (most often complemented by time averaging to improve statistical convergence and assuming ergodicity). Ideally, one would like to obtain a fully 2D wave spectrum, as it was done in smaller-scale experiments on gravity-capillary waves (Herbert *et al.*, 2010; Falcon *et al.*, 2012), but unfortunately on the scale of our experiment no similar 2D measurement techniques have been developed yet.

For a statistically steady and homogeneous state, E_ω and E_k are independent of t and $\mathbf{x}$. Several theories predict existence of scaling ranges where the spectrum has a power-law form,

$$E_\omega \propto \omega^{-\nu} \tag{6.3}$$

and

$$E_k \propto k^{-\mu}, \tag{6.4}$$

with indices ν and μ depending on a particular theory and/or the type of the scaling range (e.g. the direct cascade range or the inverse cascade range).

Weak wave turbulence theory. considers weakly nonlinear waves and assumes that these waves have random statistically-independent phases. The key ingredient of this theory is the infinite box limit. The latter is taken in such a way that the mean wave energy density in the physical space remains constant (box-size independent). Then, in the infinite box limit, the distance between the adjacent wave modes in the $\mathbf{k}$-space becomes much less than the nonlinear resonance broadening. In KAM terminology, this means that no invariant tori survive and all of the phase space is stochastic. For the wave spectrum, this approach leads to the famous Hasselmann's kinetic equation (Hasselman, 1962),

$$\dot{n}_k = 4\pi \int |W(\mathbf{k}, \mathbf{k}_1, \mathbf{k}_2, \mathbf{k}_3)|^2 \delta(\mathbf{k} + \mathbf{k}_1 - \mathbf{k}_2 - \mathbf{k}_3)\delta(\omega_k + \omega_{k_1} - \omega_{k_2} - \omega_{k_2})$$

$$\times n_k n_{k_1} n_{k_2} n_{k_3} \left(\frac{1}{n_k} + \frac{1}{n_{k_1}} - \frac{1}{n_{k_2}} - \frac{1}{n_{k_3}} \right) d\mathbf{k}_1 d\mathbf{k}_2 d\mathbf{k}_3, \tag{6.5}$$

where $n_k = E_k/(2\pi k\omega_k)$ is the wave action spectrum and $W(\mathbf{k}, \mathbf{k}_1, \mathbf{k}_2, \mathbf{k}_3)$ is the interaction coefficient given by a rather lengthy expression (Krasitskii, 1994; Zakharov, 1999; Nazarenko, 2011).

The Hasselmann equation (6.5) has an exact power-law solution

$$E_k = C_d \, g^{1/2} \epsilon^{1/3} k^{-5/2},\qquad(6.6)$$

or in the ω-domain

$$E_\omega = C'_d \, g^2 \epsilon^{1/3} \omega^{-4}.\qquad(6.7)$$

Here, ϵ is the rate of dissipation of the wave energy per unit area of the water surface and C_d and C'_d are dimensionless constants analogous to the Kolmogorov constant of the hydrodynamic turbulence. This solution is called the ZF spectrum and it describes a steady state with energy cascading through an inertial range of scales from large scales, where it is produced, to the small scales where it is dissipated by wavebreaking and viscosity. It is a direct analogue of the famous Kolmogorov spectrum describing the turbulent energy cascade in the Navier–Stokes fluids. Because of this analogy, the ZF spectrum and similar solutions in other wave systems are generally called Kolmogorov–Zakharov (KZ) spectra.

Besides the energy, the Hasselman kinetic equation (6.5) also conserves wave action. Thus, it is a dual cascade system and has an extra exact power-law solution, KZ spectrum describing an inverse cascade of the wave action from small to large scales:

$$E_k = C_i \, g^{2/3} \zeta^{1/3} k^{-7/3},\qquad(6.8)$$

or in the ω-domain

$$E_\omega = C'_i \, g^2 \zeta^{1/3} \omega^{-11/3}.\qquad(6.9)$$

Here, ζ is the rate of dissipation of the wave action per unit area of the water surface and C_i and C'_i are dimensionless constants. This spectrum was first obtained by Zakharov and Zaslavskii (1982). In the open ocean, such an inverse cascade behavior is believed to be responsible for the swell, or aged waves, whose wave length is increasing with time passed from the moment of their generation by wind. Note that initially wind inputs energy into short-wave ripples of several centimeter wavelength, as commonly seen on lakes or ponds after a fresh gust of wind. In persistent wind conditions, the maximum wind forcing gradually shifts to longer waves which creates a direct cascade range between the forcing and the (small) dissipative scales. However, even in persistent wind conditions at open seas, a broad range of long waves remains where the inverse cascade continues to act.

In our large flume, we have not yet implemented an inverse cascade state. First, it is tricky to generate short waves uniformly across the large flume area. Also, the inverse cascade is believed to be even more sensitive to the finite size effects. Indeed, as the cascade propagates to larger scales, the degree of nonlinearity is decreasing, and inevitably the resonances start disengaging in the discrete k-space at the low k region. Some evidence of generation of low frequency mode was obtained in our most recent experiments. However, it is premature to talk whether we see a genuine local inverse cascade as predicted by the weak wave turbulence theory, or the observed generation of longer waves is due to some other mechanism (e.g. an instability). Thus, we will postpone this discussion, and in the present review, we will wholly concentrate on the direct cascade settings.

Discrete and mesoscopic wave turbulence. It is important that in deriving the kinetic equation, the limit of an infinite box is taken before the limit of small nonlinearity. This means that in a however large but finite box, the wave intensity should be strong enough so that the nonlinear resonance broadening is much greater than the spacing of the k-grid (corresponding to Fourier modes in a finite rectangular box). As estimated by Nazarenko (2006), this implies a condition on the minimal surface slope γ (for definition of γ see (6.54)):

$$\gamma > 1/(kL)^{1/4}, \qquad\qquad (6.10)$$

where L is the size of the basin, which is quite a severe restriction. If this condition is not satisfied, the number of exact and quasi four-wave resonances will be drastically depleted (Kartashova, 1991, 1998; Nazarenko, 2006; L'vov and Nazarenko, 2010; Nazarenko, 2011). This can lead to a significant slowdown of the energy cascade from long to short waves and, therefore, a steeper energy spectrum. A theory of discrete and mesoscopic wave turbulence was proposed by Nazarenko (2006) and further extended in L'vov and Nazarenko (2010) and Nazarenko (2011). For very low levels of forcing, one can expect that the energy cascade will be frozen due to the deficiency of the four-wave resonances and the energy will be accumulating in the forcing scales until the modes will become sufficiently strong and the resonance broadening will reach the spacing between the neighboring modes in the discrete k-space. This will switch on quasi-resonances on and trigger an energy transfer to higher k's toward the dissipation scales. In turn, this transfer will drain the energy from the forcing scales and disengage the quasi-resonances. The wave turbulence will freeze once again, it will start accumulating at the forcing scale, and the cycle will repeat. Because of the obvious analogy, this scenario was called "sandpile" by Nazarenko (2006). For weak forcings, it is natural to assume that there is no large overshoots on the energy accumulation stage, which also means that the avalanches are weak and there is no over-draining of energy at the sandpile tip-over stage. In this case, the spectrum is weakly oscillating around a critical state where the nonlinear resonance broadening is approximately equal to the spacing of the **k**-grid. Estimating the nonlinear frequency from the kinetic equation (6.5), we have for this state

$$E_\omega = C_s \, g^{7/2} L^{-1/2} \omega^{-6}, \tag{6.11}$$

where L is the flume length and C_s is yet another order-one dimensionless constant.

The above scenario is relevant to (weakly) forced systems. In the experimental data reported in this review, we also observe sandpile spills of energy at the decay stage when the forcing is turned off. This seems to reflect the fact that in the gravity wave system there is no strict frozen turbulence stage in a sense that some resonances survive even at very low amplitudes. Thus, the energy may be trapped in a low-dimensional cluster of large-scale modes whose dynamic is quasi-periodic rather than frozen. As a consequence, the resonance broadening may also experience quasi-periodic oscillations and, when it gets large at the higher-frequency end of the cluster, wave-energy spills into the high-frequency tail may be expected.

Coherent structures. Sharp wave crests are quite common for gravity waves of sufficiently large amplitudes. The most common type of crests discussed in the literature looks like a break in the surface slope. A prototype for such structures is a sharp-crested stationary Stokes wave solution with the crest angle of 120°. Following Kadomtsev (1965), such sharp crested waves are usually associated with the Phillips spectrum. Indeed, assume that there are discontinuities occurring at isolated points. This leads to the following 1D energy spectrum in wavenumber space

$$E_k \propto k^{-3}. \tag{6.12}$$

Second, assuming that transition from the **k**-space to the ω-space should be done according to the linear wave relation $\omega = \sqrt{gk}$, we arrive at the Phillips spectrum, (Phillips, 1958),

$$E_\omega \sim g^2 \omega^{-5}. \tag{6.13}$$

An alternative way to derive the Phillips spectrum, the way it was originally done by Phillips (1958), is to assume that the gravity constant g is the only relevant dimensional physical quantity. This argument is equivalent to saying that the linear term is of the same order as the nonlinear one in the water surface equations in the Fourier space *at all wavenumbers* (since both of these terms depend only on a single physical parameter — g).

Kuznetsov (2004) questioned this picture and argued that (i) slope breaks occur on 1D lines/ridges rather than on zero-dimensional points/

peaks, and (ii) that the wave-crest is propagating with preserved shape, i.e. $\omega \propto k$ should be used instead of the linear wave relation $\omega = \sqrt{gk}$. This assumptions give $E_k \propto k^{-4}$ and respectively,

$$E_\omega \propto \omega^{-4}, \tag{6.14}$$

i.e. formally the same ω (but not k!) scaling as ZF, even though the physics behind it is completely different. Finally, it was proposed by Connaughton *et al.* (2003) that wave crest ridges may have noninteger fractal dimension D somewhere in the range $0 < D < 2$. This leads to the following 1D energy spectrum in the k-space,

$$E_k \propto k^{-3-D}. \tag{6.15}$$

Assuming, following Kuznetsov, $\omega \propto k$, we have in this case

$$E_\omega \propto \omega^{-3-D}. \tag{6.16}$$

6.2.2. *Higher order statistics of the wave field*

The spectra introduced in the previous section belong to the class of the second-order correlators. Different types of coherent and incoherent structures may lead to the same spectra. Therefore, to see an unequivocal signature of a particular kind of coherent structures or incoherent random phased field, one must consider higher-order correlators.

6.2.2.1. *Statistics of the surface height*

The most straightforward way to obtain the higher-order statistics is to measure the probability density function (PDF) of the surface elevation $\eta(\mathbf{x}, t)$ obtained from a time series measured at a fixed position $\mathbf{x}$ (e.g. from an experimental signal measured by a wire probe) or from the physical space distribution of the surface elevation at a fixed time (from an experimental or a numerical measurement). The PDF of η, denoted $P(h)$, is defined in the usual way as a probability of η to be in the range from h to $h + dh$ divided by dh, or in the symbolic form

$$P(h) = \langle \delta(h - \eta) \rangle, \tag{6.17}$$

Expression (6.17) is related to the obvious identity $\langle f(\sigma) \rangle \equiv \int f(\sigma') P(\sigma') \, d\sigma' = \int f(\sigma') \langle \delta(\sigma - \sigma') \rangle d\sigma' = \langle f(\sigma) \rangle$.

If one assumes that the waves have short correlations in the physical space, so that the waves arriving to the same point $\mathbf{x}$ from different directions are independent, then one can use the central limit theorem argument and conclude that the PDF is Gaussian,

$$P_{\text{Gauss}}(\eta) = \frac{1}{\sqrt{2\pi\sigma^2}} \exp\left(-\frac{\eta^2}{2\sigma^2}\right), \qquad (6.18)$$

where σ is the standard deviation of the surface height from the equilibrium value $\eta = 0$. This argument could be adjusted taking into account that the normal variables describing the wave amplitudes, the wave action variables, are related to the physical variables (η and the surface velocity) via a nonlinear canonical transformation. Because of the nonlinearity of this transformation, at a fixed wavenumber $\mathbf{k}$ in addition to the main mode frequency $\omega = \omega_k$ there will be secondary frequency components, which are referred to as the bound modes. On the other hand, the nonlinearity of the transformation results in non-Gaussianity of the physical variables, even if one assumes that the normal wave amplitudes are Gaussian. Tayfun distribution is a PDF for η which takes into account such non-Gaussianity and which was derived for weakly nonlinear narrowband spectra in a model taking into account the quadratic nonlinearity only (Tayfun, 1980; Socquet-Juglard *et al.*, 2005; Onorato *et al.*, 2009),

$$P_{\text{Tayfun}}(\eta) = \frac{1 - 7\sigma^2 k_*^2/8}{\sqrt{2\pi(1 + 3G + 2G^2)}} \exp\left(-\frac{G^2}{2\sigma^2 k_*^2}\right), \qquad (6.19)$$

where $G = \sqrt{1 + 2\sigma k_*^2 \eta} - 1$, σ is the standard deviation of η and k_* is a characteristic wavenumber. Tayfun distribution appears to provide reasonable fits for broadband waves too, see Onorato *et al.* (2006), Denissenko *et al.* (2007), Tayfun and Fedele (2007) and Onorato *et al.* (2009). It is often used to quantify appearance of "freak" waves whose height is significantly greater than the average wave elevation, see Onorato *et al.* (2006), Janssen (2004), Tayfun and Fedele (2007) and Onorato *et al.* (2009).

Two global parameters which could be obtained from the elevation of PDF and which are often used as indicators of deviation from Gaussianity (see e.g. Caulliez and Guerin (2012)) are the skewness $C_3 = \langle\eta^3\rangle/\langle\eta^2\rangle^{3/2}$ and the kurtosis $C_4 = \langle\eta^4\rangle/3\langle\eta^2\rangle^2 - 1$.

As a word of caution, we would like to make it clear that to date there is no rigorous theoretical justification as to why Tayfun distribution should work. Indeed, even for narrowband spectra, the contribution into the curtosis from the cubic nonlinearity, unaccounted by Tayfun distribution, is of the same order as the contribution coming from the quadratic nonlinearity; see Janssen (2009). It is quite possible that good fits of experimental data by Tayfun distribution could simply be explained by a sufficient number of fitting parameters in this distribution and by the fact that only minor "tweaking" is usually needed for curve fitting if the waves are nearly Gaussian.

Note that the bound modes are also present in strongly nonlinear coherent structures and, in particular, in the singular ones, e.g. Stokes waves with 120° crests. In fact, this statement is almost empty because by definition a structure is coherent if it consists of several Fourier modes with fixed amplitudes and phase relations. Keeping this in mind, we could also understand non-Gaussianity as a property arising from the top-bottom asymmetry in the nonlinear water waves (seen in a left-right asymmetry of the PDF of η).

6.2.2.2. *Statistics of the height increments*

To study abrupt changes in space or time of the wave field or its derivatives, it is useful to consider space and time increments of the elevation field. Such increments can be of different orders, e.g.

$$\delta_l^{(1)} = \eta(\mathbf{x}+1) - \eta(\mathbf{x}), \tag{6.20}$$

$$\delta_l^{(2)} = \eta(\mathbf{x}+1) - 2\eta(\mathbf{x}) + \eta(\mathbf{x}-1), \tag{6.21}$$

etc. (here all η's are taken at the same t), and

$$\delta_\tau^{(1)} = \eta(t+\tau) - \eta(t), \tag{6.22}$$

$$\delta_\tau^{(2)} = \eta(t+\tau) - 2\eta(t) + \eta(t-\tau), \tag{6.23}$$

etc. (here all η's are taken at the same $\mathbf{x}$).

PDFs of the above increments $P_x(\sigma)$ and $P_t(\sigma)$ are defined in the usual way as a probability of a particular increment to be in the range from σ to $\sigma + d\sigma$ divided by $d\sigma$, or in the symbolic form

$$P_x^{(j)}(\sigma) = \langle \delta(\sigma - \delta_l^{(j)}) \rangle \tag{6.24}$$

and

$$P_t^{(j)}(\sigma) = \langle \delta(\sigma - \delta_\tau^{(j)}) \rangle \tag{6.25}$$

respectively, where $j = 1, 2, \ldots$ and $\delta(x)$ is Dirac's delta function. For random phased fields, these PDFs are Gaussian, and the presence of sparse coherent structures can be detected by the deviations from Gaussianity at the PDF tails. In particular, "fatter" (slower decaying) than Gaussian tails indicate an enhanced probability of strong bursts in the signal which is called intermittency.

Let us now introduce the moments of the height increments, which are called the *structure functions* (SF's),

$$S_l^{(j)}(p) = \langle (\delta_l^{(j)})^p \rangle = \int \sigma^p P_x^{(j)}(\sigma)\, d\sigma \qquad (6.26)$$

and

$$S_\tau^{(j)}(p) = \langle (\delta_\tau^{(j)})^p \rangle = \int \sigma^p P_t^{(j)}(\sigma)\, d\sigma. \qquad (6.27)$$

In turbulence theories, the SFs asymptotically tend to scaling laws,

$$S_l^{(j)}(p) \sim l^{\xi(p)} \qquad (6.28)$$

in the limit $l \to 0$, and

$$S_\tau^{(j)}(p) \sim \tau^{\zeta(p)} \qquad (6.29)$$

in the limit $\tau \to 0$, respectively. Functions $\xi(p)$ and $\zeta(p)$ are called the SF scaling exponents, and they contain the most important information about the turbulent field coherent and incoherent components and, correspondingly, about the turbulence intermittency.

Scalings generated by waves with random phases. Now let us consider a wave field made out of modes with random phases and the energy spectrum $E_k \sim k^{-\mu}$. For random phased fields, the PDFs of the height increments are Gaussian. For Gaussian statistics, we immediately have

$$S_l^{(j)}(p) \sim l^{p(\mu-1)/2} \qquad (6.30)$$

if $\mu < 2j + 1$, otherwise $S_l^{(j)}(p) \sim l^{pj}$ because the field is j times differentiable.

Similarly, in the time domain, we have for the random-phased field with the energy spectrum $E_\omega \sim \omega^{-\nu}$:

$$S_\tau^{(j)}(p) \sim \tau^{p(\nu-1)/2} \qquad (6.31)$$

if $\nu < 2j + 1$, otherwise $S_\tau^{(j)}(p) \sim \tau^{pj}$.

Of course, the laws (6.30) and (6.31) hold only in the range of scales corresponding to the power-law ranges of the respective spectra via $l \sim 1/k$ and $\tau \sim 1/\omega$. At very small scales l and τ, the waves are either damped by viscosity or become affected/dominated by the surface tension and acquire a different spectrum.

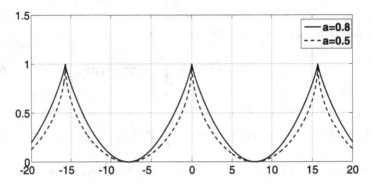

Fig. 6.1. A wave profile with singular structures of type $\eta(x) = \eta_0 - C_a |x|^a$ with $0 < a < 1$. (Plot shows $y = 1 - |\sin(0.2x)|^a$.)

Scalings generated by singular coherent structures. Let us now consider singular coherent structures and model their cross section near singularities by formula

$$\eta(x) = \eta_0 - C_a |x|^a, \tag{6.32}$$

with a singularity degree constant a such that $0 < a \leq 1$, and constants η_0 and C_a describing a reference surface elevation and the coherent structure amplitude respectively, see Fig. 6.1. For the special case of the Phillips and the Kuznetsov structures, we have $a = 1$. Note that for the scaling exponents, only small regions near the singularities give the suggested model as the profile is rather general (i.e. the specified power-law behavior provides a good fit near singularities whereas the smooth ramps away from the singularities have more arbitrary shapes, but they do not contribute much to the SFs). We will see below that structures with the additional parameter $a < 1$ also seem to be relevant to the wave turbulence in our experiments. Further, for the singularity dimension we have $D = 0$ for the Phillips and $D = 1$ for the Kuznetsov structures. In general, we will assume that the ridges of such crests may have a fractal dimension $0 \leq D < 2$.

As it was shown in Nazarenko *et al.* (2010), the singular structures of type (6.32) give the following contributions to the SFs,

$$S_l^{(j)}(p) \sim l^{pj} + N l^{2-D+ap}. \tag{6.33}$$

Note that in the limit $l \to 0$, out of the two terms on the RHS, the one with the smallest power will be dominant, and only in this limit the direct sum in (6.33) is valid. Thus, the structures of the Phillips or the Kuznetsov type, i.e. with $a = 1$, will not be seen in SFs for the first-order increments and we would have to consider $j \geq 2$. However, one should keep in mind that the finite range of excited scales makes determination of the scalings less precise for higher orders of j because of the higher number of SF points, the maximum distance between which must remain within this finite range of scales. Therefore, it is better to consider the lowest j that could allow to extract the scalings induced by the coherent structures ($j = 2$ in case of the Phillips or the Kuznetsov).

It was also shown in Nazarenko *et al.* (2010) that the singular structures of type (6.32) with $a = 1$ give rise to the following asymptotic behavior for the PDF tails, $|\sigma| \ll 2l$,

$$P_x^{(2)}(\sigma) \sim \frac{A}{l} \left[\ln \left(\frac{l}{|\sigma|} \right) + B \right], \tag{6.34}$$

where A and B are dimensionless constants which depend on the strength distribution of the singular ridges and their spatial density. For $a < 1$, the singular structures contribute to the PDF tails locally, i.e. forming a nonuniversal bump on the tail near $\sigma = C_a l^a$.

Now suppose that the wave field is bi-fractal and consists of two components: random phased modes and singular coherent structures. Avoiding the choices of j for which the field is j times differentiable, we have in this case

$$S_l^{(j)}(p) \sim l^{p(\mu-1)/2} + l^{2-D+ap}. \tag{6.35}$$

If $a < (\mu - 1)/2$ we expect to see the scaling associated with the incoherent random phased component at low p's (first term on the RHS) and the singular coherent structure scaling at high p's (second term on the RHS).

Similarly, one can consider the SFs of the time increments. Assuming, following Kuznetsov, that the coherent structures could be thought as passing a point-like probe (a wire probe in our experiment, see Sec. 6.3, Experimental setup) with constant velocity (due to shortness of the time needed for the singular ridge to pass the probe), we should obtain the time-domain scalings to be identical to the space-domain scalings obtained above, i.e. $S_\tau^{(j)}(p) \sim \tau^{2-D+ap}$.

In the case when incoherent waves and singular coherent structures are present simultaneously, we have

$$S_\tau^{(j)}(p) \sim \tau^{p(\nu-1)/2} + \tau^{2-D+ap}. \tag{6.36}$$

As before, it is understood here that the order j is chosen in such a way that the field associated with the incoherent wave component is not j times differentiable in time. For example, for spectra with $3 < \nu < 5$ (e.g. for

ZF spectrum) one should use $j \geq 2$, and for $5 < \nu < 7$ one should use $j \geq 3$, etc.

Let us discuss the relation between the statistics of the surface elevations measured at a single point (to which the Tayfun distribution is intended) and the statistics of the multi-point increments introduced in the previous section. Obviously, when the separation of the points in the increment is greater than the wave correlation length, then the multi-point PDF factorizes in terms of the one-point PDFs. Thus, in the large l limit the multi-point and the one-point PDFs contain the same information. On the other hand, for small l the elevation increments are related to the derivatives of the height elevation, e.g. $\delta_l^{(1)}$ measure the slope and $\delta_l^{(2)}$ measure the curvature. Thus, using $\delta_l^{(2)}$ allows one to pick out singular points with high curvature and magnify their effect by choosing high order p of the SFs. On the other hand, the effect of the weakly nonlinear bound modes on the statistics of such increments is expected to remain subdominant because they do not lead to the high-curvature points. Indeed, provided the nonlinearity remains weak in the scales which give dominant contribution to curvature, the bound modes will remain a perturbation at these scales, i.e. the curvature distribution will be close to the one of a Gaussian field. If, on the other hand, the curvature dominated scales are strongly nonlinear, then the perturbative bound mode approach is invalid, and we are again in the situation when the singular coherent structures are more relevant.

We can summarize that the single-point elevation PDFs are directly relevant to the rogue wave phenomenon and for which the bound modes are essential, whereas the multi-point PDFs of the elevation increments are more suited for detecting singularities associated with the wave breaking. Note, however, that single-point PDFs of *derivatives* of the surface elevation, such as the slope or curvature, are also sensitive to the singular structures and often they are used in open sea observations, see e.g. Cox and Munk (1954) and Caulliez and Guerin (2012).

6.2.3. *Wave turbulence life cycle*

Typically, both incoherent waves and coherent structures are present simultaneously and are dynamically important in wave turbulence. Their mutual interactions and transformations comprise wave turbulence life cycle which could be qualitatively understood in terms of fluxes in the wavenumber-amplitude space as described in Nazarenko *et al.* (2010) and Nazarenko (2011). We have already described the k-space fluxes associated with the KZ spectra, and we will now describe fluxes in the space of wave amplitudes.

In Choi *et al.* (2005) (see a detailed description of this approach in Nazarenko (2011)), the weak wave turbulence formalism was extended to PDF of Fourier intensities $J_k = |a_k|^2$ defined as

$$\mathcal{P}_k(J) = \langle \delta(J - |a_k|^2) \rangle. \tag{6.37}$$

Assuming as usual weak nonlinearity, random wave phases and random wave amplitudes, they derived the following PDF equation,

$$\dot{\mathcal{P}} + \partial_J F = 0, \tag{6.38}$$

where

$$F = -J(\beta \mathcal{P} + \alpha \partial_J \mathcal{P}) \tag{6.39}$$

is a probability flux in the J-space, and

$$\alpha_k = 4\pi \int |W_{23}^{01}|^2 \delta(\mathbf{k} + \mathbf{k}_1 - \mathbf{k}_2 - \mathbf{k}_3) \delta(\omega_k + \omega_{k_1} - \omega_{k_2} - \omega_{k_2})$$
$$\times n_{k_1} n_{k_2} n_{k_3} \, d\mathbf{k}_1 d\mathbf{k}_2 d\mathbf{k}_3,$$

$$\beta_k = 8\pi \int |W_{23}^{01}|^2 \delta(\mathbf{k} + \mathbf{k}_1 - \mathbf{k}_2 - \mathbf{k}_3) \delta(\omega_k + \omega_{k_1} - \omega_{k_2} - \omega_{k_2})$$
$$\times [n_{k_1}(n_{k_2} + n_{k_3}) - n_{k_2} n_{k_3}] d\mathbf{k}_1 d\mathbf{k}_2 d\mathbf{k}_3, \tag{6.40}$$

where $n_k = (L/2\pi)^2 \langle J \rangle$ and $W_{23}^{01} \equiv W(\mathbf{k}, \mathbf{k}_1, \mathbf{k}_2, \mathbf{k}_3)$ is the interaction coefficient.

Equation (6.38) has a solution with $F = 0$, $\mathcal{P}_k(J) = \frac{1}{\langle J \rangle} e^{-J/\langle J \rangle}$, which is called Rayleigh distribution and which corresponds to a Gaussian wave field. In addition, there are solutions for a constant J-flux, $F = \text{const} \neq 0$. They have power-law tails and, therefore, correspond to intermittent wave turbulence. At the tail of the PDF, $J \gg \langle J \rangle = (2\pi/L)^2 n_k$, the solution can be represented as series in $\langle J \rangle / J$,

$$\mathcal{P}_k(J) = -F/(J\beta) - \alpha F/(\beta J)^2 + \cdots. \tag{6.41}$$

It was speculated that such a flux in the amplitude space can be physically generated by the wave breaking events. On the other hand, we remember that the KZ state corresponds to the energy flux through wavenumbers k. Thus, it is natural to consider a combined flux which has both k and J components and, thereby, clarify the picture of the wave turbulence cycle which involves both random waves and coherent structures which can interact and get transformed into each other. The relevant quantity whose flux has both k and J components is the 2D energy density in the (k, J)-space,

$$\mathcal{E}_{k,J} = 2\pi k \omega_k J \mathcal{P}_k(J), \tag{6.42}$$

for which we can obtain from (6.38) the following equation,

$$\dot{\mathcal{E}}_{k,J} + \partial_k \mathcal{F}_k + \partial_J \mathcal{F}_J = 0, \tag{6.43}$$

with the k and J flux components, respectively,

$$\mathcal{F}_k = -2\pi \int_0^k k' \omega_{k'} F_{k'} \, dk' \quad \text{and} \quad \mathcal{F}_J = 2\pi k \omega_k F_k.$$

Importantly, the intermittent solution (6.41) corresponds to negative J-flux, from large to small amplitudes, and not in the opposite direction as one could naively expect based on the picture that the wave breaking occurs and dissipates turbulence when amplitudes become large. However, a more careful look reveals that the wave breaking not only dissipates energy but also returns part of it into weak incoherent ripples. This can be understood by considering the fluxes on the (k, J)-plane, see Fig. 6.2. Let us force turbulence by generating weak waves at low k's — region marked as "source" in Fig. 6.2. The energy cascade will proceed from the forcing region to higher k predominantly along the curve $J(k) = \langle J_k \rangle = (2\pi/L)^2 n_k$. For example, for the ZF state this curve is

$$J(k) = J_{\mathrm{ZF}}(k) \sim \epsilon^{1/3} k^{-4},$$

which immediately follows from (6.6).

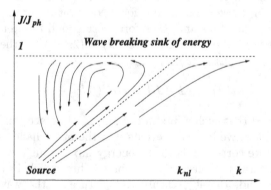

Fig. 6.2.   Wave turbulence fluxes in the (k, J)-plane. The amplitude J is normalized to the Phillips spectrum, $J_{\mathrm{Ph}} \sim g^{1/2} k^{-9/2}$. The ZF spectrum and the Phillips spectrum are shown by the dashed horizontal and dashed inclined lines, respectively.

When amplitudes become large, all k-modes become correlated (i.e. we observe occurrence of the coherent structures). Namely, in the (k, J)-plane, these modes are concentrated at the line $J = J_{\text{Ph}}$, where

$$J_{\text{Ph}} \sim g^{1/2} k^{-9/2},$$

see Fig. 6.2 (we normalize J by J_{Ph} in this figure, hence this line is horizontal). Subscript Ph here stands for "Phillips" to emphasize that it corresponds to the Phillips scaling for which, as follows from (6.12), the linear and the nonlinear timescales are of the same order, which in turn implies presence of coherent structures whose Fourier components have correlated phases.

Around some scale k_{nl}, where the Phillips and ZF curves intersect, the weak wave turbulence description breaks down because the nonlinear term becomes of the same order as the linear one. At this point, the phases get correlated, which arises in the form of coherent structures in the x-space. Such coherent structures are made of a broad range of Fourier modes which are correlated and for each of whom the linear and nonlinear terms are in balance. This is generally called a critical balance state, and it corresponds to the Phillips spectrum shown by the horizontal dashed line in Fig. 6.2. The critical balance state appears to be attracting. Indeed, if the linear term for some k was greater than the nonlinear term, then this wave would quickly de-correlate from the rest of the modes. If, on the other hand, the nonlinear term gets larger than the linear one at some k, then the inertial forces on a fluid element would get larger than the gravity force, and this fluid element would separate from the surface and exit the coherent structure. However, such sea spray and foam do form occasionally via wave breaking which provides the main mechanism of the wave energy dissipation (above the horizontal dashed line in Fig. 6.2). Thus, on the (k, J)-plane, the flux turns at k_{nl} and goes back to lower k's along the Ph curve, with some energy lost to the regions above the Ph curve via wave breaking. Occasionally, the coherent structures lose their coherence, due to the energy loss, to the sea spray and foam thus causing a corresponding reduction in nonlinearity. Additional mechanisms that can promote de-correlation of coherent structures are due to their mutual interactions and due to interactions with the incoherent component. On the (k, J)-plane, this corresponds to turning of the flux down below the Ph curve and toward the ZF curve. This closes the cycle of the wave turbulence, in which the energy cascade of the random phased waves leads to creation of coherent structures, which in turn, break down with their energy partially dissipated in whitecapping and partially returned into the incoherent random phased component. The exact partition of the energy dissipated vs. the energy returned into the random waves is not known, but it is natural to think that these parts are of the same order of magnitude.

The last part of the wave turbulence cycle is crucial for understanding intermittency. Indeed, this part corresponds to a flux in the negative J-direction, which, as we mentioned, corresponds to the fat tails of the PDF of the Fourier modes. Of course, such a flux is not constant and, therefore, one should not literally apply the $1/J$ solution obtained for this idealized case. However, we will use this solution for order-of-magnitude physical estimates below. Note that the J-flux is negative only at the low k's whereas at higher k's the J-flux is positive and one expects depleted PDF tails with respect to Rayleigh distribution.

Let us now obtain some order-of-magnitude qualitative predictions from the wave turbulence cycle picture presented above. Taking into account the definition of the energy flux in the k space, ϵ_k, from (6.43) we have for the energy spectrum $E_k = 2\pi k\omega_k n_k = (L/2\pi)^2 \int \mathcal{E}_{k,J} dJ$,

$$\dot{E}_k = -\partial_k \epsilon_k = 2\pi k\omega_k (L/2\pi)^2 \int_0^{J_{\mathrm{Ph}}} J\partial_J F_k dJ, \qquad (6.44)$$

where we took into account the cut-off at $J = J_{\mathrm{Ph}}$ related to the fact that for $J > J_{\mathrm{Ph}}$ the nonlinearity is stronger than the linear terms which means severe damping via wave breaking (i.e. the gravity force is not able to keep the fluid particles attached to the surface). This leads to the following estimate of the relationship between the (k-integrated) J- and the (J-integrated) k-fluxes,

$$F_k \sim \frac{\partial_k \epsilon_k}{2\pi k\omega_k n_{\mathrm{Ph}}}. \qquad (6.45)$$

Thus, the intermittent tail of the PDF (6.41) becomes

$$\mathcal{P}_k(J) \approx \frac{F_k}{J\beta_k} \sim \frac{\partial_k \epsilon_k}{2\pi k\omega_k \beta J n_{\mathrm{Ph}}} \sim \frac{n_k}{J n_{\mathrm{Ph}}}. \qquad (6.46)$$

Here, we used the fact that in the kinetic equation

$$\dot{n}_k = \alpha_k - \beta_k n_k \qquad (6.47)$$

the two terms on the RHS are of the same order.

The PDF tail (6.46) gives the following contribution to the moments of the Fourier amplitudes

$$M_k^{(p),\mathrm{tail}} = (L/2\pi)^{2p} \int_0^{J_{\mathrm{Ph}}} J^p \mathcal{P}_k(J)\, dJ \sim \frac{n_k}{p} n_{\mathrm{Ph}}^{(p-1)}. \qquad (6.48)$$

For example, for the ZF states $n_k = n_{\mathrm{ZF}}$, we have

$$M_k^{(p),\mathrm{tail}} \sim \frac{1}{p}\epsilon^{1/3} g^{(p-1)/2} k^{1/2-9p/2}. \qquad (6.49)$$

On the other hand, the PDFs core has a Rayleigh (exponential) shape,

$$\mathcal{P}_k(J) = \frac{(L/2\pi)^2}{n_k} e^{-(L/2\pi)^2 J/n_k},$$

which gives the following contribution to the moments,

$$M_k^{(p),\text{core}} = p! n_k^p. \tag{6.50}$$

For the ratio of the tail and core contributions we have

$$M_k^{(p),\text{tail}}/M_k^{(p),\text{core}} \sim \frac{1}{pp!}(n_{\text{Ph}}/n_k)^{(p-1)}. \tag{6.51}$$

In particular, for the ZF state

$$M_k^{(p),\text{tail}}/M_k^{(p),\text{core}} \sim \frac{1}{pp!}\left(\frac{k}{k_*}\right)^{(1-p)/2}, \tag{6.52}$$

where $k_* = g\epsilon^{-2/3}$.

One can see that for fixed ϵ the core will dominate when $p \to \infty$. However, for fixed $p > 1$, the PDF tail will dominate in the moments as $\epsilon \to 0$. This is a rather unexpected result saying that the weaker the wave turbulence is, the more visible will be the k-space intermittency.

6.2.4. *Non-stationary wave turbulence*

All previous considerations assume the stationary case of wave turbulence, while nonstationary processes, turbulence formation and decay are also important for practice and theory. Evolving wave turbulence is the least studied and most poorly understood object. In this review, we will report new experimental data on the nonstationary behavior preceding formation of the steady state and on the decaying regime — when the wave makers are switched off. We will see several puzzling effects for which we will suggest possible explanations and outline open problems. Clearly, most of the work on identifying the mechanisms in the evolving wave turbulence still remains to be done.

Two important time scales will be important for the subsequent discussion: the characteristic times of the nonlinear dynamical and the kinetic evolutions. The dynamical time scale τ_D is the characteristic time of the nonlinear deterministic dynamical equations for the water waves. It can simply be obtained based on the knowledge that the leading order process for the surface gravity waves is four-wave and, therefore, the nonlinear time is inversely proportional to the square of the wave amplitudes. The rest of the expression can be reconstructed by matching physical dimensions from using the wave frequency ω and the gravity constant g:

$$\tau_D \sim \frac{g^2}{\omega^5 \eta^2}. \tag{6.53}$$

This allows us to introduce a convenient measure of nonlinearity of the waves,

$$\gamma = k\eta \sim \frac{1}{\sqrt{\tau_D \omega}}, \tag{6.54}$$

which can be interpreted as a typical slope of the water surface. The kinetic time scale is the characteristic time of the kinetic equation (6.5). It can also be obtained from the dimensional analysis based on the fact that, according to the kinetic equation, it must be proportional to the inverse square of the wave spectrum, which means that it is proportional to the inverse fourth power of the wave amplitude. The rest is reconstructed dimensionally:

$$\tau_K \sim \frac{g^4}{\omega^9 \eta^4} = \tau_D^2 \omega = \gamma^{-2} \tau_D. \tag{6.55}$$

Let us estimate τ_D and τ_K for our typical experiment with $\eta \sim 5\,\text{cm}$ and $\omega \sim \omega_f = 2\pi\text{s}^{-1}$ (1 Hz forcing). We have $\tau_D \sim 4\,\text{s}$ and $\tau_K \sim 100\,\text{s}$. This corresponds to the nonlinearity parameter $\gamma \sim 0.2$.

Note, however, that once the forcing is switched off the kinetic time scale rapidly increases due to the rapid dependence of τ_K on η. Also, the time scales are very sensitive to the value of the energy containing frequency ω.

6.2.5. *Self-similar solutions of the kinetic equation and formation of the steady state*

Let us suppose that the kinetic equation of the weak wave turbulence (6.5) is valid from the very first moment after the wave maker is switched on, and let us discuss how it would predict formation of the steady state spectrum. Appropriate description of such a dynamical phase is given by self-similar solutions of the kinetic equation (6.5), see e.g. Zakharov *et al.* (1992) and Nazarenko (2011).

There are several types of self-similar evolutions. To identify one for the gravity waves, it is important to know that this is a finite capacity system, i.e. only finite amount of wave energy is necessary to fill up the high-frequency tail of the ZF spectrum for whatever large inertial intervals. Thus, no matter how high is the dissipation frequency, the propagating front of the spectrum will reach from the forcing frequency to the dissipation frequency in a finite time which is of order of the characteristic time of the kinetic equation (6.55). It is interesting that the spectrum left behind such a propagating front is not going to be the KZ spectrum (ZF spectrum in the case of the gravity waves), but a power-law spectrum with a steeper exponent. The KZ spectrum in this case will form as a reflection wave propagating toward lower frequencies after the front hits the high-frequency end of the inertial range, see e.g. Nazarenko (2011). In wave turbulence, this kind of self-similar behavior of the finite capacity systems was discovered first in the example of MHD waves by Galtier *et al.* (2000), and later was also found in a model of hydrodynamic turbulence in Connaughton and Nazarenko (2004).

However, as we will see later, our experimental results do not show any detectable front propagation toward the high frequencies during the stage of the steady state formation. Possibly, this is because the initial phases are not yet quite random and the kinetic equation does not yet fully describe the averaged wave behavior. On the other hand, the evolution time scale at the formation stage seems to be consistent with the kinetic time scale (6.55). We thus conclude that the kinetic equation is at least partially valid for the nonstationary early evolution, which implies that at least some phase randomization necessary for the statistical regime has occurred. We attribute such a rapid phase randomization to the billiard-like mixing due to the wave reflections from the walls.

6.2.6. *Decay of the wave turbulence via the energy cascade*

Now, we will estimate characteristics of the wave turbulence in the decaying regime, when its forcing is switched off. Again, we will first assume the weak turbulence regime described by the kinetic equation (6.5). The forward cascade of the gravity wave turbulence is a finite capacity system, as was explained in the previous section. This means that only a small amount of energy supply is necessary to maintain the ZF spectrum, and in the decaying stage this supply is provided by the lowest frequencies which contain most of the energy. Thus, in the decaying stage, we should expect the ZF spectrum in the inertial range whose overall amplitude is gradually decreasing as the energy is slowly leaking from the system. Note that this situation is typical not only for the finite capacity wave turbulence systems, but also for the strong hydrodynamic turbulence which is also of the finite capacity type. That is why the famous Kolmogorov spectrum is observed, e.g., in a free stream behind a grid, even though, strictly speaking, it is in a decaying state.

Let us find the energy decay law. Because most of the wave energy resides at the largest scales near the forcing frequency ω_f, the wave energy density per unit area of the water surface $\mathcal{E}$ is related with the energy spectrum at the forcing scale E_{ω_f} as $\mathcal{E} = \int_{\omega_f} E_\omega d\omega \sim E_{\omega_f}\omega_f$. Thus, for the the energy dissipation rate $\epsilon = -\dot{\mathcal{E}} \sim -\dot{E}_{\omega_f}\omega_f$. Substituting ϵ from the ZF spectrum (6.7) taken at ω_f, we have

$$\dot{E}_{\omega_f} \sim -\epsilon/\omega_f \sim -E_\omega^3 \, g^{-6}\omega_f^{11}. \qquad (6.56)$$

Solving this equation gives

$$E_{\omega_f} \sim \frac{g^3}{\omega_f^{11/2}}t^{-1/2}. \qquad (6.57)$$

Thus, the prediction of the weak turbulence theory is that the peak of the energy spectrum (which is at ω_f, i.e. where the forcing was before it was switched off) should decay as an inverse square root of time. We will see that our experimental data agree with this prediction but only at the initial decay stage. Thus, for the long-time decay, let us explore another possible mechanism of the energy dissipation, which could take over from the cascade mechanism when it becomes inefficient after a quick initial drop in the wave amplitudes.

6.2.7. *Decay of the wave energy via wall friction*

As we will see later, our experimental data for the decay of the spectrum agrees with the power-law decay derived in the previous section, but only initially. On the other hand, for large times, it could fit quite well with an exponential decay law. This situation is puzzling because the exponential decay could indicate a linear dissipation mechanism which is dominant over the cascade mechanism assumed in the previous section. In fact, if such a mechanism could act parallel to the cascade process in itself it would not contradict the weak turbulence theory. Here, we will explore a possibility that after a short transient after stopping the forcing, the largest (energy containing) scales are dissipated directly via the skin friction at the flume walls. At the same time, these energy containing scales could still serve as an energy reservoir for the ZF cascade toward the higher frequencies. The latter would serve as an additional and perhaps sub-dominant dissipation mechanism in this case.

The simplest estimate could be made based on the well-known textbook problem about the oscillating laminar flow with velocity amplitude u and frequency ω near a flat boundary. This layer has an exponential velocity profile with thickness $\delta \sim \sqrt{2\nu/\omega}$ and, therefore, the energy dissipation rate in this layer is $2\nu\delta^{-2}u^2V \sim \omega u^2 V$, where $V = 4L\delta H$ is the total volume of the boundary layer, with L being the flume size and $H \sim g/\omega^2$ being the characteristic depth of the wave motion. Dividing the energy dissipation rate in the layer by the total wave energy $u^2 L^2 H$, we get the exponent of the exponential decay of the wave energy due to the wall friction:

$$\sigma_\nu \sim \frac{4\sqrt{2\omega\nu}}{L}. \tag{6.58}$$

Substituting here $\nu = 10^{-6} \, \text{m}^2/\text{s}$, $\omega \sim \omega_f = 2\pi \, \text{s}^{-1}$ (1 Hz forcing) and $L \sim 10 \, \text{m}$, we get $\sigma_\nu \sim 1/(15 \, \text{min})$. As we will see, this estimate is consistent with our experimental data for the decay rate although with some underestimation (by a factor of ~ 3).

Now, we can estimate the amplitude η_c corresponding to the crossover from the dominant cascade to the friction dissipation mechanism. It occurs when $\tau_K \sigma_\nu \sim 1$, which, taking into account (6.55), gives $\eta_c \sim 3 \, \text{cm}$. Thus, we see that after a rather quick decay of η from 5 cm to just 3 cm, we should expect that the cascade dominated dissipation of energy should be replaced by the wall friction mechanism. For the crossover time, we have respectively $t_c \sim \tau_K \ln(5/3) \sim 500$ s, which is consistent with our experimental results. Because of the singular character of the $t^{-1/2}$ law, at $t = 0$, the crossover time is almost insensitive to the initial wave intensity, which is confirmed in our experiments.

Note, however, that the above estimates are rather rough because one of the four sides of the flume is not flat as it contains the wavemaker, and because in some places the boundary layer may become turbulent as the average Reynolds number near the wall is about 200 and may get higher during splashes. Also, long waves may experience a bottom friction (the large bottom area may somewhat compensate for the small velocity near the bottom). These additional factors may contribute to further increase of the skin friction dissipation.

6.3. Experimental Setup

The experiments were conducted in a rectangular tank with dimensions $12 \times 6 \times 1.5$ meters filled with water up to the depth of 0.9 m, see Fig. 6.3. The gravity waves were excited by a piston-type wavemaker. The wavemaker consists of eight vertical paddles of width 0.75 m covering the full span of one short side of the tank. An amplitude, frequency and phase can be set for each panel independently which, in particular, allow to control directional distribution of generated waves. A motion controller is used to program parameters of the generated wave field by specifying its amplitude and a number of wavevectors (given by a set of frequencies and directions). In the experiments described here, the wavemaker generated a superposition of

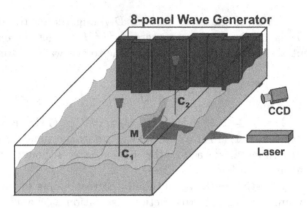

Fig. 6.3. The experimental setup: M, the first surface mirror; C_1 and C_2, the capacitance wire probes and CCD, the digital camera.

two waves of equal amplitude with frequencies $f_1 = 0.99\,\text{Hz}$ and $f_2 = 1.14\,\text{Hz}$ (the wavelengths are 1.59 and 1.2 m correspondingly). The wavevector k_1 was perpendicular to the plane of the wavemaker and k_2 was at the angle $7°$ to k_1. It is assumed that energy dissipation is low and the waves undergo multiple reflections from the flume walls, interact with others and form a chaotic wave field homogeneous in the central area of the flume. The main control parameter was an oscillation amplitude of the wavemaker, by varying it we study the dependence of the spectrum and PDFs on the average wave intensity.

Our choice of the two-mode forcing configuration, the mode wavenumbers and directions were motivated by searching for an optimum where the forcing is narrowband enough to allow a large inertial range, but at the same time leads to a quick phase mixing and isotropization due to the wall reflections, and does not produce peaks due to the resonances with the flume eigenmodes or harmonics.

Two capacitance wire probes were used to measure the wave elevation as a function of time $\eta(t)$ in two fixed points in the central part of the flume as it is shown in Fig. 6.3. The distance between the probes was 2 m. Signals from the probes were amplified and digitized by an analog-to-digital converter (NI6035) controlled by the LabView and stored in a PC. Typical signal acquisition parameters were as follows: the bandwidth — 32 Hz and the recording time — 2000 s. The wire probes were calibrated before the measurements in the same tank with a stationary water surface.

In addition to the measurements of the time dependence, in the present work, we introduced a new technique, similar to that described by Mukto *et al.* (2007) which allows us to measure the dependence of the surface elevation on the space coordinate along a line. For this, we used a vertical

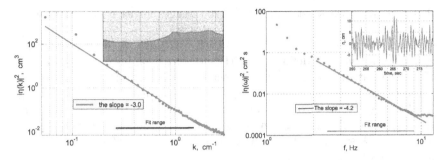

Fig. 6.4.   Left plot: The spectrum in k-domain. The inset shows an image of air–water interface. Right plot: The spectrum in ω-domain. The inset shows the corresponding function $\eta(t)$. Both spectra were measured at the wave nonlinearity $\gamma = 0.2$.

cross section image of the air–water interface. The upper layer of water was colored by a fluorescent dye Rhodamine 6G. The water–air interface area was illuminated from below by a narrow light sheet from a pulsed Yag laser (power 120 mW, wavelength 532 nm), see Fig. 6.3. The images were captured by a 1.3 Mpixel digital camera (Basler, A622f) synchronously with laser pulses at the sampling frequency from 8 to 24 Hz. The image size was 900 × 1200 mm with the resolution 0.93 mm/pixel and 0.90 mm/pixel in the vertical and the horizontal directions, respectively. Typically, we collected five sets of images, each set consisting of 240 frames. The time interval between the sets was 5 min. As an example, we present an inset with instantaneous wave profile, see Fig. 6.4.

The data from the capacitance probes were acquired continuously and in parallel with the images during this time. The measurements were done at fixed excitation parameters. The measurement procedure included setting the amplitude of the wavemaker oscillations, waiting for a transient time interval, 20–30 min, and recording the signals during 35 min.

The stored data were processed using the Matlab. The wire probe data were filtered by a band-pass filter within 0.01–20 Hz frequency band. The image sets were processed using standard binarization and the boundary detection procedures from the Matlab Image Processing Toolbox. Detected air–water boundaries were stored as a set of arrays, $\eta(x)$, for a following statistical analysis. The images where the boundary was not a single valued function of x or when it had significant jumps ($|\delta\eta(x)|/\delta x > 4$) were deleted. A proportion of such images did not exceeded 3% for highest wave amplitude regimes. To calculate spectra from the wire probes we used the Welch algorithm with the Hamming window and the averaging performed over 1000 spectral estimates for each signal record. The k-spectra were calculated for each array of boundaries (one array from each image) and then averaged over a set from up to 1200 images for each stationary wave field.

As a characteristic of the averaged wave strength, we used a nonlinearity parameter which is defined in (6.54) and has a meaning of the mean slope of the wave at the energy containing scale, $\gamma = k_m A$, where A is the rms of η and k_m is the wavenumber corresponding to the maximum of the energy spectrum $|\eta_\omega|^2$. In all our experiments, k_m was approximately the same and located in the forcing range, $k_m \approx 5.2\,\mathrm{m}^{-1}$ that corresponds to the wavelength $\lambda \approx 1.2$ m. In this experiment, the range of the nonlinearity parameter was $0.1 < \gamma < 0.25$.

6.4. Experimental Results

We will start by describing our previous results on stationary wave turbulence reported in Denissenko *et al.* (2007), Lukaschuk *et al.* (2009) and Nazarenko *et al.* (2010). We will follow that by describing our new results on evolving wave turbulence.

6.4.1. *Spectra*

Typical energy spectra in the ω and k domains are shown in Fig. 6.4. For the ω-spectra, we usually have about one decade of the fitting range which, according to the dispersion relation, should correspond to the two decades for the k-spectra. In reality, the k-spectra have shorter scaling range which is limited, on the low-k side, by the width of laser sheet, 1.2 m, and on the high-k end by insufficient vertical resolution of the images and limited statistics. In addition, the scaling ranges getting narrower for the flume runs with weaker forcing.

The slopes of the energy spectra in the ω- and k-domains as functions of the wave field intensity are shown in Fig. 6.5. We see that for both ω- and k-spectra the slopes are steeper for the weaker wave fields with respect to the stronger ones. One can see that at low wave intensities the data scatter and uncertainty are much greater than for stronger wave turbulence.

In Fig. 6.6, we show the graph of k-slope versus ω-slope for the energy spectra measured in the same experiments with the laser sheet and the capacitance wire techniques, respectively. Filled circle and diamond data points correspond to two different orientations of the laser sheet — perpendicular to the wavemaker and inclined at 20° to it respectively. Small differences between these two data sets demonstrate good isotropy. The solid line is where the ω- and the k-slopes are related via the linear dispersion relation, i.e. $\nu = 2\mu - 1$. As we see, the experimental data deviate

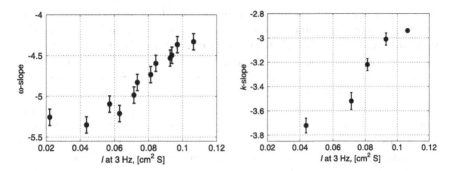

Fig. 6.5. Slopes of the k- and ω-spectra as functions of the wave intensity I defined as the value of the ω-spectrum of η measured at the frequency 3 Hz, i.e. $I = \langle |\eta_\omega|^2 \rangle_{\omega=3\,\mathrm{Hz}}$.

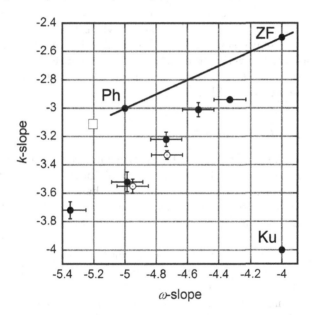

Fig. 6.6. The k-slope vs. ω-slope. See the main text for the description.

from this line, which could be due to a direct influence of long waves onto shorter ones (leading to a doppler shift), finite depth effects or/and presence of coherent structures.

We also put the points corresponding to the theoretical predictions: Phillips (Ph), Zakharov-Filonenko (ZF) and Kuznetsov (Ku) spectra. We see that both Ph and Ku points are rather far from the experimental data, whereas ZF point is more in agreement with the experiment: the plot suggests that if one would perform experiments at even higher amplitudes then the experimental data could have crossed the ZF point. This may seem surprising as one could expect the ZF theory to work better for weaker waves. However, the finite flume size effects are more important for weaker waves because weakly nonlinear wave packets preserve their integrity while traveling across the flume and, upon reflection, create standing waves whose structure is determined by the flume's shape (in the Fourier space this corresponds to a regular grid of eigenmodes whose spectral broadening is much less than the intermode distance and whose resonant interaction among themselves, therefore, is depleted). This is the most likely explanation for why there is a significant deviation from the ZF spectrum at smaller wave amplitudes. Note that the simultaneous measurement of the k- and ω-spectra allows one to resolve the uncertainty of the previous results reporting on the the ω-spectra only. Namely, we are now able to differentiate between the ZF and Ku states which have undistinguishable ω-slopes but different k-slopes. Now, it is clear that the ZF spectrum is more consistent with the experimental data than the Ku spectrum. However, as we will see, singular structures of the Ku type do seem to show up in the scalings of the high order SFs.

6.4.2. *Probability density functions*

In this section, we will focus on the experimental run with spectra $E_k \sim k^{-3.02}$ and $E_\omega \sim \omega^{-4.2}$. Because for each of these spectra both k and ω slopes are steeper than -3 but shallower than -5, the second-order increments will be most appropriate. Experimental PDFs of the $j = 2$ time and space height increments are shown in Figs. 6.7(a) and 6.7(b), respectively. For the space increments, there are clear deviations from Gaussianity at the PDF tails indicating intermittency due to coherent structures. For the time increments, the deviations from Gaussianity is weaker, which could be due to slow propagation speed of the coherent structures leading to their less infrequent occurrence in the t-domain in comparison with the x-domain. Both the t- and the x-domain PDF's are asymmetric, with dominant negative increments, which reflects the fact that breaks occur at wave crests rather than troughs.

Asymmetry with respect to the crests and the troughs is also seen on the one-point PDF of the surface elevation shown for the same run in Fig. 6.8. A fatter tail on the right corresponds to the typical wave structure where the crests deviate from the mean elevation more than the troughs. As we see in Fig. 6.8, the PDF can be well fitted by Tayfun distribution (6.19) with skewness 0.4 and kurtosis 3.3.

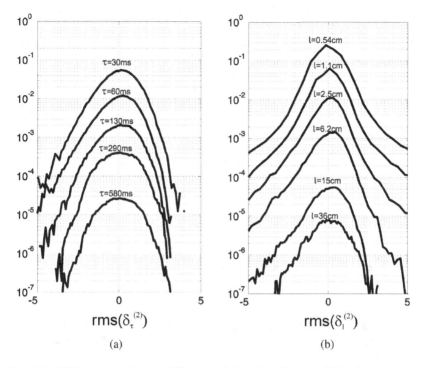

Fig. 6.7. PDFs of second-order differences: (a) in the t-domain, (b) in the x-domain.

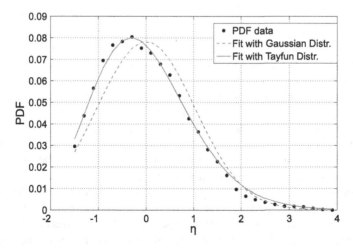

Fig. 6.8. PDFs of the wave elevation η (scaled with standard deviation) compared to the Tayfun distribution (solid line) and the Gaussian distribution (dashed line).

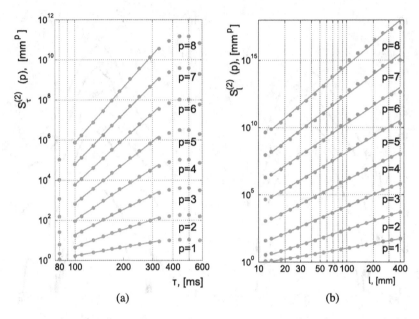

Fig. 6.9.   The elevation SF's (moments) from the 2nd to 8th order ($p = 2, \ldots, 8$): (a) in the t-domain, $S_\tau^{(2)}(p)$, and (b) in the x-domain, $S_l^{(2)}(p)$, as the functions of τ and l, respectively.

6.4.3. *Structure functions*

In our data on the SFs, both $S_\tau^{(2)}(p)$ as a function of τ and $S_l^{(2)}(p)$ as a function of l exhibit clear power-law scalings in the range of scales corresponding to the gravity waves for all p at least up to 8 (see Fig. 6.9). The SF exponents for the time and space domains are shown in Figs. 6.10(a) and 6.10(b), respectively. Straight lines on these graphs represent the ZF scaling (solid line), scaling of waves with the spectrum as measured in the experiment (dash line) and the fit of the high-p behavior with a scaling corresponding to singular coherent structures (dash-dot line). For the time domain, the scaling at low p is close to the ZF scaling, this is surprisingly more consistent than with the scaling calculated from the actual measured spectrum. For an infinite scaling range the $p = 2$ point must, of course, lie exactly on the value corresponding to the spectrum, $(\nu - 1)$, irrespective of the presence or absence of the phase correlations. Thus, we attribute the observed discrepancy to the finiteness of the scaling range. Furthermore, the fit of the high p dependence indicates the presence of singular coherent structures with $D = 1, a = 1.05$ that is very close to the Ku's $D = 1, a = 1$.

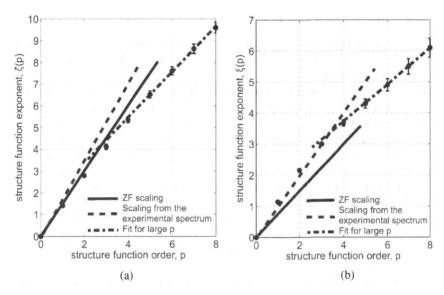

Fig. 6.10. SFs scaling exponents: (a) $\zeta(p)$ in the t-domain and (b) $\xi(p)$ in the k-domain.

For the space-domain, at low p, there is an agreement with the scaling of the random phased waves having the actual measured spectrum, and less agreement with the random phased waves having the ZF spectrum. This is not surprising since the scaling range in k is greater than in ω and, therefore, there is a better agreement between the spectrum and the SF exponent for $p = 2$. More importantly, we see again the dominance of the random phased waves in the low-order SFs, and the dominance of coherent breaks in the high-order SFs. The fit at high p's gives the dimension and the singularity parameter of the breaks $D = 1.3$ and $a = 1/2$, respectively. We see that the breaks appear to be more singular and "spiky" than the Ku-type breaks ($a = 1$). Visually, we observed numerous occurrences of these kinds of spiky wave breaks, which are not propagating (or propagating very slowly) and producing vertical splashes. These kinds of structures should be probable in isotropic wave fields due to the collision of counter-propagating waves, which in our flume appear due to multiple wave reflections from the walls. The slow propagation speed of such breaks means their seldom crossing through the capacitance probe even if there is a large number of them in the x-domain (i.e. more than the Ku-type breaks). This could explain why the Ku-breaks show up in the SF scalings in the t-domain, whereas more singular spiky structures are seen in the x-domain.

6.4.4. *Statistics of the Fourier modes*

The results of this section are relevant to the theory of the Fourier mode statistics and the wave turbulence cycle described in Sec. 6.2.3. However, our results are rather preliminary and they do not allow to fully test the theoretical predictions. First, the theoretical picture was developed for weak waves where random waves dominate over the coherent structures in most of the inertial range, whereas the experimental runs selected in the present work corresponded to larger excitations where the rms amplitude and wavebreaking amplitude are not different by orders of magnitude. Second, like the classical weak wave turbulence, the theory corresponds to an infinite system, whereas the finite size effects are likely to be important in our flume. Recall that the finite size effects are strongest at low amplitudes, and it is impossible to implement weak wave turbulence and eliminate the finite size effects simultaneously in our flume. Third, our statistical data are not sufficient for the single Fourier modes and, as a result, the PDF tails are rather noisy.

Nonetheless, the results on Fourier space PDFs are quite interesting. PDFs of ω-modes, $J_\omega = |\eta_\omega|^2$, as measured by the wire probes, are presented in Fig. 6.11. Here, the averaging was done over modes in the frequency range from 5 to 7 Hz via band-pass filtering. One can see that the PDF core can be fitted with an exponential function which corresponds to Gaussian statistics. At the tail, one can see a significant deviation from the exponential fit corresponding to intermittency. As predicted by the theory, the tail follows a power-law (with cut-off), but with a different index, -3

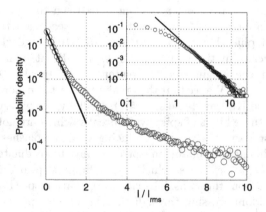

Fig. 6.11.   Normalized PDF of the spectral intensity $I = |\eta_w|^2$ band-pass filtered with the frequency window ± 1 Hz centered at $f = 6$ Hz, measured at wave nonlinearity $\gamma = k_0 \eta_{rms} \approx 0.16$ (k_0 is the wave vector at the maximum spectral power). The inset shows the same plot in log–log coordinates.

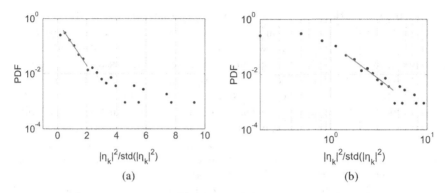

Fig. 6.12. The PDF of the k-mode centered at $k = 54.3\,\mathrm{rad/m}$, filtered within the window $\pm10.8\,\mathrm{rad/m}$ and normalized by its standard deviation. (a) A semilog plot with the exponential fit $y \propto 10^{-1.6x}$. (b) A log–log plot with the power fit $y \propto x^{-2.5}$.

rather than -1. This deviation in the power-law index could be due to one of the reasons mentioned in the beginning of this subsection.

PDFs of k-modes, $J_k = |\eta_k|^2$, are shown in Fig. 6.12. Here, the averaging was done over three adjacent modes. Again, we see a Gaussian PDF core and a power-law tail, now with index -2.5, which is also different from the theoretical index -1. The fact that the PDF tail for the k-modes decays slower than for the ω-modes ($J^{-2.5}$ vs. J^{-3}) is consistent with our conclusion (which we have made based on the statistics of the x- and t-domain increments) that turbulence shows stronger intermittency in space than in time. Recall that the power-law tail of the PDF correspond to the production of the random waves by breaking coherent structures.

6.4.5. *Energy spectrum dependence on the energy dissipation rate*

In this section, we will report on a study of the energy spectrum dependence on the energy dissipation rate previously reported in Denissenko *et al.* (2007). Such a dependence and the other statistical properties of the energy dissipation rate in wave turbulence are considered in detail by Falcon *et al.* (2012); see also references therein on the previous works on this subject. They observe a linear scaling $E_k \propto \epsilon$, which is in a striking disagreement with the weak wave turbulence predictions $E_k \propto \epsilon^{1/3}$ for the gravity waves and $E_k \propto \epsilon^{1/2}$ for the capillary waves. The way this group has measured the energy dissipation rate is via a direct reading of work produced by the wave maker. However, it is clear that a large part of work produced by the wave maker goes into the bulk of the fluid volume in the form of turbulent vorticity and not the irrotational surface waves. Indeed, the velocity field associated with the waves is exponentially decaying with the fluid depth, whereas the wave maker motion is horizontal and depth independent. Naturally, hydrodynamic turbulence has its own mechanism of dissipation, e.g. for Kolmogorov isotropic homogeneous turbulence one has $E_k \propto \epsilon^{2/3}$, and more complicated behavior should be expected when the inhomogeneous/anisotropic character of turbulence in the vortex wake and the boundary layer around the wave maker is considered. Moreover, observed $E_k \propto \epsilon$ dependence is suspiciously suggestive that the energy dissipation on the wave maker could even be driven by a simple linear process, e.g. wall friction which is discussed in Sec. 6.2.7. In other words, it is not clear how much of the energy produced by the wave maker actually goes into the wave turbulence and, therefore, it is not surprising that the weak turbulence scalings are not observed when ϵ is assumed to be equal (or proportional) to the wave maker produced work.

In our work, we measured the energy dissipation in wave turbulence in a different way. Namely, the value of ϵ was found directly from the decay rate of the wave energy measured immediately after switching off the wave maker (by fitting the initial dependence of energy on time by an exponential function). For different wave intensities, the decay rates were constant within a time interval of at least 500 periods of the energy containing waves (i.e. at the forcing scales). The results are shown in Fig. 6.13 in a

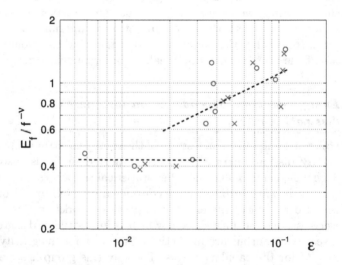

Fig. 6.13.   Dependence of the spectral intensity $E_f/f^{-\nu}$ on the energy dissipating rate. Circles and crosses correspond to the signals measured simultaneously in two different points separated by 0.4 m.

log–log plot of $[E_f/f^{-\nu}]_{f=3\,\mathrm{Hz}}$ (the steady-state energy spectrum at a fixed frequency of 3 Hz) vs. ϵ. Despite a significant scatter of the data points, which is mainly due to the limited time of the fit interval and the chaotic nature of signals, one can see two different regimes. At higher dissipation rates, $E_{f=3\,\mathrm{Hz}}$ is an increasing function of ϵ. For comparison, we show the slope 1/3 corresponding to the weak turbulence prediction, which appears to be mildly consistent with the data. At low excitations, the 3 Hz spectrum is nearly constant independent of ϵ, which agrees with predictions of the discrete wave turbulence. In this regime, the wave spectrum saturates at an ϵ-independent level and the excesses of energy are released downscale via sandpile avalanches, the later being stronger or/and more frequent for larger values of ϵ, see Nazarenko (2006), L'vov and Nazarenko (2010) and Nazarenko (2011).

It is clear, however, that the results presented in Fig. 6.13 require further improvements, and the work is under way to obtain significantly greater amount of data in order to improve the statistics.

6.4.6. *Evolving wave turbulence*

Now we will report results of our new experiments specially devoted to studying the evolution of wave turbulence.

6.4.6.1. *Experimental procedures*

To study nonstationary regimes of wave turbulence, we performed a series of new experiments where each continuous run included, along with a stationary part, transitional regimes both at the beginning, when the wave maker was just switched on, and at the end, after the wave maker was turned off. The measurements were done using both the wire-probe and the laser-sheet techniques described in Sec. 6.3 above. Two capacitance probes measured elevations continuously as a function of time at two points separated by a distance of about 2.5 m. The fluorescent imaging technique was used to record the movies of the vertical cross-sectional profiles along the line perpendicular to the wave maker plane.

Characteristic parameters for the set of the experiments considered here are summarized in Fig. 6.14. In each experiment, the wave maker was started at some oscillation amplitude G that was kept constant during the rise and the stationary phases, after which the wave maker was turned off to observe the decay phase. Thus, each experiment included three phases of the wave turbulence evolution — rising waves, statistically stationary wave turbulence and its decay. We have ordered our experiments in the table in the order of increase of the forcing defined by the wave maker

	Wave maker amplitude a.u.	Stationary Wave height RMS, cm	Coef. of Nonlinearity	k - Slope	ω - Slope
1	0.2	2.0	0.09	−5.7	−6.68
2	0.25	2.7	0.125	−3.85	−5.38
3	0.3	2.9	0.134	−3.66	−5.4
4	0.35	3.3	0.150	−3.58	−4.91
5	0.4	3.3	0.15	−3.53	−5.03
6	0.45	3.9	0.18	−3.46	−4.88
7	0.5	4.8	0.225	−3.13	−4.69
8	0.55	4.6	0.21	−2.97	−4.56
9	0.6	5.2	0.24	−2.92	−4.55

Fig. 6.14.   Characteristics of experiments measured during stationary stage.

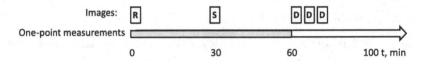

Fig. 6.15.   Flowchart of the typical data set recording for a fixed amplitude of the wave maker. R, S and D mark 120-second continuous image sets taken at the rise, stationary and decay phases.

amplitudes, see column 1. The second column in the table shows the RMS of the wave elevation which is proportional to the coefficient of nonlinearity introduced before. Figure 6.14 also includes the exponents of k- and the ω-spectra measured in the stationary time intervals, they show quite a similar dependence on wave intensity as in our previous experiments on the stationary wave turbulence described in the previous section.

Figure 6.15 shows a flowchart of data recording for one experiment. Data from the wire probes was recorded continuously from $t = 0$, when the wave maker was switched on, until the end of the observed wave decay phase (where just a few low frequency modes have survived). Typical duration of one wire probe record was about 100 min. Due to the computer memory limitations, the movies of the wave profiles were recorded in several short intervals 120 sec long each. The first movie, "R", starts at $t_0 = 0$ and covers the initial part of the rising phase. The second one, "S", starts at $t = 30$ min when the wave field has reached a stationary state. The 3rd movie, "D", starts at the moment the wave maker is turned off at $t = t_f$, and the following two D-sets are acquired at $t_f + 300$s and $t_f + 600$s to cover two further advanced decay stages.

6.4.6.2. *Rising waves — formation of the wave turbulence cascade*

In Sec. 6.2.4, we estimated the dynamics and the kinetic characteristic time scales for the evolving gravity wave turbulence. Based on those estimations for the pumping frequency of about 1 Hz and wave nonlinearity of 0.2, one can expect the characteristic time scales of order $\tau_D \sim 4$s and $\tau_K \sim 100$s. From our previous gravity wave experiments in the same flume, we know that the stationary regimes are reached after at least half an hour transient. To capture nonstationarity, we use a relatively short averaging time for finding the standard deviation of the wave amplitudes compared with the averaging time intervals used for the stationary phase. Applied to the band-pass filtered wave elevation signals, such an approach allows to observe time delays in development of the spectral amplitudes caused by the energy propagation from the forcing to the dissipation scale. Figure 6.16 shows the time evolution of the filtered wave amplitudes, A_i, for four spectral bands, $i = 1, \ldots, 4$, centered at frequencies $1, 3, 5$ and 7 Hz, respectively. In all of these cases, we used the same band-pass filter of bandwidth 1 Hz. The presented time interval, 1200 s, covers both the rising and the stationary regimes. Dashed lines show the stationary levels of $A_i(t)$; they serve as an eye guide for the distance to the stationary level. As we can see, the

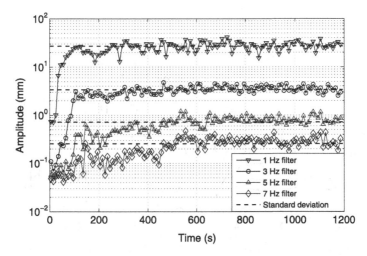

Fig. 6.16. Evolution of the Fourier spectrum at several different frequencies for the experiment with $G = 0.25$.

low frequency components, 1 and 3 Hz, reach their stationary levels after about 100 s which is fairly close to the estimate of the kinetic time. Higher frequency components rise with a delay and first reach their stationary levels about 50 s later, but this follows a further adjustment interval of about 600 s. One can notice that these four spectral components do not start to grow simultaneously. Instead, there is a kind of a front propagation with characteristic time of 50–100 s commensurable with the kinematic time. An additional feature in the rising waves is an overshooting observed for all the components: after a fast initial increase of the spectral amplitudes up to its stationary level, they drop up to 2 times and then return back to the stationary level. As it was mentioned, adjustment of the spectral amplitudes at higher frequencies is a much slower process which lasts about 400 s.

Due to a substantial averaging time, 100 s, the frequency-domain filtered components are inadequate for resolving processes with time scales shorter than 100 s. Much better resolution can be achieved using movies with spatial wave profiles $\eta(t, x)$ analyzed in the k-domain. Figure 6.17 shows the k-domain spectral components at $1/\lambda = 2, 20, 80, 160$ and $320\,\mathrm{m^{-1}}$. In the optical fluorescence method, each frame in the recorded movie represents a profile snapshot $\eta(x, t_i)$ taken at instant t_i, where t_i is sampled with the frequency 24 Hz. Spatial k-domain components were obtained from FFT transform in the x-domain and then averaged over 1 s (24 spectra). This enables a temporal zoom into the rising stage.

Now, the spectral energy front propagation becomes more evident, such that one can see 20 s delay between the wavelengths 0.5 m and 5 cm

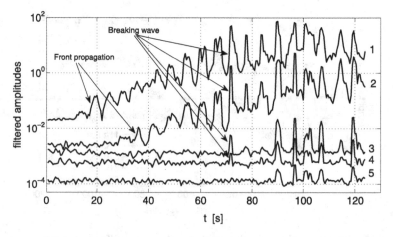

Fig. 6.17. The k-domain spectral amplitudes. Upper plot k-domain spectral amplitudes fluctuations for the experiment with $G = 0.35$. $1/\lambda = 2, 20, 80, 320\,\mathrm{m^{-1}}$.

(corresponding frequencies are 1.8 and 5.6 Hz). Moreover, we can see what was really happening to the short wavelength (high frequency) spectral components. They show a "fake" growth due to appearance of breaking waves whose energy synchronously spreads over a wide spectral interval and contributes to the high-frequency components in the time domain. Therefore, the first time when 5 and 7 Hz components arise in Fig. 6.16 should be associated with the breaking events whereas the "adjustment" interval of 500–600 s may be considered as an actual time required for the energy to get transferred from 1 Hz mode, which corresponds to the pumping energy scale, down to 7 Hz, which is at the far end of the gravity wave scale. It is also interesting that the wave breaks in Fig. 6.17 appear with a certain degree of periodicity with the period of about 4.5 s which is close to the period of the forcing. Thus, we could suggest that in the beginning, before a significant wave mixing has been developed, the forcing provokes breaking, probably in resonance with the nearest standing eigenmode of the flume.

6.4.6.3. *Stationary wave turbulence*

Rising waves reach a statistically stationary turbulent state after about 30 min after the wave maker starts. Our previous experiments (Denissenko *et al.*, 2007; Lukaschuk *et al.*, 2009; Nazarenko *et al.*, 2010) showed that the slope of the stationary 1D k-spectra increases from -3.8 to -3 when the averaged wave intensity grows from weak, with the nonlinearity coefficient less than 0.1, to strong, with the nonlinearity coefficient of about 0.25 (stronger amplitudes are impossible due to wave breaking). Now when we study evolving wave turbulence, we can apply a similar approach and observe temporal fluctuations of the spectral amplitude. Spectral amplitudes in the k-domain are the most convenient objects due to the relatively short averaging times. Temporal fluctuations of four spectral modes, $2, 20, 80$ and $320\,\mathrm{m}^{-1}$, are shown in the top plot of Fig. 6.18. The modes 2 and $20\,\mathrm{m}^{-1}$ are at the opposite sides of our gravity wave k-interval. The mode $80\,\mathrm{m}^{-1}$ is in the middle of gravity-capillary range and $320\,\mathrm{m}^{-1}$ is in the capillary range. Large synchronous fluctuations of the spectral mode amplitudes are related to the breaking wave events which are characterized by a wide coherent spectrum and showing themselves as simultaneous peaks appearing at all the sampled frequencies. Besides these breaking events, nothing significant has reached the waves at the capillary range. On the other hand, the gravity-capillary range has both the breaking (coherent) and the weak (incoherent) wave components. The bottom plot shows fluctuations of the k-slope. There is a visible correlation between the breaking waves and the flattening of the slope. Therefore, the increase of the mean spectral exponent when the forcing is increased can be explained

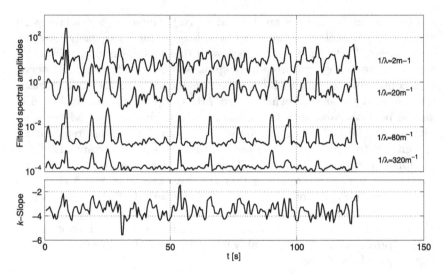

Fig. 6.18.   Stationary regime of wave turbulence, G = 0.35. Upper plot: time fluctuations of four wave k-modes at $1/\lambda = 2, 20, 80, 320\,\mathrm{m}^{-1}$. Lower plot: time dependence of the k-spectral slope measured over the gravity range.

by an increase in the frequency of the breaking events, at least when the wave intensity is quite significant.

6.4.6.4. *Decaying wave turbulence*

The decaying phase of wave turbulence starts when the wave maker is switched off. The latter does not stop instantly and it continues to move for few seconds which causes a small uncertainty in the decay start time.

In Secs. 6.2.6 and 6.2.7, we discussed two different mechanisms of the wave energy decay. The wave turbulence theory predicts for the spectral components a power-law decay $\propto t^{-1/2}$, see (6.57). The wall friction is predicted to prevail after passing a crossover point estimated at 8 min and leads to an exponential decay in time. Figure 6.19 presents the time decay of the main ω-spectrum component at 1 Hz filtered with the 0.4 Hz bandwidth for three different experiments, 3, 5 and 6, with different pumping energies, see Fig. 6.14. The main plot in Fig. 6.19 is semi-logarithmic and the insert in log-log coordinates is such that the difference between the power and the exponential decays is evident. For small time, the decay is consistent with the theoretical power-law $\propto t^{-1/2}$, whereas for large time the decay is clearly exponential, with a rate $\sigma_\nu \sim 1/(4.5\,\mathrm{min})$ which agrees with (but somewhat faster than) the order-of-magnitude theoretical estimate $\sigma_\nu \sim 1/(15\,\mathrm{min})$. As we mentioned before, somewhat faster dissipation could be due to the

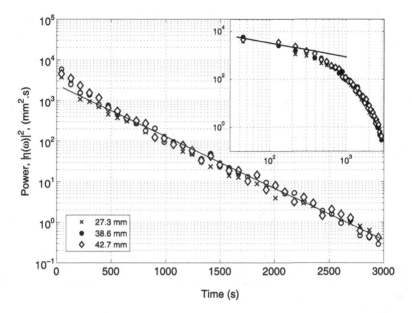

Fig. 6.19.   Decay of the main peak of the ω-spectra of the wave elevations as a function of time measured at a fixed point and filtered between 0.8 and 1.2 Hz (the energy containing range). The boundaries were rectangular. The main plot is in lin–log coordinates, and the inset is a log–log plot.

additional friction with the bottom or/and enhanced dissipation in the flume corners. The crossover between the power-law and the exponential decays occurs at about 500 s which agrees with the theoretical prediction.

The crossover amplitude is the same for any initial amplitude, and the crossover time is almost the same. Hence, the decay curves for the three different experiments collapse onto each on average.

To see how the wave turbulence decay is affected by the geometry, we have also changed the shape of the flume to trapezoidal by inclining one of its sides which is opposite to the wave maker by 30° relative to its original 90° position. A comparative plot for the main (energy containing) peak decay in the experiments with the inclined and the straight back walls is shown in Fig. 6.20. One can see that the dissipation rate is increased in the trapezoidal flume in both the initial (power-law) and the large-time (exponential) stages and the crossover point is slightly shifted to the left. The increased decay rate in the initial stage is probably due to the fact that the linear eigenmodes in the trapezoidal flume are such that the four-wave resonances are less suppressed by the finite-size effects than in the rectangular flume, and the resulting energy cascade is faster. On the other

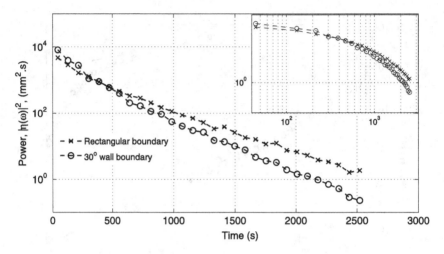

Fig. 6.20. Comparison between the rectangular and trapezoidal boundary conditions for the decay of the main peak from the ω-spectra of wave elevations as a function of time, measured at a point and filtered between 0.8 and 1.2 Hz. The main plot is in lin–log coordinates, and the inset is a log–log plot.

hand, faster long-time decay indicates that the wall friction may also be greater in the trapezoidal flume.

The exponent of the power-law decay is about -0.5 for both the rectangular and the trapezoidal flume shapes, which is consistent with the theoretical wave turbulence prediction (6.57).

It is interesting that the nonlinear wave evolution remains, and is in fact highly nontrivial, at the exponential stage of decay. This may seem paradoxical given the fact that the wall friction effect causing the exponential decay is linear. However, the wall friction is mostly acting on the large (energy containing) scales, whereas the smaller scales may still be nonlinear. Indeed, the nonlinearity is known to increase along the KZ cascade toward higher wave numbers, and one should expect a similar behavior deviating from pure KZ, e.g. due to finite size effects. To demonstrate nontrivial nonlinear effects, we present in Fig. 6.21 a black and white plot of the ω-spectra as a function of time and ω for an experiment with the standard deviation of the elevation being 27.9 mm at the stationary phase. We see that after passing the crossover point, in the exponential decay region, the spectral amplitudes start to oscillate in time with a period of about 700 s. These oscillations occur synchronously for different frequencies, which can be seen in Fig. 6.21 showing time dependence of normalized frequency spectra. A plausible explanation of such an oscillatory behavior is that, because of the finite flume size, there exists an active

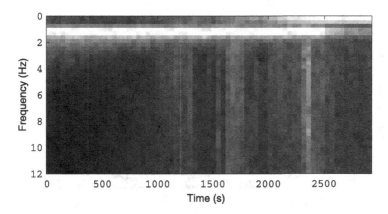

Fig. 6.21. Black and white plot of the ω-spectra as a function of time and ω for an experiment with the standard deviation of the elevation being 27.9 mm at the stationary phase. The spectra are normalized at each t; lighter color means higher value of the spectrum.

cluster of discrete wave modes, consisting of one or more coupled resonant (or quasi-resonant) quartets. The cluster is semi-insulated in a sense that its wave vectors are not too small to be strongly dissipated by the friction and they are not resonantly connected to the small-scale modes which could drain their energy via cascade. Such an insulated low-dimensional system could experience quasi-periodic behavior in a way an isolated integrable wave quartet would. Yet, the degree of "insulation" in the cluster could vary in time: in the oscillation phase, when the most small-scale modes in the cluster are excited, they could leak some energy into even smaller scales beyond the the cluster in a cascade-like fashion. Possibly, it is this energy spills into the high-ω tail that we see in Fig. 6.21. It is important to reiterate that these nontrivial nonlinear oscillations are not significant for the decay of the total energy: they are not visible on the decay of the main energy peak shown in Fig. 6.19. As we said before, the main peak at this stage decays mainly due to friction — a linear mechanism.

To further check the idea that the observed late-time oscillations at the high-ω tail are related due to the finite-size effects and the wave mode discreteness, we have studied these oscillations in the runs characterized by different wave amplitudes at the stationary phase. In Fig. 6.22, we show decay of the band-pass filtered (between 4 and 7 Hz) wave elevation for experiments with different initial wave intensities (i.e. with different wave amplitudes at the steady state before the start of the decaying phase). One can see that the oscillations are most pronounced for the weakest case and completely absent for the strongest case, which is consistent with the

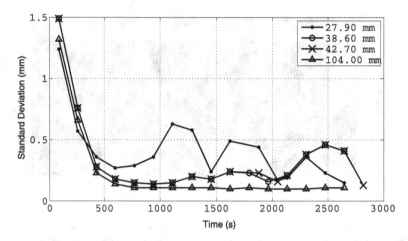

Fig. 6.22.   Decay of standard deviation of wave elevation as a function of time, measured at a point and filtered between 4 and 7 Hz. The configuration of the boundaries was rectangular.

explanation in terms of the finite-size effects because these effects arise only for sufficiently weak waves. Remarkably, for different initial wave intensities the oscillations have the similar periods and are even aligned in phase. This suggests that starting from different initial intensities the system quickly decays to an intensity at which further energy cascade to higher frequencies is blocked by the discreteness effects. Since this blocking intensity is determined only by the structure of the linear eigenmodes (i.e. by the flume shape) it is the same in all of these experiments and, therefore, the energy remaining in the discrete cluster is the same. Thus, because the nonlinear oscillations in the cluster should have the same period and, considering the fact that the initial decay of energy before trapping it in the cluster is small, the phases of the nonlinear oscillations could also be aligned.

6.5. Summary and Discussion

We have presented an overview of the results of the seven years of research conducted on the study of gravity wave turbulence at the Deep flume facility at the University of Hull. We have complemented our review with an extensive theoretical part explaining both basics of the wave turbulence theory and the new developments in this field including the higher-order wave statistics, description of the breaking coherent structures and their role in the wave turbulence life cycle, characterization of the finite size effects leading to discrete and mesoscopic wave turbulence. We also evaluate the role of the wall friction in dissipating the wave energy. In the experimental part, we start by overviewing our previous results which are mostly (apart from the runs with measuring ϵ) concerned with the steady wave turbulence states. We performed measurements of the gravity wave turbulence statistics in both x- and t-domains, which allowed us to differentiate between the states which have the same ω-spectra but different k-spectra, particularly between the ZF and the Ku states. The exponents of both ω- and k-spectra show dependence on the amplitude of the forcing. We see a certain evidence favoring ZF theory at larger wave amplitudes. However, the ZF state is not fully achievable because the two fundamental limits of weak wave turbulence, weak nonlinearity and infinite box, are impossible to implement simultaneously even in such a large flume as ours. Theories dealing solely with random weakly nonlinear waves or solely with coherent wave crests of a particular type cannot fully explain our experimental results. Instead, there is a strong evidence for coexisting and interacting random and coherent wave components, as it was speculated before based on the open-sea data by Kudryavtsev *et al.* (2003). The random waves are captured by the PDF cores and the low-order SFs, whereas the coherent wave crests leave their imprints on the PDF tails and on the high-order SFs. The singular wave crests themselves consist of structures of different shapes: numerous nonpropagating spikes/splashes (which show in the x-domain SFs) and propagating Ku-type breaks (seen in the t-domain SFs). We suggested a plausible scenario for the dynamics and mutual interactions of the coexisting random-phased and coherent wave components based on a wave turbulence life cycle. In this scenario, the coherent structures arise from the energy cascade within the random wave component when the nonlinearity becomes strong at some (small) scale along the cascade. Such coherent waves break and partially dissipate their energy, whereas the other part of their energy goes into weak ripples, i.e. it is returned into incoherent waves with a wide range of frequencies. Based on this picture, we made qualitative theoretical estimates for the scalings of the moments of the Fourier modes. However, our present data is insufficient for finding these moments, and longer experimental runs will be needed in future to accumulate required statistics.

The finite flume size leads to the prevalence of splash-type breaks over the progressively propagating breaks, because reflections from the walls produce counter-propagating waves which collide. A prototype for these splashes could be a standing wave in which at some limiting amplitude the wave nodes break symmetrically and form an upward jet. Naturally, the situation is different at open seas where the progressive wave breaks are dominant. However, our technique based on SF scalings can be useful in future for analyzing the open sea data in terms of characterizing the wave breaking events and their interactions with the random wave component. Recently, SFs were measured in the open seas studies by Mironov *et al.* (2012), and we hope that the findings of our work would help to develop such techniques further. It is also clear that such random-coherent interactions are the key to understanding the wave turbulence life cycle, and their description should become a priority in the future for the WT theory.

Discussion of the inverse cascade was not included into our review apart from the WT theoretical prediction for such a state and a brief mention of an ongoing experimental work in our lab implementing an inverse cascade setup. This is a very important subject for future studies keeping in mind that in the open seas it is the inverse cascade that is responsible for the energy transfer from the small centimeter-long scales generated by wind initially, to longer waves within the gravity range. At later times, provided the wind is steady, the effective forcing scale may move to smaller wavenumbers clearing a range for the direct cascade. However, such steady winds are rare, and the inverse cascade is probably more important than the direct one in the most open sea conditions. Our diagnostic techniques based on the SFs will certainly be useful for characterizing the underlying dynamics in the inverse cascade too.

We would also like to highlight our result on the dependence of the wave energy spectrum E_k on the dissipation rate ϵ. We measured ϵ directly by observing the wave energy decay immediately after switching off the forcing, and our results are consistent with the wave turbulence predictions for stronger wave fields and with the discrete turbulence predictions for weak wave fields. The other paper in this book, Falcon *et al.* (2012), is specially devoted to the statistics of the energy dissipation, and they report $E_k \propto \epsilon$ which is in disagreement with the weak wave turbulence theory. We attribute this to their identification of ϵ as the power dissipated by the wave maker. In our opinion, this identification is questionable because it is not clear how much of the wave maker power actually goes into waves and how much of it goes into hydrodynamic turbulence or directly dissipated by the wall friction. On the other hand, our data on ϵ is very scattered and further work is underway to improve statistics and to reach more conclusive results.

In our new data devoted to evolving wave turbulence, we have considered the rising stage (between switching the forcing on and reaching the steady state), the steady stage and the decaying stage (after switching the forcing off). In the rising stage, we see a front in the k-space propagating toward higher k's at the timescale consistent with the kinetic equation. On the background of this slow rise, we see peaks synchronously appearing across a broad range of wavenumbers with a periodicity of about $4.5\,\mathrm{s}$, which could be interpreted as breaking of a standing wave mode which is in resonance with the forcing.

In the steady state, the synchronous peaks corresponding to wave breaking are also clearly seen, although no longer periodic. We detect a clear correlation between flattening of the wave spectra and occurrence of the breaking peaks.

The decaying stage displays two different types of behavior. Early decay proceeds in agreement with the wave turbulence predictions based on the kinetic equation: the energy decays as $t^{-1/2}$. This is followed by a longer time interval when the energy decays exponentially, which is consistent with our estimate of the decay rate caused by the wall friction. Despite the monotonous decay of the total energy, at this stage, we observe quasi-periodic synchronous nonlinear oscillations across the high-frequency tail of the spectrum. We attribute these oscillations to a possibility that at intermediate scales the energy is trapped into a low-dimensional cluster of resonant waves. Energy migrates quasi-periodically within such a cluster, and when it reaches the shortest waves of the cluster, it may spill into the high-frequency tail beyond the cluster. This behavior is in agreement with the previously suggested scenarios for the discrete and mesoscopic wave turbulence.

We also studied wave turbulence decay in a trapezoidal flume (when one of its walls was inclined). Such decay appears to be faster than for the rectangular flume at both the power-law and the exponential decay stages, which we interpret as a possible increase of efficiency of the four-wave resonances and an enhanced wall friction for the trapezoidal shape.

In summary, the observed behavior of the gravity wave turbulence is extremely rich and its description requires understanding an intricate interplay of the well-understood wave turbulence processes and more poorly understood effects of the coherent structures, finite size effects, the wall friction and other types of interaction with the boundaries and the wave maker. We described some solid results which we firmly understand, presented plausible interpretations for more uncertain and preliminary results, and outlined the problems which must be addressed in future.

Bibliography

Caulliez G, Guerin C-A. 2012. Higher-order statistical analysis of short wind-waves. *J. Geophys. Res.* 117: C06002.

Choi Y, Lvov Y, Nazarenko S, Pokorni B. 2005. Anomalous probability of large amplitudes in wave turbulence. *Phys. Lett. A* 339: 361–369.

Connaughton C, Nazarenko S, Newell AC. 2003. Dimensional analysis and weak turbulence. *Physica D* 184: 86–97.

Connaughton C, Nazarenko S. 2004. Warm cascades and anomalous scaling in a diffusion model of turbulence. *Phys. Rev. Lett.* 92: 044501.

Cox C, Munk W. 1954. Statistics from the sea surface derived from the sun glitter. *J. Marine Res.* 13: 198227.

Denissenko P, Lukaschuk S, Nazarenko S. 2007. Gravity wave turbulence in a laboratory flume. *Phys. Rev. Lett.* 99: 014501.

Falcon E, Laroche C, Fauve S. 2012. Fluctuations of the energy flux in wave turbulence. *this volume.*

Galtier S, Nazarenko SV, Newell AC, Pouquet A. 2000. A weak turbulence theory for incompressible mhd. *J. Plasma Phys.* 63: 447–488.

Hasselman K. 1962. Anomalous probability of large amplitudes in wave turbulence. *J. Fluid Mech.* 12: 481.

Herbert E, Mordant N, Falcon E. 2010. Observation of gravity-capillary wave turbulence. *Phys. Rev. Lett.* 105: 144502.

Janssen PAEM. 2004. *The Interaction of Ocean Waves and Wind.* Cambridge: Cambridge University Press.

Janssen, PAEM. 2009. On some consequences of the canonical transformation in the Hamiltonian theory of water waves. *J. Fluid Mech.* 637: 1–44.

Kadomtsev BB. 1965. *Plasma Turbulence.* New York: Academic Press.

Kartashova EA. 1991. On properties of weakly nonlinear wave interactions in resonators. *Physica D* 54: 125–134.

Kartashova EA. 1998. Wave resonances in systems with discrete spectra. In *Nonlinear Waves and Weak Turbulence*, Advances in the Mathematical Sciences, pp. 95–129.

Krasitskii VP. 1994. On reduced equations in the hamiltonian theory of weakly nonlinear surface-waves. *J. Fluid Mech.* 272: 1–20.

Kudryavtsev V, Hauser D, Caudal G, Chapron G. 2003. A semiempirical model of the normalized radar cross-section of the sea surface. *J. Geophys. Res* 108: 8054.

Kuznetsov EA. 2004. Turbulence spectra generated by singularities. *JETP Lett.* 80: 83–89.

Lukaschuk S, Nazarenko S, McLelland S, Denissenko P. 2009. Gravity wave turbulence in wave tanks: Space and time statistics. *Phys. Rev. Lett.* 103: 044501.

L'vov VS, Nazarenko SV. 2010. Discrete and mesoscopic regimes of finite-size wave turbulence. *Phys. Rev. E* 82: 056322.

Mironov AS, Kosnik MV, Dulov VA, Hauser D, Guerin C-A. 2012. Statistical characterization of short wind waves from stereo images of the sea surface. Submitted to *J. Geophys. Res.*

Mukto MA, Atmane MA, Loewen MR. 2007. A particle-image based wave profile measurement technique. *Exp. Fluids* 42: 131–142.

Nazarenko SV. 2006. Sandpile behavior in discrete water-wave turbulence. *J. Stat. Mech.* L02002.

Nazarenko S, Lukaschuk S, McLelland S, Denissenko P. 2010. Statistics of surface gravity wave turbulence in the space and time domains. *J. Fluid Mech.* 642: 395–420.

Nazarenko SV. 2011. *Wave Turbulence*. Berlin: Springer.

Onorato M, Osborne A, Serio M, Cavaleri L, Brandini C, Stansberg CT. 2006. Extreme waves, modulational instability and second order theory: Wave flume experiments on irregular waves. *Eur. J. Mech. B/Fluids* 25: 586–601.

Onorato M, Cavaleri L, Fouques S, Gramstad O, Janssen PAEM, Monmaliu J, Osborne AR, Pakozdi C, Serio M, Stansberg CT, Toeffoli A, Trulsen K. 2009. Statistical properties of mechanically generated surface gravity waves: A laboratory experiment in a three-dimensional wave basin. *J. Fluid Mech.* 627: 235–257.

Phillips OM. 1958. The equilibrium range in the spectrum of wind generated waves. *J. Fluid Mech.* 4: 426–434.

Shrira VI, Annenkov SY. 2012. Towards a new picture of wave turbulence. *this volume.*

Socquet-Juglard H, Dysthe K, Trulsen K, Krogstad HE, Liu J. 2005. Distribution of surface gravity waves during spectral changes. *J. Fluid Mech.* 542: 195–216.

Tayfun MA. 1980. Distribution of surface gravity waves during spectral changes. *J. Geophys. Res.* 85: 1548–1552.

Tayfun MA, Fedele F. 2007. Wave-height distributions and nonlinear effects. *Ocean Eng.* 34: 1631–1649.

Zakharov VE, Filonenko NN. 1967. Weak turbulence of capillary waves. *J. Appl. Mech. Tech. Phys.* 4: 506–515.

Zakharov VE, Zaslavskii MM. 1982. *Izv. Acad. Nauk SSSR, Atm. Ocean Phys.* 18: 970.

Zakharov VE, Lvov VS, Falkovich GE. 1992. *Kolmogorov Spectra of Turbulence*. Berlin: Springer-Verlag.

Zakharov V. 1999. Statistical theory of gravity and capillary waves on the surface of a finite-depth fluid. *Eur. J. Mech. B* 18: 327–344.

Chapter 7

Towards a New Picture of Wave Turbulence

V. I. Shrira and S. Y. Annenkov

Department of Mathematics,
EPSAM, Keele University,
Keele ST5 5BG, UK

The existing wave turbulence paradigm assumes, in particular, a proximity to stationarity and spatial homogeneity of the wave field under consideration, which is often violated in nature. In this review, we outline a new theoretical approach free from these restrictive assumptions. We present and discuss a novel approach to the direct numerical simulation (DNS) of random wind wave fields, based on the Zakharov equation. This approach can be used in situations when the statistical theory is not applicable. It can be also used to provide an independent corroboration to the wave turbulence theory. We review the experimental evidence indicating that often wave fields evolve much faster than the existing theory predicts, we then employ the DNS to study the fast evolution of wave fields far from equilibrium. These fields evolve on the dynamic timescale typical of coherent wave processes, rather than on the kinetic timescale typical of broadband random wave fields. This phenomenon of "fast evolution", its causes, the specifics of its mathematical description and implications are the main focus of the review. We also present a generalisation of the existing statistical theory of wave turbulence, based on the new generalised kinetic equation. This equation is free of the assumption of quasi-stationarity and can account for both the fast and kinetic-scale processes in wave fields. Throughout the review water waves are used as a typical example of weakly nonlinear wave field, however, the new approach is equally valid for any random broadband weakly nonlinear wave field evolving due to nonlinear resonant interactions, either triad or quartic; the specificity of wave field is only in the form of the interaction coefficients.

Contents

7.1. Introduction

The challenge of describing the evolution of broad-band random nonlinear wave fields crops up in many very diverse areas of physical sciences. In spite of diversity of physical situations and types of interacting waves, there is a common conceptual and mathematical core which is the main subject of the theory of "wave turbulence" (e.g. Zakharov *et al.*, 1992; Nazarenko, 2011). Although the wave (or weak) turbulence theory has reached a venerable age of 80-plus, it continues to evolve. In this review, we focus upon three recent developments which are chosen partly because we ourselves initiated and actively pursued these developments and because we genuinely believe them to be game changing.

First, in Sec. 7.2 we discuss the paradigm shift concerned with direct numerical simulation (DNS) of wave turbulence. For most of its history, the wave turbulence theory lacked ways of verification and direct examination of the basic underlying assumptions. The numerous attempts to employ DNS based upon off-the-shelf numerical codes for the primitive equations have had limited success because of the inherent limitations. Here, we discuss a novel approach, the only one specially designed for DNS of long-term evolution of random weakly nonlinear wave fields.

Second, in Sec. 7.3 we employ the DNS approach to study the recently discovered phenomenon of "fast evolution" of wave turbulence. This phenomenon refers to situations where random wave fields evolve on the dynamic time scale typical of coherent wave processes, rather than on the kinetic timescale typical of broadband random wave fields. We focus on the causes of fast evolution and specifics of its mathematical description. The implications of this phenomenon are also briefly outlined.

In Sec. 7.4, we discuss the generalization of the existing statistical theory of wave turbulence, based on the new kinetic equation, which is not based on the assumption of quasi-stationarity and can account for fast processes in wave fields far from equilibrium.

The survey is not aimed at presenting a comprehensive established picture of the developments in the chosen areas, since such a picture does not exist yet; rather, we attempt to outline the vectors of new developments. The survey is based primarily on our own published and unpublished results and will discuss both the results and outstanding questions.

Without much loss of generality, we present these developments using wind waves as a convenient master example. Wind waves represent probably the most common and well-studied example of wave turbulence, and wind-wave modeling is of tangible practical importance for a variety of human activities. The quality of wave forecasts is essential for the reliability of shipping, development of offshore resources, prevention of human life loss at sea, and local and global weather prediction (Young, 1999; Lavrenov, 2003; Janssen, 2004; Babanin, 2011).

7.2. DNS of Evolution of Wave Turbulence

7.2.1. *Background*

Within the framework of the classical weak turbulence paradigm, the wave field represents a continuum of interacting weakly nonlinear waves (e.g. Zakharov *et al.*, 1992; Nazarenko, 2011). The word "turbulence" in the term not only indicates intrinsic randomness of such a field, but also stresses its affinity with the classical hydrodynamic turbulence: the interactions are primarily local in the Fourier space and there are cascades of energy and/or other invariants. In contrast to the classical turbulence, the basic interacting elements (waves) are weakly nonlinear, which enables one to develop a consistent asymptotic theory. In the linear limit, the wave field is just a superposition of free harmonic waves of the type $a_{\mathbf{k}}e^{i(\mathbf{k}\cdot\mathbf{x}-\omega t)}$, which are solutions of the primitive equations. Here, $\mathbf{k}$ is the wavevector, $\mathbf{x}$ is the spatial coordinate, $\omega(\mathbf{k})$ is the linear dispersion relation and $a_{\mathbf{k}}$ is the amplitude of the harmonic component with wavevector $\mathbf{k}$. In the chosen context $\mathbf{k} = \{k_x, k_y\}$ and $\mathbf{x} = \{x, y\}$ are two-dimensional vectors and $\omega(\mathbf{k}) = \sqrt{g|\mathbf{k}|}$. The weakly nonlinear waves interact via quartic resonant interactions

$$\mathbf{k}_0 + \mathbf{k}_1 = \mathbf{k}_2 + \mathbf{k}_3, \quad \omega_0 + \omega_1 = \omega_2 + \omega_3. \tag{7.1}$$

Thus, all waves are coupled, but the coupling is weak in the sense that wave amplitudes change only slightly over the wave period. Smallness of nonlinearity is characterized by a small parameter ε, which in the water wave context can be interpreted as characteristic wave steepness. In a more general interpretation, applicable to most of the wave turbulence cases, it is the ratio of characteristic maximal orbital velocity to the wave phase velocity.

Evolution of a random wave field could be adequately described only in terms of ensemble averaged quantities. In contrast to the hydrodynamic turbulence, a regular asymptotic procedure based on smallness of ε and a number of additional assumptions, which we will discuss in more detail later, makes it possible to decouple equations for statistical moments of the field (e.g. Zakharov *et al.*, 1992; Nazarenko, 2011). In this way, a closed equation describing evolution of the second statistical moments of the field has been derived, first by Peierls (1929) for phonons in crystals, then, starting from the sixties, for many other wave fields (Litvak, 1960; Zakharov *et al.*, 1992; Nazarenko, 2011) including water waves (Hasselmann, 1962; Janssen, 2004). Such an equation is referred to as the kinetic equation (hereinafter KE).

Over much of its history, the science of wave turbulence was primarily a theoretical construct with a relatively weak experimental and DNS support. The basic question — to what extent the existing theory captures the actual behavior of physical systems — remains open to a large degree. In particular, a detailed analysis of applicability of various statistical assumptions at the foundations of the wave turbulence theory is still lacking. Thus, an independent corroboration, which would be able to establish the range of validity of the theory, is essential. Such a corroboration can be provided with DNS of a random wave field, based upon an algorithm that would be free of statistical assumptions. The second basic question on how a random wave field evolves when the existing theoretical models are not applicable also could be greatly advanced by DNS.

However, developing a DNS of the evolution of random wave fields proved to be less straightforward and more challenging than it might seem at first glance. For the reasons we discuss below, the existing off-the-shelf numerical approaches as a rule do not work. There are several fundamental difficulties. First, the primitive equations (e.g. equations of hydrodynamics) should be integrated with high accuracy over quite large time intervals, much larger than those required in simulations of classical turbulence. The necessary times are at least $O(\varepsilon^{-4})$ of characteristic wave periods for the media with prevailing quartet interaction, as it is the case for water waves, where ε is small by the definition of weak turbulence. Moreover, a random wave field is a continuum of waves, and it is not *a priori* clear how to perform the unavoidable discretization. It is crucial to include into simulations a very large number of modes, involved in both resonant and nonresonant interactions.

In the water wave context, although there exists a large variety of algorithms developed to simulate wave evolution, none of them proved to be particularly well suited to the challenge. Many different groups employed some modifications of the existing algorithms, although with a limited success (Willemsen, 2001; Tanaka, 2001; Onorato *et al.*, 2002; Dyachenko *et al.*, 2004; Yokoyama, 2004; Korotkevich *et al.*, 2008). These works are mostly based on the amplitude expansions of the original primitive hydrodynamic equations and the subsequent integration of the resulting equations using spectral methods, employing efficient realizations of fast Fourier transform (FFT). Using high-order spectral method, Tanaka (2001) studied short-term evolution of gravity wave spectra, while Onorato *et al.* (2002) were able to demonstrate the emergence of the power-like energy spectrum corresponding to direct cascade. Willemsen (2001), Dyachenko *et al.* (2004) and Yokoyama (2004) employed a numerical model based on dynamical Hamiltonian equations in physical space, with calculations involving FFT to make use of the convolution form of the equations. With this numerical method, it was possible to trace the wave field evolution at larger timescale, demonstrating some of the effects predicted by the weak turbulence theory: frequency downshift, angular spreading and the start of the formation of the spectrum corresponding to the inverse cascade. However, these computations requiring very substantial computational resources were limited to the timescale of $O(10^3)$ characteristic wave periods, and still contained certain artifacts of discreteness that could not be eliminated (Korotkevich *et al.*, 2008).

The key difficulty in applying Fourier transform-based spectral methods is that the vicinity of each resonance has to be sufficiently well resolved in the wavevector space, in order to capture resonance interactions of a continuous wave field. When this condition is violated, the mode interaction weakens and can even disappear completely. Such a regime is called *frozen turbulence* (Pushkarev and Zakharov, 2000; Lvov *et al.*, 2006). In principle, it is possible to resolve wave resonances in the wavevector space using a sufficiently fine mesh. However, in the generic situation for all regimes with the inverse cascade, the wave spectra spread towards smaller and smaller wavenumbers. Thus, any *a priori* chosen fine mesh becomes not fine enough, and with time any code based upon such a mesh ceases to be applicable. The origin of the difficulties lies in the fact that the commonly employed FFT techniques require the discretization of a wave field onto a regular grid of Fourier harmonics. Regular grids are known to bring undesirable artifacts in behavior of wave turbulence (Pushkarev and Zakharov, 2000; Lvov *et al.*, 2006). In our context, it is especially important that the number of both exact and approximate resonances in a regular grid of harmonics is very limited, a grid too large to be realistic being necessary for a successful modeling of weak turbulence. Moreover, a regular grid is, in fact, an unnatural representation of a continuous wave field with waves of very different scales. Even if the grid is sufficiently dense for short waves to form a rich enough system of resonances in the high-frequency part of the spectrum, the scarcity of resonances in the low-frequency part is inevitable. Thus, all the FFT-based spectral approaches face the same difficulty, which confines their range of applicability to not too broad spectra and relatively short spans of evolution.

Here, we outline a novel approach specially designed for simulations of wave kinetics aimed to overcome these limitations. The approach is applicable to most of weakly nonlinear nearly conservative wave fields. Its starting point is the integrodifferential Zakharov equation and, hence, we are speaking about DNS within the framework of the Zakharov equation. The equation is derived by performing expansion in wave amplitudes in the Hamiltonian with the desired accuracy. Small dissipation and forcing are also taken into account without destroying the Hamiltonian structure in the lower orders of the perturbation scheme. The variables are diagonalized and canonically transformed in such a way that all bound waves are eliminated, which in itself provides a quite considerable gain in efficiency. (Below, we consider the procedure for water waves as a typical example.) The key point of the new numerical approach to the DNS of statistical ensembles of weakly nonlinear waves is that it is not restricted to regular grids of harmonics. Instead, the wave field is represented as an ensemble of a moderately large number of finite-size wave packets with random phases, linked by a dense grid of approximate resonances, which represent the interactions of individual harmonics within wave packets. The key idea is that by allowing and facilitating approximate resonant interactions, we can faithfully capture the dynamics of a wave field with a huge number of simultaneous interactions, employing a moderate number of dynamical modes.

7.2.2. *The Zakharov equation. Water waves example*

Although the derivation of the Zakharov equation is well known in various physical contexts (e.g. Zakharov *et al.*, 1992; Krasitskii, 1994; Janssen, 2004; Nazarenko, 2011), it makes sense to repeat here its key points for water waves. We confine our consideration to potential waves at the surface of an ideal incompressible fluid of infinite depth. The coordinate system has the origin located at the undisturbed water surface, the vertical axis z oriented upward and the horizontal axes x, y. Let $z = \zeta(\mathbf{x}, t)$ specify the free surface, and let $\varphi(\mathbf{x}, z, t)$ be the velocity potential, with $\psi(\mathbf{x}, t) = \varphi(\mathbf{x}, \zeta(\mathbf{x}, t), t)$ being the flow potential at the surface. Then, the governing equations can be written in the Hamiltonian form

$$\frac{\partial \zeta(\mathbf{x}, t)}{\partial t} = \frac{\delta H}{\delta \psi(\mathbf{x}, t)}, \tag{7.2a}$$

$$\frac{\partial \psi(\mathbf{x}, t)}{\partial t} = -\frac{\delta H}{\delta \zeta(\mathbf{x}, t)}, \tag{7.2b}$$

where δ denotes the operator of functional differentiation, and the Hamiltonian H is the total energy of the system.

Equations (2a) and (7.2b) can be rewritten, to a desired order of accuracy, in the form

$$i\frac{\partial b(\mathbf{k})}{\partial t} = \frac{\delta H}{\delta b^*(\mathbf{k})}, \tag{7.3}$$

where the complex variable $b(\mathbf{k})$ is linked to the Fourier-transformed primitive physical variables $\zeta(\mathbf{k}, t)$ and $\psi(\mathbf{k}, t)$ through an integral-power series (Krasitskii, 1994)

$$b(\mathbf{k}) = \frac{1}{\sqrt{2}} \left\{ \sqrt{\frac{\omega(\mathbf{k})}{k}} \zeta(\mathbf{k}) + i\sqrt{\frac{k}{\omega(\mathbf{k})}} \psi(\mathbf{k}) \right\} + O(\varepsilon), \tag{7.4}$$

asterisk means complex conjugation, $\omega(\mathbf{k}) = [gq(\mathbf{k})]^{1/2}$ is the linear dispersion relation, $q(\mathbf{k}) = |\mathbf{k}| = k$ for infinite depth. Without the loss of generality, we set $g = 1$, with the corresponding change of length scale. The Hamiltonian H is a functional of $b(\mathbf{k})$, $b^*(\mathbf{k})$, and can be written in the form of a series in powers of these variables.

The specific choice of the coefficients in the series (7.4) enables *reduction* of the Hamiltonian, in other words, the essential simplification of H, which now contains resonant terms only. All the details of the derivation can be found in Krasitskii (1994).

For the case of pure gravity waves, the reduced Hamiltonian has the form

$$H = \int \omega_0 b_0 b_0^* \, \mathrm{d}\mathbf{k}_0 + \frac{1}{2} \int T_{0123} b_0^* b_1^* b_2 b_3 \delta_{0+1-2-3} \, \mathrm{d}\mathbf{k}_{123}$$

$$+ \frac{1}{2} \int W_{01234} (b_0^* b_1^* b_2 b_3 b_4 + b_0 b_1 b_2^* b_3^* b_4^*) \delta_{0+1-2-3-4} \, \mathrm{d}\mathbf{k}_{1234} + \dots, \quad (7.5)$$

so that the corresponding reduced equation, to the fifth order in ε, is

$$\mathrm{i}\frac{\partial b_0}{\partial t} = \omega_0 b_0 + \int T_{0123} b_1^* b_2 b_3 \delta_{0+1-2-3} \, \mathrm{d}\mathbf{k}_{123}$$

$$+ \int W_{01234} b_1^* b_2 b_3 b_4 \delta_{0+1-2-3-4} \, \mathrm{d}\mathbf{k}_{1234}$$

$$+ \frac{3}{2} \int W_{43210} b_1^* b_2^* b_3 b_4 \delta_{0+1+2-3-4} \, \mathrm{d}\mathbf{k}_{1234}. \quad (7.6)$$

Here, the compact notation used designates arguments by indices, e.g. $\omega_0 = \omega(\mathbf{k})$, $T_{0123} = T(\mathbf{k}, \mathbf{k}_1, \mathbf{k}_2, \mathbf{k}_3)$, $\delta_{0+1-2-3} = \delta(\mathbf{k} + \mathbf{k}_1 - \mathbf{k}_2 - \mathbf{k}_3)$. Equation (7.6), being a generalization of the original result of Zakharov (1968) to the next order in ε, is known as the five-wave Zakharov equation for gravity waves. The interaction coefficients of the reduced equation T and W are listed in Krasitskii (1994). This equation is quite general and could be derived for almost any weakly nonlinear dispersive wave field, provided that it is primarily conservative and triad interactions are prohibited. The specifics of each type of waves is accumulated in the expressions for the coefficients. Examples for waves of various nature governed by the Zakharov equation differing only in the interaction coefficients are scattered in the literature (e.g. Zakharov *et al.*, 1992; Nazarenko, 2011). When triad interactions are permitted, the Zakharov equation is much easier to derive (e.g. Zakharov *et al.*, 1992) since, for most applications, only quadratic terms have to be retained. To take into account weak dissipative effects and forcing, we allow frequency in the linear term $\omega_0 b_0$ in (7.7) to have a small imaginary correction $\mathrm{i}\gamma_0$, that is, in the explicit presentation the linear term takes the form $(\omega_0 + \mathrm{i}\gamma_0)b_0$. Throughout the paper, we will mostly consider the version of the Zakharov equation without quintet interactions

$$\mathrm{i}\frac{\partial b_0}{\partial t} = (\omega_0 + \mathrm{i}\gamma_0)b_0 + \int T_{0123} b_1^* b_2 b_3 \delta_{0+1-2-3} \, \mathrm{d}\mathbf{k}_{123}. \quad (7.7)$$

It is important to note that the variables $b(\mathbf{k})$ are not physical wave amplitudes, but appear as a result of the canonical transformation (7.4). The $O(\varepsilon)$ terms due to nonresonant three-wave interactions must be taken into account when the solutions of the Zakharov equation are transformed back to physical space; more precisely the series (7.4) has to be inverted. The corresponding formulae could be found in Krasitskii (1994). In deep water, the difference between $b(\mathbf{k})$ and their physical-space counterparts is small (Zakharov, 1999). The statistical theory for water waves is most naturally formulated in terms of $b(\mathbf{k})$, i.e. in the same canonically transformed space.

7.2.3. *Numerical algorithm for a deterministic wave system*

The Zakharov equation, originally derived in the water wave context in 1968 (Zakharov, 1968), was long considered to be inconvenient and expensive to solve numerically (e.g. Trulsen and Dysthe, 1996). Nevertheless, it has important advantages from the point of view of numerics, due to the elimination of nonresonant interaction terms by canonical transformation, and complete freedom in the choice of a computational grid. Since there is no need to perform a Fourier transform during time stepping, a regular grid is not required.

An algorithm based on the Zakharov equation was proposed in Annenkov and Shrira (2001) and successfully applied to the study of a number of physical problems with a relatively small number of degrees of freedom. The essence of the method is in the efficient computational strategy, where all the coefficients T and W are computed by a preprocessing routine and stored in a way that facilitates all subsequent operations of integration in time.

To simulate the Zakharov equation (7.7) or (7.6), it is convenient to write it in a slightly different form by change of variable

$$b(\mathbf{k}, t) = B(\mathbf{k}, t) \exp\left[-i\omega(\mathbf{k})t\right],$$

which in the conservative case with five-wave interactions dropped leads to

$$i\frac{\partial B_0}{\partial t} = \int T_{0123} B_1^* B_2 B_3 e^{i(\omega_0 + \omega_1 - \omega_2 - \omega_3)t} \delta_{0+1-2-3} \, d\mathbf{k}_{123}. \tag{7.8}$$

The new variable B does not contain fast dependence on time. For the integration in time, the discretized version of (7.8) is used, performing calculations with the efficient use of the results of the preprocessing.

The algorithm was first devised for deterministic wave systems, where an initial state of the wave field is given as a set of discrete complex amplitudes. In other words, a wave field is *assumed* to be discretized.

However, the problem of representing a continuous wave field as a set of discrete wave harmonics is far from being straightforward, as discussed below.

7.2.4. *Numerical simulation of random wave fields*

At first glance, the numerical simulation of a random wave field looks deceptively simple. Upon the discretization of the field, the resulting set of ordinary differential equations can be integrated numerically using a standard scheme, for example the Runge–Kutta method. Then, the results are ensemble averaged over a chosen number of realizations. The nontrivial point here is the discretization. To capture the evolution of a continuous system by a discrete system, we have to emulate somehow the continuum of waves involved in resonant, approximately resonant and nonresonant interactions. It is not *a priori* clear how to retain the properties of the continuum in a discrete system, and how to represent correctly the role played by each type of resonances. To clarify these points, and to outline the basic idea of algorithm construction in the generic case, we first consider a "toy model" consisting of just four wavepackets.

7.2.4.1. *A toy model: resonant interaction of four wavepackets*

In order to examine the basic questions on the role of resonant and nonresonant interactions, it is instructive to examine the simplest possible model with a nontrivial evolution of the spectrum. This will also help in understanding what is needed for an adequate DNS method. We recall that the method based on the Zakharov equation is not restricted to regular grids, so that we are free to consider a wave system with any distribution of interacting harmonics.

The most elementary model is, evidently, a single resonant quartet

$$\mathbf{k}_0 + \mathbf{k}_1 = \mathbf{k}_2 + \mathbf{k}_3, \quad \omega_0 + \omega_1 = \omega_2 + \omega_3, \tag{7.9}$$

an example is shown in Fig. 7.1. The KE for this particular model can be obtained either by discretization of the standard KE (7.13) (its derivation is discussed below), or derived directly from the Zakharov equation by using the random phase approximation (Kadomtsev, 1982). Both approaches lead to the equation for the quartet in terms of wave action $n(\mathbf{k})$,

$$\frac{\partial n_0}{\partial t} = 8\pi T_{0123}^2 \left[n_2 n_3 (n_0 + n_1) - n_0 n_1 (n_2 + n_3) \right], \tag{7.10}$$

where the right-hand side contains a dimensional constant ('phases correlation time'), which arises in the derivation based on the random phase approximation (Kadomtsev, 1982) and is set to unity according to the

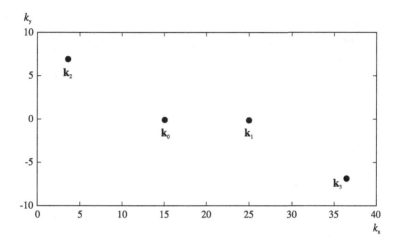

Fig. 7.1.   Resonant quartet (7.9) in Fourier space. Dots show position of harmonics.

requirement that the 'discrete' limit of the continuous KE for strongly localized wave packets and the discrete equation for these packets coincide (Annenkov and Shrira, 2006a).

Numerical solution of (7.10) for quartet (7.9), for a set of initial conditions

$$b_0(0) = 1.5 \times 10^{-2}, \quad b_1(0) = 1.0 \times 10^{-2},$$
$$b_2(0) = 0.8 \times 10^{-2}, \quad b_3(0) = 0.5 \times 10^{-2}, \tag{7.11}$$

$n_i(0) = b_i(0)^2$, is shown in Fig. 7.2 (dashed curves). Here and below in this section, we prefer to plot and discuss the evolution of amplitudes $|b(t)|$, instead of wave packet intensities $n(t) = |b(t)|^2$. For this discrete wave system, the solution to the Zakharov equation does not tend to the Rayleigh–Jeans equilibrium despite the fact that the system is conservative (no forcing or dissipation terms are included). This striking discrepancy is due to the presence of additional integrals relating the amplitudes of interacting packets (the Manley–Rowe integrals). In order to reproduce the solution of the KE (7.10) by DNS, the most direct way would be, at first glance, to build an ensemble of realizations of the same quartet, each realization having a different set of random initial phases, with subsequent averaging. This was performed by Stiassnie and Shemer (2005), who noted that the result was qualitatively different from the evolution obtained with the KE (7.10).

Since the correct procedure of simulation should take into account approximate resonant interactions as well, let us first examine what happens

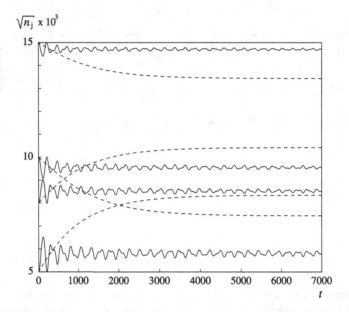

Fig. 7.2.   Dashed curves: numerical solution of the KE for the single resonant quartet (7.9), for initial conditions (7.11). Solid curves: ensemble averaged evolution for 10,000 quartets (7.12), with $\Delta = 0.6$ and the same initial conditions.

when we introduce a frequency mismatch into each realization, so that each quartet has the form

$$\hat{\mathbf{k}}_0 + \hat{\mathbf{k}}_1 = \hat{\mathbf{k}}_2 + \hat{\mathbf{k}}_3, \quad \hat{\omega}_0 + \hat{\omega}_1 = \hat{\omega}_2 + \hat{\omega}_3 + \hat{\Omega}, \qquad (7.12)$$

where $\hat{\mathbf{k}}_i = \mathbf{k}_i + \mathbf{K}_i$, $\hat{\omega}_i = \omega(\hat{\mathbf{k}}_i)$, $\mathbf{K}_i$ is a random vector, $|\mathbf{K}_i| \leq \Delta$, $\hat{\Omega}$ is the frequency mismatch, and Δ is a parameter. For each quartet (7.12), evolution is computed with the Zakharov equation (7.8), with subsequent ensemble averaging. The result is shown in Fig. 7.2 (solid curves) and is qualitatively similar to the picture observed by Stiassnie and Shemer (2005). The amplitudes of the harmonics exhibit fast, gradually diminishing, oscillations. Such a scenario of evolution radically differs from the corresponding solution of the KE: there is not the slightest resemblance. Clearly, an alternative procedure is needed, which (i) would take into account both exactly resonant and approximately resonant interactions, for each realization, (ii) would not depend on the position of additional harmonics in the Fourier space, or on other parameters.

7.2.4.2. *The role of the exactly resonant, approximately resonant, and nonresonant interactions. Idea of clusters*

The most natural way to build a wave system that would satisfy the conditions formulated above, at the same time still corresponding to the case of four resonantly interacting waves, is to consider four interacting *wave packets of finite width* in the Fourier space.

Let us model each packet as a "cluster" of random-phase harmonics, such that the sum of their amplitudes squared is equal to the total intensity of a packet. In Fig. 7.3(a), each wave packet is represented as a cluster of five harmonics $\mathbf{k}_j$, $\mathbf{k}_j \pm \mathbf{d}_x$, $\mathbf{k}_j \pm \mathbf{d}_y$, where $\mathbf{d}_x = \Delta \kappa_x$, $\mathbf{d}_y = \Delta \kappa_y$, and $\kappa_x = (1, 0)$, $\kappa_y = (0, 1)$ are the unit wave vectors. We will refer to the parameter Δ as the *cluster size* in the $\mathbf{k}$-space. The adopted procedure of constructing clusters includes a parallel translation of the original resonant quartet by $\mathbf{d}_x$, $\mathbf{d}_y$, the resulting quartets being in approximate resonance, due to the nonlinearity of $\omega(\mathbf{k})$ in the resonance conditions. Simultaneously, in each pair of clusters there are close pairs of harmonics that form approximately resonant quartets. Besides that, in each cluster, a central harmonic interacts with its sidebands, in a Benjamin–Feir type interaction. In this way, in the considered "minimal" configuration, evolution of a single resonant quartet in the KE corresponds to evolution of 181 coupled quartets in the dynamical equation.

Figure 7.3(b) shows the evolution of packet amplitudes with time obtained by the numerical integration of the Zakharov equation for the wave system specified in Fig. 7.3(a). Here and below, time is measured in characteristic wave periods. Within each cluster, the initial amplitudes of individual harmonics were chosen randomly, but subject to the condition that the total packet intensity is equal to the prescribed initial value, given by (7.11). In this and subsequent simulations, this distribution of amplitudes within each cluster was kept the same for all realizations, and only the random phases of harmonics were different for each realization. It was also verified that if all the amplitudes are random as well, this does not affect the results, provided that the initial conditions for the amplitudes of the clusters are satisfied. The evolution almost coincides with that obtained with the KE, except at the initial stage where the timescale appears to be somewhat different. The problem of timescales will be discussed separately below; here, it is essential to note that the numerical solution in Fig. 7.3(b) is virtually independent of the way the clusters were chosen (i.e. the cluster size Δ, the number and positions of harmonics within the clusters). Specifically, the numerical solution for the example shown in Figs. 7.3(a) and 7.3(b) was found to be independent of Δ in the range $0.05 \leq \Delta \leq 2.0$. It is, however, important to choose a

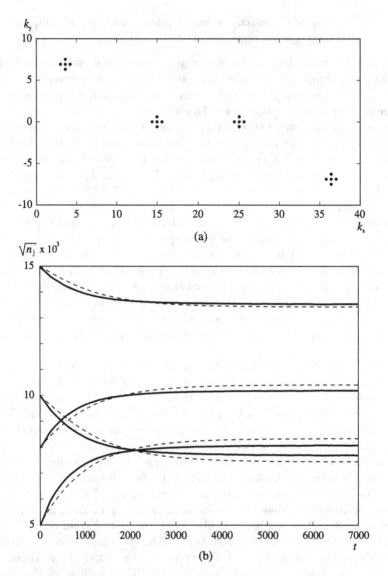

Fig. 7.3. (a) Resonant quartet (7.9): an example of cluster representation, with the cluster size $\Delta = 0.6$. (b) Comparison of the DNS with the KE solution. Solid curves: evolution of amplitudes of four wave packets (7.9) for initial conditions (7.11), averaged over 50,000 realizations. Each realization corresponds to a particular choice of random initial phases. Dashed curves: numerical solution of the KE.

configuration of resonances that is most structurally stable, in a certain sense, with respect to the change of amplitude.

Similar properties are also obtained for much more complex systems consisting of numerous (up to 1000) localized wave packets.

Simulations of a few simple wave systems show that the interactions that are not exactly resonant (i.e. with a small but finite mismatch) play a crucial role in the DNS of the evolution of the statistical properties. An adequate account for both resonant and near-resonant interactions is achieved with the configuration based on the representation of waves by clusters of harmonics. Physically, such a representation corresponds to the interaction of wave packets of finite width in Fourier space. It is essential that with only the exact resonances, the DNS of a wave system cannot be even qualitatively similar to the solution of the KE. On the other hand, simultaneously taking into account exact and approximate resonances, even in the most crude way (say, representing a wave packet with just two harmonics) leads to a qualitative similarity, at least. Note that the presence of *exact resonances* is not essential, i.e. they do not play any special role. A wave field can be adequately described by taking into account approximate resonances only. Inclusion of interactions that are far from being resonant (with frequency mismatch considerably larger than $O(\varepsilon^2)$) does not have any discernable effect on the evolution.

The issue of quantitative comparison is more subtle. Before comparing the DNS with anything, the employed DNS should first be validated internally. The extensive testing we carried out showed remarkable robustness of the DNS results obtained by employing different moderately crude representations of wave packets, i.e. models of the clusters. The specific configuration used in Annenkov and Shrira (2006a) and reproduced here has been found to provide the best compromise: an economical way of taking into account the near-resonant interactions without a noticeable dependence on parameters. In this way, we have got a relatively simple and robust tool for the quantitative modeling of statistical evolution of wave ensembles made of finite number of wave packets. We reiterate that the elementary interaction in the kinetic description of nonlinear field evolution is not an interaction of four waves but an ensemble average of a multitude of near-resonant interactions between harmonics of four packets. This understanding is crucial for extending the approach to generic continuous wave fields.

7.2.4.3. *DNS of a continuous wave field. Numerical algorithm*

To extend the approach outlined above for DNS of a generic continuous wave field, it is necessary to work out a procedure of discretization, representing a continuous field in terms of a large number of harmonics coupled by exactly and approximately resonant interactions. We note that, as shown above, exact resonances do not play any special role in the field evolution (Annenkov and Shrira, 2006a). In other words, we have shown how to model a wave field made of a number of *a priori* prescribed packets; now, we discuss how to represent a generic continuous field by a tractable number of wave packets.

In the Zakharov equation, the resonance condition on the wave vectors should be exact, as prescribed by the presence of the δ-function. However, for a numerical treatment the equation needs to be discretized, with Kronecker delta replacing the Dirac δ-function, and the complex amplitude $b(\mathbf{k}, t)$ must be replaced by a set of N discrete variables. In such a discrete system, the number of resonant and approximately resonant nonlinear wave interactions will be proportional to N, perhaps with a large coefficient, but it will never approach the $O(N^2)$ number of interactions in a continuous wave field, where every degree of freedom is expected to interact with every other degree of freedom. Besides that, there will always be a large number of wave pairs involved in just one near-resonant interaction, with amplitudes linked by the Manley–Rowe integrals.

Therefore, the resulting discrete wave system has properties that are different from those of a continuous wave field, both quantitatively and qualitatively. This means that in order to model a continuous wave field correctly, we need to work out, instead of discretization, the concept of *coarse-graining* of the wave field, which would retain its nonlinear interaction properties.

To that end, we build in Fourier space a grid consisting of $\sim 5 \times 10^3$–10^4 wave packets, coupled through exact and approximate resonant interactions. A wave packet, centered at $\mathbf{k}_0$, is characterized by one amplitude and one phase, but has a finite size in Fourier space, and is allowed to enter into nonlinear interactions with other wavepackets, provided that the wavevector mismatch $|\Delta \mathbf{k}| = |\mathbf{k}_0 + \mathbf{k}_1 - \mathbf{k}_2 - \mathbf{k}_3|$ does not exceed a certain threshold which we call the *coarse-graining parameter*. Thus, the usual resonance condition $\mathbf{k}_0 + \mathbf{k}_1 - \mathbf{k}_2 - \mathbf{k}_3 = 0$ is relaxed.

The specific parameters of the scheme employed in the simulations below are as follows. The grid is logarithmic in the wavenumber k (161 points within a span $0.13 < k < 2.12\,\mathrm{m}^{-1}$) and regular in the angle θ (31 point within $-\pi/3 \leq \theta \leq \pi/3$). Within the coarse-graining approach, a quartet of grid points is assumed to be in approximate resonance if its wavevector and frequency mismatch satisfies a pair of conditions $\Delta \omega / \omega_{\min} < \lambda_\omega$, $|\Delta \mathbf{k}| / k_{\min} < \lambda_k \bar{\omega} / \omega_{\min}$, where $\Delta \omega$ and $|\Delta \mathbf{k}|$ are the frequency and wavenumber mismatch in the quartet, $\omega_{\min}$ and $k_{\min}$ are the minimum values of frequency and wavenumber in the quartet, $\bar{\omega}$ is the mean frequency, and λ_ω and λ_k are detuning parameters, chosen to ensure that the total number of resonances is $O(N^2)$, where N is the number of grid points. In the simulations to follow, $N = 4991$, $\lambda_\omega = \lambda_k = 0.01$; results were verified to be nondependent on specific values of the detuning parameters in a wide range of λ_ω, λ_k. Forcing in the Zakharov equation is presumed to be known, and can be prescribed by any given function of $\mathbf{k}, t$. In the simulations below, it is calculated using an empirical formula (Hsiao and Shemdin, 1983). In order to model developed sea waves in the relatively narrow computational domain of the DNS, forcing can be applied to the high-frequency part of the spectrum only, which means that for dominant waves and waves of comparable scales, the wind input is exactly balanced by dissipation. In the simulations below, forcing is confined to the range $1.0 < k < 1.29 \, \mathrm{m}^{-1}$, and the most energetic part of the wave field evolves entirely due to nonlinear interactions, which is typical of developed sea waves (Badulin *et al.*, 2007; Banner and Young, 1994). Strong dissipation is applied to $k > 1.62 \, \mathrm{m}^{-1}$. The ensemble averaging is performed over 30 realizations.

Thus, we have got a robust universal tool for quantitative modeling of statistical evolution of continuous wave fields. An example of DNS of typical wave field evolution exhibiting formation of direct and inverse cascades is shown in Fig. 7.4.

Obviously, no DNS algorithm can compete in efficiency and scale range with numerical models based on the statistical theory. However, the DNS approach is free from any assumptions on which the statistical theory is based, and can be used in various contexts for independent verification of the validity of these assumptions, and for modeling of the phenomena not captured by the existing statistical theory. In the next sections, it will be employed for a detailed analysis of the phenomenon of "fast" field evolution.

7.3. "Fast" Evolution of Wave Turbulence. DNS Study

7.3.1. *Background*

The major part of the existing understanding of wave turbulence comes from the analysis of wave KEs. It is fair to say that the reduction of the wave turbulence problem to the analysis of the KE and revealing the affinity of wave turbulence cascades to those in hydrodynamic turbulence are among the major achievements of the 20th century physical sciences (Zakharov *et al.*, 1992; Nazarenko, 2011). Let us now briefly summarize the existing state of knowledge on wave turbulence, for definiteness focussing on wind water waves. In terms of wave action density spectrum $n(\mathbf{k}, t)$, where $\mathbf{k}$ is

Advances in Wave Turbulence

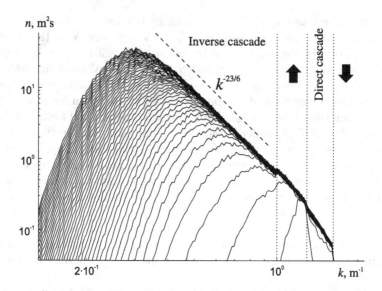

Fig. 7.4. Time evolution of wave action spectrum under constant forcing, in steps of approximately 400 periods of the spectral peak. Regions of forcing and dissipation are delimited by dotted vertical lines and shown with up and down arrows, respectively. The theoretical asymptotic slope of the spectrum is indicated by dashed line. The forcing is due to 10 m/s wind calculated using parametrization (Hsiao and Shemdin, 1983) and confined to the range $1.0 < k < 1.29 \, \text{m}^{-1}$; the range is indicated by two vertical dotted lines and a upward arrow.

the wavevector and t is the "slow" temporal variable, the KE has the form

$$\frac{\partial n(\mathbf{k}, x, t)}{\partial t} = S_{\text{input}} + S_{\text{diss}} + S_{nl}. \tag{7.13}$$

Here S_{input}, S_{diss} and S_{nl} describe contributions to rate of change in action density due to a variety of physical processes grouped as input, dissipation and nonlinear interactions, respectively. The inclusion of processes resulting in input and dissipation ensures the existence of cascades (energy and wave action in our case), but we will not dwell upon the physics of input and dissipation (see e.g. Komen *et al.*, 1994; Janssen, 2004; Babanin, 2011). We just note that although sources and sinks are not well separated in Fourier space, we still have, at least for dominant wind waves, an analogue of inertial interval, since the role of nonlinear interactions (S_{nl} term) is far more important (Badulin *et al.*, 2005). The prevailing nonlinear interaction term S_{nl} represents a summation with appropriate weights over all resonant quartets (7.1), in which the chosen wave vector, say, $\mathbf{k} \equiv \mathbf{k}_0$ is involved, and is given by the following integral over a surface in six-dimensional space

$$S_{nl} = 4\pi \int T^2_{0123} f_{0123} \delta_{0+1-2-3} \delta(\omega_0 + \omega_1 - \omega_2 - \omega_3) \, dk_{123}, \qquad (7.14)$$

where

$$f_{0123} = n_2 n_3 (n_0 + n_1) - n_0 n_1 (n_2 + n_3), \quad n_i \equiv n(\mathbf{k}_i),$$
$$\delta_{0+1-2-3} \equiv \delta(\mathbf{k}_0 + \mathbf{k}_1 - \mathbf{k}_2 - \mathbf{k}_3), \qquad (7.15)$$

and the interaction kernel T_{0123} for water waves is given by an explicit formula (e.g. Krasitskii (1994)). Numerical evaluation of this integral, often referred to as the Boltzmann integral, is rather complicated; it took more than two decades to develop a reliable and efficient code (see van Vledder (2006) and Benoit (2005) for a review).

We summarize the fundamentals of the established picture of wave field evolution due to the quartet interactions as follows: (i) Waves are weakly nonlinear, and their characteristic wave steepness ε (a measure of nonlinearity) is small (Komen *et al.*, 1994; Janssen, 2004); (ii) Due to the dominant (quartet) nonlinear interactions "individual" wave amplitudes evolve on the $O(\varepsilon^{-2})$ timescale, where the characteristic wave period provides the unit scale; (iii) Random wind-wave fields are described by ensemble averaged quantities, then the $O(\varepsilon^{-2})$ dynamics of the individual waves averages out and the evolution of wave energy spectra occurs on the $O(\varepsilon^{-4})$ timescale and is described by the KE; (iv) The KE prescribes the scaling of energy fluxes as $O(\varepsilon^6)$, and the existence of Kolmogorov–Zakharov (KZ) cascades of energy and wave action that gives rise to the development of power-like wave spectra (Komen *et al.*, 1994; Zakharov *et al.*, 1992; Janssen, 2004; Badulin *et al.*, 2005, 2007; Nazarenko, 2011). This picture is supported experimentally (Komen *et al.*, 1994; Janssen, 2004; Badulin *et al.*, 2007) and by numerical simulations (e.g. Tanaka, 2001; Onorato *et al.*, 2002; Dyachenko *et al.*, 2004, Annenkov and Shrira, 2006b); (v) If we consider wave fields evolving due to triad interactions, then individual waves evolve on the $O(\varepsilon^{-1})$ time scale while evolution of wave energy spectra occurs on the $O(\varepsilon^{-2})$ timescale.

The novel phenomenon we focus upon here is concerned with random wave fields evolving on a faster time scale than follows from the KE (7.13). At first glance, there is no room for a faster evolution, since the well established KE prescribes quite rigidly the "kinetic" time scale, that is $O(\varepsilon^{-4})$ and $O(\varepsilon^{-2})$, for quartic and triad interactions, respectively. However, the derivation of the KE assumes that the wave fields are close to stationarity, and that there is spatial homogeneity of the environment. In nature, these assumptions are often violated and the classical theory cannot be applied. For example in the wind wave context, gusts and sudden changes of wind speed or direction often occur in real oceanic conditions. Meanwhile, there is experimental evidence that the adjustment of the wave fields to sudden changes of wind direction is much faster than predicted by the KE (van Vledder and Holthuijsen, 1993; Young, 2006), and that a sharp change of wind speed can also cause a much faster field evolution (Waseda *et al.*, 2001). In the most detailed and well-documented laboratory study of the effects of a sudden wind increase (Autard, 1995), it was found that the wave field adjustment occurs on the timescale of only a few dozen wave periods, which is incompatible with the $O(\varepsilon^{-4})$ timescale estimate.

On the theoretical side, it was noted that the field evolution of an arbitrary initial wave field not in the near-stationary state can occur on the dynamic $O(\varepsilon^{-2})$ timescale during a short initial stage (Janssen, 2003); the conjecture was supported by the numerical simulations of the Dysthe equation (Dysthe *et al.*, 2003). However, a systematic study of wave turbulence beyond the classical KE paradigm has never been attempted. Thus, it is necessary to work out a statistical theory that is able to describe the evolution of nonstationary wave fields, for example, those driven out of equilibrium by an external perturbation or also fields developing rapidly evolving fronts in $\mathbf{k}, t$ space under stationary conditions.

In this section, we present a survey of our recent results on the evolution of water wave spectra in the cases where the standard statistical theory based upon the classical KE cannot be used. We start with a brief analysis of the laboratory observations by Autard (1995), which demonstrate the phenomenon of fast evolution. Then, to establish the main features of the evolution, we employ the DNS described in the previous section. In Sec. 7.4, of the review, we derive the new generalized kinetic equation (GKE), which is free from the stationarity assumption and captures the fast evolution phenomenon.

7.3.2. *Laboratory observations*

In the experiments (Autard, 1995; Caulliez, 2008, private communication), performed at the large (40 m long, 2.6 m wide and 0.9 m deep) IRPHE-Luminy wind-wave tank with 1.5 m high air recirculation tunnel, the spatial development of wind waves subjected to an abrupt increase of wind was studied. Wind increase, by approximately 50% over a distance of 2 m, was created by narrowing the wind tunnel cross section using a false ceiling (Fig. 7.5(a)), located at the distance 19 m from the entrance. Wind speed was measured by the Pitot tube. Waves were measured by capacitance gauges placed at a sequence of different fetches for four values of initial wind speed (3, 4, 5 and 6 m/s). At the start of wind increase, peak frequency was between 2.5 and 4 Hz (wavelength 15–25 cm). With the account of drift

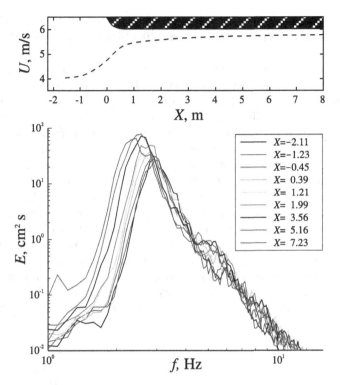

Fig. 7.5. Experiments on wind waves subjected to an abrupt increase of wind. (a) Scheme of the experiment with false ceiling, and wind dependence on fetch for initial wind speed $U = 4$ m/s. (b) Evolution of energy spectra with fetch for $U = 4$ m/s.

current, the waves of scales under consideration fall into the gravity range. Energy-frequency wave spectra were obtained at various locations along the wind acceleration profile for four initial wind speeds (Fig. 7.5(b)). The wind and wave field evolution are referred to the initial conditions observed just 1 m ahead of the false ceiling edge, and are investigated as functions of fetch (the relative distance from the edge position). Prior to the change of forcing, dominant waves were still growing, but appeared to be in a local equilibrium with the wind, characterized by slow increase of energy, almost constant steepness, and slow frequency downshift. At first, the increase of wind does not significantly manifest itself in the spectrum, the effect of wind increase being felt at high wave frequencies only, where waves are forced by the wind directly. Then, the disturbance propagates to the spectral peak, and after a short delay wave spectral energy and especially steepness grow quickly with fetch, with a considerable increase over just about a dozen wavelengths. Meanwhile, the maxima of the spectra stay approximately at the same frequency (the downshift stops) for about 2 m, after which the spectral peak undergoes a rapid downshift which then gradually slows down with fetch.

The focus of our analysis is on the behavior of growth rates of the spectral slope. For these growth rates, KE predicts a strict ε^6 scaling (corresponding to $O(\varepsilon^{-4})$ evolution timescale), provided that the growth is caused by the nonlinearity of the wave field, rather than by direct wind action. In the experiments, we observe the transition from a state where waves were in a local equilibrium with the wind, to a state when they are again in a local equilibrium with the new wind conditions, so that growth rates are small at the start and finish of the transition, and have maximum at a certain fetch. We study these maximal growth rates for the frequencies in the range $f_p < f < 2f_p$ (i.e. for frequencies higher than the frequency of the spectral peak f_p, but sufficiently small to neglect the capillarity effects). After the increase of wind, the maximal growth rates for all frequencies lower than 4 Hz are much higher than those expected due to linear growth, which allows us to attribute them to nonlinear growth caused by the propagation of the disturbance due to change of wind to lower frequencies. In Fig. 7.6, these growth rates, plotted with the ε^{-4}-

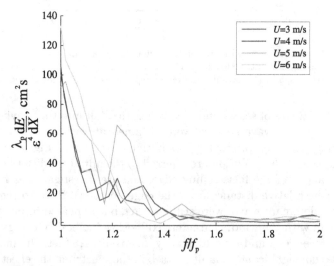

Fig. 7.6.   Maximal values of absolute spatial growth rates of energy density $dE/dX \cdot \lambda_p$ in the transition zone, as functions of frequency relative to the peak, for different wind speeds. Here λ_p is the characteristic wavelength of the spectral peak. Growth rates are normalized by ε^4, where ε is the new equilibrium value of wave steepness after the change of wind, measured at fetch $X = 1.2$ m. Time is normalized by the period of the spectral peak.

scaling, nearly collapse onto a single curve, indicating evolution on $O(\varepsilon^{-4})$ timescale and being in sharp contrast with the KE predictions. To allow for an accurate measurement of the exponent, these experiments need more ensemble averaging and longer times of observation. However, in the ε^{-4} vs. ε^{-6} dichotomy even these preliminary observations leave very little room for doubt.

7.3.3. *DNS of fast evolution*

7.3.3.1. *Evolution timescales*

To study timescales of field evolution by DNS, we begin with the toy model made of four wave packets introduced in Sec. 7.2.4.1. In the toy wave systems examined so far, the evolution is quite simple. In the absence of any nonconservative effects, the system tends to a certain equilibrium state specified by the initial conditions. We have seen that this equilibrium state is obtained both by the DNS and by the numerical solution of the KE, with the same initial conditions. Now, we will perform a thorough comparison of the evolution timescales. The initial state of the toy wave system is far from equilibrium; will the timescale of evolution obtained with the DNS be different from the strict ε^{-4} scaling implied by the classical KE?

In Fig. 7.3, we have plotted the evolution of the toy model obtained with the KE and the DNS, for a specific characteristic steepness of the waves, close to 0.1. Although the curves were pretty close, a certain discrepancy in the timescales could be seen. Moreover, the timescales of the evolution are found to have different amplitude dependence. Repeating the numerical simulations for half-amplitudes, taking the initial values of b_i as

$$\hat{b}_i(0) = 0.5b_i(0), \quad i = 0, 1, 2, 3 \tag{7.16}$$

we arrive at a somewhat different picture: both sets of curves are presented in Fig. 7.7. While the KE solution timescale is proportional to ε^{-4}, the DNS shows a ε^{-2} dependence. This can be seen in Fig. 7.8, where the evolution of amplitudes multiplied by 0.5, 1.0 and 2.0 is plotted versus time normalized by ε^{-2}. Again, no essential dependence on the cluster size was found, provided that $0.05 < \Delta < 1.0$. Thus, for small amplitudes the evolution is much faster according to DNS than within the framework of the KE. Such a behavior was found to be typical of all the systems made of finite number of localized packets: in any conservative wave system that is initially far from equilibrium, practically all the evolution towards the equilibrium occurs on the "fast" ε^{-2} timescale, contrary to the KE predictions.

This observation has two important consequences. First, the fast adjustment towards equilibrium is not resolved by the KE and, in fact,

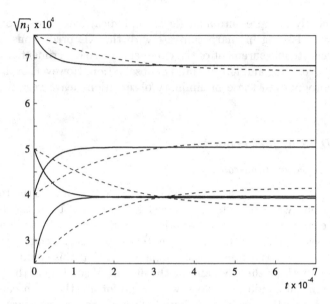

Fig. 7.7.   Evolution of amplitudes of four wave packets, obtained with the DNS (solid curves), with averaging over 50,000 realizations, and with numerical solution of the KE (dashed curves). Time is measured in characteristic wave periods. Compared to Fig. 7.3(b), initial amplitudes of all wave packets are multiplied by 1/2.

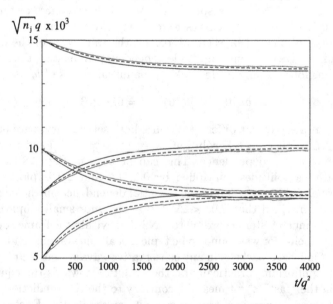

Fig. 7.8.   Evolution of amplitudes of four wave packets (7.9), normalized by q^{-1}, for $q = 1$ (solid curves), $q = 0.5$ (dashed curves), $q = 2$ (dotted curves), with timescales normalised by q^2. The averaging is over 50,000 realizations.

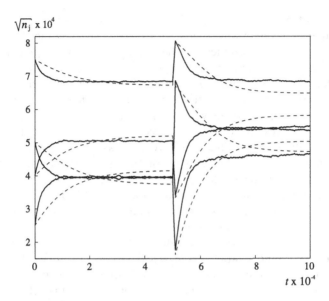

$\sqrt{n_j} \times 10^4$

Fig. 7.9. Evolution of the resonant quartet for the same initial conditions as in Fig. 7.7, with rapidly varying forcing. After relaxation of the conservative wave system to equilibrium, nonconservative effects are turned on at $t = 5 \times 10^4$, with $\gamma_0 = 1 \times 10^{-4}$, $\gamma_1 = 2.5 \times 10^{-4}$, $\gamma_2 = \gamma_3 = -3.5 \times 10^{-4}$, and then switched off at $t = 5.1 \times 10^4$. Time is in characteristic wave periods. Solid curves: DNS, with averaging over 1000 realizations, dashed curves: numerical solution of the KE. Time is measured in characteristic wave periods.

could be interpreted as a modification of the initial conditions for field evolution governed by the KE at a later stage. Second, reaction to any rapidly varying forcing would be incorrectly described by the KE, since the system will evolve to equilibrium with a different timescale. An example of such a scenario is presented in Fig. 7.9.

Moreover, since the "dynamical" wave field reacts to a fast external forcing on a different timescale than the "statistical" one, this reaction can be different. Shorter reaction timescale means that the wave field can have sufficient time to adapt to an external forcing, which is instantaneous in terms of the KE. As we will discuss below, these conclusions hold not only for any number of localized wave packets but also for continuous wind wave fields.

7.3.3.2. *Evolution of continuous wind wave spectra after an instant change of forcing*

The response of random wave fields to sharp change of forcing was examined in the context of wind waves by Annenkov and Shrira (2009), and the present chapter largely follows this work. To examine the effect of sharp wind change on random wave fields, we focus on wave nonlinear interactions and do not aim to reproduce all the complexity of wave generation and dissipation. Instead, preserving the basic physics, we choose a simplified description: wave breaking is not explicitly included, and the forcing we plug-in into the Zakharov equation is wind input minus dissipation. Forcing is applied to the high-frequency part of the spectrum only, which means that for dominant waves and waves of comparable scales, wind input is exactly balanced by dissipation. Thus, we have the classical transparency interval for this range of scales: the most energetic part of the wave field evolves entirely due to nonlinear interactions, which is typical of developed sea waves (Banner and Young, 1994; Badulin *et al.*, 2007). In order to quantify the dependence of wave evolution on nonlinearity, a wide range of wave steepness is used in the simulations.

We start with low-intensity white noise, and then run the model for several thousand characteristic wave periods, to allow the spectrum to develop under the constant wind. Thus, prior to the change of forcing, the developed wave spectrum is nearly stationary, and its evolution is fairly described by large-time asymptotics of the KE (Badulin *et al.*, 2005). In particular, the wave action spectral density of the dominant waves, $n(k_p)$, where k_p is the wavenumber of the spectral peak, grows slowly: $n(k_p) \sim t^{23/11}$, the peak position gradually shifts to low wavenumbers as $k_p \sim t^{-6/11}$ due to the inverse KZ cascade of wave action, and the dominant wave steepness ε (defined as $\varepsilon = \sqrt{2E}k_p/2\pi$, where $E = \int \omega n_k \, dk$ is the wave energy) slowly decreases. Several values of wind speed in the range 8–16 m/s were used in different runs. In all runs, evolution of the wave spectra continued until the spectral peak, which was slowly moving towards large scales, reached the value $k_p \approx 0.45 k_f$, where $k_f = 1.0$ is the lower boundary of forcing. Then, at time T_0 (Table 7.1), the wind was instantaneously increased to 16 or 24 m/s, so that the spectrum started to grow from the quasi-equilibrium value $n_0(k)$. An example of the evolution of the wave action spectrum for Run 3 of the numerical model (Table 7.1) is shown in Fig. 7.10(a), and the corresponding evolution of wave steepness and spectral peak in Figs. 7.10(b) and 7.10(c), respectively. After the sharp increase in wind at $t \simeq 6500$ characteristic peak wave periods, the steepness rises almost instantly, and reaches its new "equilibrium value" in a few hundred wave periods. During this short period, the disturbance of the spectrum "propagates" from the forcing region to the spectral peak on a timescale of tens of wave periods, so that energy at all wavenumbers on the high-frequency side of the peak starts to grow almost simultaneously, leading to a fast increase in the steepness. Meanwhile, the downshift of the spectral peak stops for a few hundred wave periods, and then resumes

Table 7.1. Runs of the numerical model. U_1 and U_2 — wind speeds before and after the change of forcing at time T_0; ε_1 and ε_2 — equilibrium values of wave steepness; k_p — wavenumber of the spectral peak at the change of forcing.

Run	U_1 (m/s)	U_2 (m/s)	ε_1	ε_2	T_0 (periods)	k_p (m^{-1})
1	8	16	0.099	0.174	15280	0.45
2	8	24	0.099	0.238	15280	0.45
3	10	16	0.118	0.173	6455	0.51
4	10	24	0.119	0.231	6455	0.51
5	12	16	0.139	0.169	5348	0.46
6	12	24	0.138	0.233	5348	0.46
7	16	24	0.170	0.217	4011	0.40

at an increased rate, gradually slowing down back to the KE asymptotics (Fig. 7.10(c)).

Our primary interest is in the growth of waves on the spectral slope, i.e. between the peak and forcing region, during the adjustment process. The response of several sample spectral bands to the wind change is illustrated in Fig. 7.11(a), shown as a function of time in terms of the normalized wave action $(n(k) - n_0(k))/(n_f(k) - n_0(k))$, where $n(k)$ is the wave action at time t, and $n_0(k)$ and $n_f(k)$ are the quasi-equilibrium values of the wave action before and after the wind increase. For all spectral bands, the growth rates have a peak in the first few hundred wave periods after T_0, and then gradually decrease, as the spectrum slowly adjusts to the new forcing.

We introduce two characteristic values of the growth rate: the "maximal" rate $M(k)$ (defined as the maximum, for each k, of $dn(k)/dt$ over 1000 wave periods after the change of wind), and the "average" rate $m(k)$ defined as the mean growth rate during 3/4 of the total change between $n_0(k)$ and $n_f(k)$. The dependence of $M(k)$ and $m(k)$ on k is shown in Fig. 7.11(b). There is no net forcing or dissipation for $k < 1$, so that all spectral growth illustrated in Fig. 7.11 is entirely due to nonlinear interactions. Dependence of the growth rate on time and wavenumber resembles the results of simulations of the KE (Young and van Agthoven, 1997); however, we are primarily interested in its dependence on nonlinearity ε (the KE predicts strict ε^6 scaling). Here, nonlinearity can be characterized by two parameters, namely the "equilibrium" values of ε before and after the wind change (ε_1 and ε_2, respectively, see Table 7.1). In all runs, dependence of $m(k)$ on ε_1 was weak, and virtually no dependence of $M(k)$ on ε_1 was found, but both $m(k)$ and $M(k)$ strongly depend on ε_2. To quantify this dependence, we choose the integrals of $m(k)$ and $M(k)$ with respect to k on the spectral slope as characteristic values, denote them as $\bar{m}$ and $\bar{M}$, and

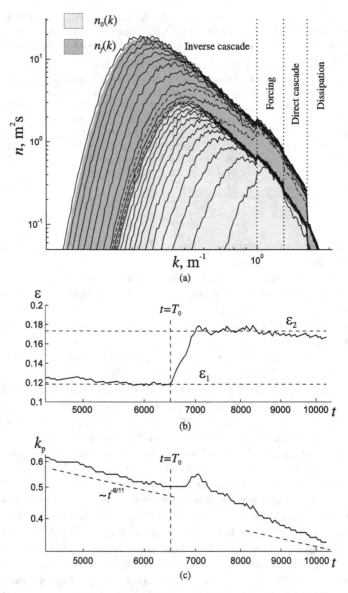

Fig. 7.10.   An example of DNS of random wave field evolution under an abrupt increase of wind (Run 3). (a) Time evolution of wave action spectrum, in steps of approximately 400 periods of the spectral peak (intermediate curves are dashed). Regions of forcing and dissipation are delimited by dotted vertical lines. "Equilibrium" spectra $n_0(k)$ and $n_f(k)$ are shown by shading. (b) Evolution of wave steepness ε. "Equilibrium" values of steepness before and after the increase of forcing (which occurs at $t = T_0$) are marked by dashed horizontal lines. (c) Evolution of spectral peak k_p. Theoretical (according to the large-time KE asymptotics) downshift rate $k_p \sim t^{-6/11}$ is shown by dashed line.

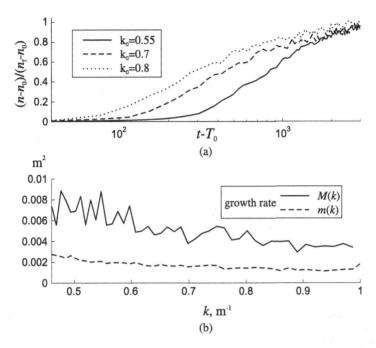

Fig. 7.11. Response of the wave action spectrum to the sharp increase of forcing (Run 3). (a) Response of sample spectral bands (integrated from k_0 to $1.1k_0$). Values of the integrated spectrum $n(k)$ are normalized by the "equilibrium" values $n_0(k)$ and $n_f(k)$. (b) Maximal and average growth rates ($M(k)$ and $m(k)$, respectively, defined in the text) as functions of wavenumber.

plot them for all runs versus ε_2, on a logarithmic scale (Fig. 7.12). Both $\bar{m}$ and $\bar{M}$ are scaled with ε^λ, where λ is close to 4 (least-squares fits yield 4.3 and 3.5). This scaling is in sharp contrast with the ε^6 scaling of the KE, and corresponds to the $O(\varepsilon^{-2})$ evolution timescale.

The DNS of wave fields subjected to an abrupt change of forcing shows that the nonlinear evolution of random wave fields occurs on the fast (dynamic) timescale. This is similar to the fast evolution of model spectra at a short initial stage found earlier by two different methods (Dysthe *et al.*, 2003; Annenkov and Shrira, 2006b), since the initialization of a spectrum by random initial phases is another example of a strong perturbation.[1] This implies that wave modeling in common situations of rapidly changing

---

[1]Such an initialization is a strong perturbation of a wave field, since in the process of natural evolution wave phases are not completely random and the corresponding correlator is nonzero ensuring the wave nonlinear evolution. By imposing the total randomness of phases, we interfere in a very crude way with the field natural evolution.

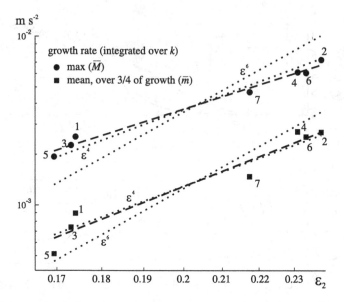

Fig. 7.12. Integrated (from $k = 0.5$ to $k = 1.0$) maximal growth rates $\bar{M}$ and mean growth rates $\bar{m}$ vs. wave steepness ε_2, on a logarithmic scale, with least-squares fits (dashed lines) for the scaling ε^λ ($\lambda = 3.5$ for $\bar{M}$ and $\lambda = 4.3$ for $\bar{m}$). Theoretical "dynamic" ε^4 and "kinetic" ε^6 scalings are also shown (dotted lines). Numbers identify numerical runs (Table 1).

or gusty winds, or in the presence of spatial inhomogeneities, should be radically revised. In particular, the situation with a rapidly fluctuating (gusty) forcing was examined with DNS in Annenkov and Shrira (2011), where it was found that although the KE is invalid at any instant, it still captures wave evolution averaged over a time scale far exceeding the gusts. Since our study is based upon the Zakharov equation, in which all specific properties of waves are contained merely in the interaction coefficients (Zakharov *et al.*, 1992; Nazarenko, 2011), the conclusion about the scaling of wave field evolution is not confined to wind waves, but applies to all random weakly nonlinear dispersive wave fields whenever there is a sharp and strong external perturbation.

7.3.3.3. *Emergence of fronts in the wave spectrum under constant forcing*

We have shown (in the context of wind wave evolution) that strong perturbations evolve towards equilibrium much faster than is allowed by the KE: on the dynamic $O(\varepsilon^{-2})$ timescale rather than the $O(\varepsilon^{-4})$ kinetic

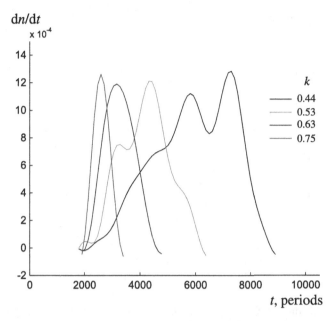

Fig. 7.13.   Rate of wave action increase dn/dt vs time (smoothed by a spline) during the passing of the spectral front for different wavenumbers, for wind speed $10\,\text{m/s}$.

timescale. The perturbations were assumed to be due to external factors, such as rapid change of forcing or initial artificial randomization of phases.

However, the phenomenon of fast evolution of random wave field is not confined to the situations when external factors are rapidly changing. Even in the idealized generic situation of perfectly uniform and steady conditions, the downshifting wave spectra always develop fronts. At these fronts, the classic KE is not applicable. Now, we will presents results of the DNS study of the evolution of the spectrum at the front.

During the passage of the spectral front through a certain point in Fourier space, wave spectral density sharply increases. In Fig. 7.13, we show the rate of this increase vs. time for a few arbitrarily picked wavenumbers, as the front passes through while moving to larger scales. For each wavenumber, the rate has a well-pronounced maximum. For all the wavenumbers for a wave spectrum developing under constant wind forcing the maxima appear to be approximately the same. In Fig. 7.14, the values of these maxima, normalized by ε^4, are plotted for different constant wind speeds. All the values collapse onto a single curve, demonstrating the $O(\varepsilon^4)$ scaling of the growth rate. Thus, even under constant forcing the spectral evolution at the front occurs on the dynamic $O(\varepsilon^{-2})$ timescale,

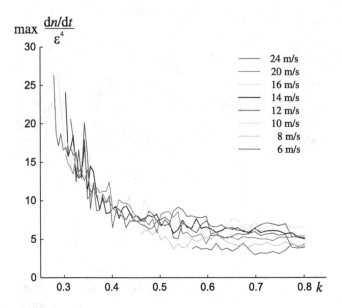

Fig. 7.14. Maxima of dn/dt vs. k for different wind speeds, normalized by ε^4 (wave steepness at the passing of the spectral front for a given k).

at a sharp contrast with the $O(\varepsilon^{-4})$ timescale predicted by the KE. Thus, the KE is inapplicable at the spectral front. Here, the analogy with gas dynamics could be helpful: the KE describes macroscopic evolution of the field, playing the same role as the Navier–Stokes equations. In gas dynamics, wide classes of initial conditions of the Navier–Stokes equations lead to the formation of shocks. However, often the inner structure of shocks cannot be described by the Navier–Stokes equations and might require going beyond the continuous mechanics paradigm. In wave kinetics, the spectral fronts are analogues of the gas dynamics shocks with the inapplicability of the KE being similar to the failure of the Navier–Stokes equations in gas dynamics.

7.3.4. *Concluding remarks*

Thus, the reviewed examples of DNS simulations demonstrated that the fast evolution of random wave field is a generic phenomenon which always takes place: it occurs both when there is an external perturbation of any kind (including spatial nonuniformity or nonstationarity) and in perfectly uniform and stationary environment caused by intrinsic mechanisms. Here, we confined our consideration of exogenous and endogenous fast evolution to just a single example of each. Among many other situations of genuine interest, which were left aside and still await a dedicated study, we list a few exogenous ones: field interaction with spatial nonuniformities, with bodies, interaction of two (or more) systems of waves, e.g. wind waves with swell, and an endogenous one: formation of direct cascade by a field distribution initially localized in **k**-space. The rich variety of situations and scenarios leading to fast evolution in other physical contexts have not been studied yet.

7.4. Generalization of Statistical Theory. Fast and Slow Evolution of Wave Turbulence

7.4.1. *Derivation of the GKE*

Now, let us turn to analytical statistical description of a wave system in terms of the correlation functions of the field $b(\mathbf{k}, t)$. The classical derivation procedure, described in detail in, e.g., Zakharov *et al.* (1992), Janssen (2004) and Nazarenko (2011), uses (7.7) as the starting point and leads to the KE [(7.13) and (7.14)], namely

$$\frac{\partial n(\mathbf{k}, x, t)}{\partial t} = 4\pi \int T_{0123}^2 f_{0123} \delta_{0+1-2-3} \delta(\omega_0 + \omega_1 - \omega_2 - \omega_3) \, \mathrm{d}\mathbf{k}_{123}, \qquad (7.17)$$

plus input and dissipation terms, which will be henceforth omitted.

First, we briefly review the most common way of derivation of the KE (7.17), pointing out the key approximations and assumptions involved. The wave field is taken to be statistically homogeneous in space, as implied by the form of the correlator

$$\langle b(\mathbf{k}, t) b^*(\mathbf{k}_1, t) \rangle = n(\mathbf{k}) \delta(\mathbf{k} - \mathbf{k}_1) \equiv n_{\mathbf{k}} \delta_{01}. \qquad (7.18)$$

In the zeroth approximation in ε, the free-wave field has Gaussian statistics, for which all odd-order correlators vanish, and the fourth-order correlator decomposes into products of pair correlators,

$$J_{0123}^{(0)} \delta_{0+1-2-3} = \langle b_0^* b_1^* b_2 b_3 \rangle = n_0 n_1 \left(\delta_{0-2} \delta_{1-3} + \delta_{0-3} \delta_{1-2} \right). \qquad (7.19)$$

In the next approximation in ε, correlations due to resonant nonlinear interactions result in deviations from strict Gaussiniaty and manifest themselves in a nonzero fourth-order cumulant $J_{0123}^{(1)}$. The essential hypothesis is that the wave field is presumed to remain quasi-Gaussian over the timescale of evolution, that is, for all times the cumulant $J_{0123}^{(1)}$ should remain small compared to $J_{0123}^{(0)}$.

Multiplying the Zakharov equation (7.7) by b_j^* and performing statistical averaging, we calculate $\partial n_0/\partial t$,

$$\frac{\partial n_0}{\partial t} = 2\mathrm{Im}\int T_{0123} J_{0123}^{(1)} \delta_{0+1-2-3}\,\mathrm{d}\mathbf{k}_{123}. \qquad (7.20)$$

The cumulant $J_{0123}^{(1)}$ is specified by an evolution equation containing on the right-hand side the sixth-order correlator I_{012345}, which by invoking the quasi-Gaussianity assumption is replaced by the corresponding free-field Gaussian correlator $I_{012345}^{(0)}$ representable in terms of the products of pair correlators and cumulant $J_{0123}^{(1)}$. Neglecting the cumulant, which is small compared to the pairs of correlators by virtue of the quasi-Gaussianity, we have

$$\left(\mathrm{i}\frac{\partial}{\partial t} + \Delta\omega\right) J_{0123}^{(1)} = -2T_{0123}f_{0123}, \qquad (7.21)$$

where $\Delta\omega = \omega_0 + \omega_1 - \omega_2 - \omega_3$, and f_{0123} is specified by (7.15). It is usually assumed that n_0 and, hence, f_{0123} depends on slow time μt, such that $\mu/\Delta\omega \ll 1$. Then the closed equation for the evolution of $n(t)$ in its classical form (7.17) is obtained (see e.g. Zakharov *et al.* (1992)) by making the following steps. Employing the assumed wide separation of scales and neglecting the oscillating terms $\sim \mathrm{e}^{-\mathrm{i}\Delta\omega t}$ in the full solution to (7.21), an approximate solution to (7.21) for large t that depends only on the slow time scale

$$J_{0123}^{(1)}(t) = -\frac{2T_{0123}}{\Delta\omega}f_{0123}, \qquad (7.22)$$

is substituted into (7.20). Here this solution is understood in terms of generalized functions

$$J_{0123}^{(1)}(t) = -2T_{0123}\left[\frac{P}{\Delta\omega} + \mathrm{i}\pi\delta(\Delta\omega)\right]f_{0123}(t), \qquad (7.23)$$

where P stands for "principal value". Note, it is commonly assumed that $\mu \sim \varepsilon^4$ while $\Delta\omega$ for four-wave interactions is $O(\varepsilon^2)$. Thus, indeed $\mu/\Delta\omega \sim \varepsilon^2 \ll 1$, and the asymptotic derivation is valid as long as our interest is confined to slow $O(\varepsilon^{-4})$ evolution.

The fact that (7.17) is obtained as a large-time limit of the theory requires a more detailed discussion. If we drop the assumption that $\mu/\Delta\omega \ll 1$, i.e. include faster variability of statistical moments of wave field, we, strictly speaking, should use instead of (7.22) the exact solution to (7.21) in the form

$$J_{0123}^{(1)}(t) = 2\mathrm{i}T_{0123} \int_0^t \mathrm{e}^{-\mathrm{i}\Delta\omega(\tau-t)} f_{0123} \, \mathrm{d}\tau + J_{0123}^{(1)}(0)\mathrm{e}^{\mathrm{i}\Delta\omega t}. \qquad (7.24)$$

The point about large-time limit was highlighted by Janssen (2003), who instead of using (7.24) pulled f_{0123} out of the integral in (7.24) and set $J_{0123}^{(1)}(0) = 0$. The resulting modification of the KE has the form

$$\frac{\partial n_0}{\partial t} = 4 \int T_{0123}^2 f_{0123} \delta_{0+1-2-3} R_i(\Delta\omega, t) \, \mathrm{d}k_{123}, \qquad (7.25)$$

where $\Delta\omega = \omega_0 + \omega_1 - \omega_2 - \omega_3$, and

$$R_i(\Delta\omega, t) = \frac{\sin(\Delta\omega t)}{\Delta\omega}.$$

Then, in the limit $t \to \infty$, (7.25) gives

$$\lim_{t\to\infty} R_i(\Delta\omega, t) = \pi\delta(\Delta\omega),$$

leading to (7.17), while for small times, although still large compared to characteristic wave periods (Janssen, 2003)

$$\lim_{t\to 0} R_i(\Delta\omega, t) = t.$$

It is easy to see that while for large times the characteristic time of the evolution is proportional to ε^{-4}, in accordance with the well-known property of (7.17), for small times it is scaled as ε^{-2}.

It is also worth noting that while the classical version of the KE (7.17) creates a false impression that the spectral evolution is caused by exactly resonant interactions only, the modified equation (7.25) via the explicit presence of function $R_i(\Delta\omega, t)$ highlights the importance of nonresonant interactions in two different ways. First, it emphasizes the fact that the classical delta function in (7.17) is a specific large time limit: for any finite time the width of $R_i(\Delta\omega, t)$ remains finite and, thus, implies the importance of near-resonant interactions for any time. Second, the fact that at small times there is a fast spectral evolution entirely due to near-resonant interactions suggests a special significance of the early stages of evolution. The latter point at first glance looks strange, since the initial moment could be chosen arbitrarily.

Unfortunately, the range of applicability of Janssen's equation (7.25) is not clear. Indeed, its derivation is based upon the implicit assumption that $\mu/\Delta\omega \ll 1$. As we already mentioned, in generic situations, $\Delta\omega$ for four-wave interactions is $O(\varepsilon^2)$, which requires $\mu \ll \varepsilon^2$ and, thus, contradicts the small time scaling $\mu \sim \varepsilon^2$ implied by (7.25). Hence, in generic situations

(7.25) is not self-consistent at initial stages of evolution which this equation was aimed to describe.

In generic situations, there is no small parameter to be utilized, and in order to obtain a KE valid at all stages of evolution, including the initial one, one should use the closed form solution (7.24) without any further simplifications. The resulting GKE derived in Annenkov and Shrira (2006a) reads

$$\frac{\partial n_0}{\partial t} = 4\mathrm{Re} \int \left\{ T_{0123}^2 \left[\int_0^t e^{-i\Delta\omega(\tau-t)} f_{0123}\, d\tau \right] - \frac{i}{2} T_{0123} J_{0123}^{(1)}(0) e^{i\Delta\omega t} \right\}$$
$$\times \delta_{0+1-2-3}\, d\mathbf{k}_{123}. \tag{7.26}$$

Thus, in the general setting, the evolution of spectral density n depends not only on the initial distribution of n, but also (to a degree) on the initial distribution of phases accumulated into single characteristics $J_{0123}^{(1)}(0)$. A zero value of $J_{0123}^{(1)}(0)$ corresponds to situations where the wave field is initially free, so that the wave components are not correlated, and waves begin to interact only after $t = 0$. Thus, this type of initial conditions is quite special. We will refer to it as "cold start". Nonzero $J_{0123}^{(1)}(0)$ corresponds to generic initial conditions where the wave field has been evolving for a sufficiently long time, being governed by the same equations, and small but nonzero correlators have emerged due to nonlinear interactions. The effect of nonzero $J_{0123}^{(1)}(0)$ decays with time, due to the factor $e^{i\Delta\omega t}$ and integration in $\mathbf{k}$-space.

A further refinement of the closure (7.24) was proposed in Gramstad and Stiassnie (2012). It takes into account the cumulant $J_{0123}^{(1)}$, which is neglected in (7.21). Then, the analogue of equation (7.21) for $J_{0123}^{(1)}$ takes the form

$$\left(i\frac{\partial}{\partial t} + \Delta\Omega \right) J_{0123}^{(1)} = 2T_{0123} f_{0123}, \tag{7.27}$$

where

$$\Delta\Omega = \omega_0 + \omega_1 - \omega_2 - \omega_3 + 2 \int n_4 [T_{0404} + T_{1414} - T_{2424} - T_{3434}]\, d\mathbf{k}_4.$$

Its solution is given by formula (7.24) with $\Delta\omega$ replaced by $\Delta\Omega$. Correspondingly, the resulting KE is the GKE (7.26) with $\Delta\omega$ replaced by $\Delta\Omega$. At present, it is not clear whether the modified closure (7.27) brings any advantages compared to (7.21), which we have adopted, but it provides an interesting possibility that remains to be explored. Thus, apart from the classical KE (7.17), we have three different KEs which to various degrees capture the phenomenon of fast field evolution. We focus our further discussion on the GKE (7.26).

7.4.2. *Properties of the GKE*

7.4.2.1. *Initial stage of evolution*

At large times, Eq. (7.26) tends to the classical KE (7.17). Let us now consider evolution at small times in more detail.

"Cold start" implies $J_{0123}^{(1)}(0) = 0$ and hence the second term in (7.26) vanishes. Then the GKE reads

$$\frac{\partial n_0}{\partial t} = 4 \int T_{0123}^2 \left[\int_0^t \cos[\Delta\omega(\tau - t)] f_{0123} \, d\tau \right] \delta_{0+1-2-3} \, d\mathbf{k}_{123}.$$

It is easy to see that at $t = 0$

$$\frac{\partial n_0}{\partial t}\bigg|_{t=0} = 0, \quad \frac{\partial^2 n_0}{\partial t^2}\bigg|_{t=0} = -4 \int T_{0123}^2 f_{0123} \delta_{0+1-2-3} \, d\mathbf{k}_{123} = \alpha. \quad (7.28)$$

It is also straightforward to find higher even-order derivatives of n_0, all odd-order derivatives being zero at $t = 0$. Then, for small t $n_0 = n_0(0) + \alpha t^2/2$, where α is a constant specified by (7.28). Since $n \sim \varepsilon^2$ and the right-hand side is $\sim n^3 \sim \varepsilon^6$, the timescale of initial evolution is $O(\varepsilon^{-2})$.

For generic initial conditions with nonzero $J_{0123}^{(1)}(0)$ the second term on the right-hand side in (7.26) should be taken into account, which adds

$$-2 \operatorname{Re} \int [iT_{0123} J_{0123}^{(1)}(0) e^{i\Delta\omega t}] \delta_{0+1-2-3} \, d\mathbf{k}_{123}|_{t=0}$$

$$= 2 \operatorname{Im} \int T_{0123} J_{0123}^{(1)}(0) \delta_{0+1-2-3} \, d\mathbf{k}_{123}.$$

By virtue of (7.23) the characteristic value of $J_{0123}^{(1)}(0)$ resulting from long-term evolution is $\sim n^3 \sim \varepsilon^6$ which implies $\frac{\partial n_0}{\partial t}|_{t=0} \sim \varepsilon^6$, and, hence, the classic kinetic $O(\varepsilon^{-4})$ timescale. Thus, in agreement with the DNS and available experimental data the GKE predicts $O(\varepsilon^{-2})$ fast initial evolution and $O(\varepsilon^{-4})$ evolution at large times.

7.4.2.2. *Conservation laws*

Conservation laws determine the key characteristics of wave turbulence specific for a system under consideration. Crucially, they prescribe the type and direction of fluxes in the weakly turbulent cascades (Zakharov *et al.*, 1992; Nazarenko, 2011). Therefore, it is of prime importance to establish what are the invariants of the GKE and related equations aimed at describing the fast and slow evolutions.

Similar to the classical KE (7.17), the GKE conserves wave action N, momentum Π and energy E. The proof for wave action and momentum conservation is nearly identical to that in Zakharov *et al.* (1992). For example, the conservation of the total wave action $N = \int n(\mathbf{k})d\mathbf{k}$ is easy to see from the integration of the GKE (7.26) over the entire $\mathbf{k}$-space

$$
\begin{aligned}
\frac{\partial N}{\partial t} &= \int \frac{\partial n}{\partial t} d\mathbf{k} \\
&= 4\mathrm{Re} \int \left\{ T_{0123}^2 \left[\int_0^t e^{-i\Delta\omega(\tau-t)} f_{0123}\, d\tau \right] - \frac{i}{2} T_{0123} J_{0123}^{(1)}(0) e^{i\Delta\omega t} \right\} \\
&\quad \times \delta_{0+1-2-3}\, d\mathbf{k}_{0123}.
\end{aligned}
$$

The integrand changes sign after the interchange $\mathbf{k}_0 \leftrightarrow \mathbf{k}_2$, $\mathbf{k}_1 \leftrightarrow \mathbf{k}_3$ of the dummy integration variables. Conservation of momentum $\Pi = \int \mathbf{k} n(\mathbf{k})d\mathbf{k}$ is easily seen after the similar interchange of the dummy integration variables, taking into account delta function in the integrand.

Conservation of total energy E requires some discussion. For the classical KE (7.17) the conserved quantity is not the total energy but the leading quadratic term in the Hamiltonian, $E_0 = \int \omega(\mathbf{k})n(\mathbf{k})d\mathbf{k}$. Its conservation is ensured by the delta function in frequencies in (7.17). The discrepancy between the full energy and the Hamiltonian corresponding to the Zakharov equation

$$
H = \int \omega_0 b_0 b_0^* \, d\mathbf{k}_0 + \frac{1}{2} \int T_{0123} b_0^* b_1^* b_2 b_3 \delta_{0+1-2-3}\, d\mathbf{k}_{123}, \tag{7.29}
$$

has not got any explanation in the literature. Meanwhile, the GKE does not conserve E_0, but instead it conserves a quantity E explicitly dependent on time,

$$
E = \int \omega_0 n_0 \, d\mathbf{k}_0 - \mathrm{Im} \int T_{0123}^2 I(t) \delta_{0+1-2-3}\, d\mathbf{k}_{0123},
$$
$$
I(t) = e^{i\Delta\omega t} \int_0^t f_{0123} e^{-i\Delta\omega\tau}\, d\tau. \tag{7.30}
$$

Indeed,

$$
\frac{dE}{dt} = \int \omega_0 \dot{n}_0 \, d\mathbf{k}_0 - \mathrm{Re} \int T_{0123}^2 \Delta\omega I(t) \delta_{0+1-2-3}\, d\mathbf{k}_{0123}, \tag{7.31}
$$

which is zero by (7.26). Alternatively, on ensemble averaging of (7.29), we find

$$\bar{H} = \int \omega_0 n_0 \, \mathrm{d}\mathbf{k}_0 + \frac{1}{2} \int T_{0123} \langle b_0^* b_1^* b_2 b_3 \rangle \delta_{0+1-2-3} \, \mathrm{d}\mathbf{k}_{0123}, \qquad (7.32)$$

where

$$\langle b_0^* b_1^* b_2 b_3 \rangle \delta_{0+1-2-3} = n_0 n_1 \left(\delta_{0-2} \delta_{1-3} + \delta_{0-3} \delta_{1-2} \right) + J_{0123}^{(1)} + o(\varepsilon^6).$$

Then, using (7.24)

$$\bar{H} = \int \omega_0 n_0 \, \mathrm{d}\mathbf{k}_0 - \mathrm{Im} \int T_{0123}^2 I(t) \delta_{0+1-2-3} \, \mathrm{d}\mathbf{k}_{0123}.$$

Thus, within the framework of the GKE, we have the same three conservations laws (action, momentum and energy) as with the classical KE, the only difference being a higher order correction to the energy integral. Correspondingly, we can expect qualitative similarity of field evolution within the framework of the GKE and classical KE. It is easy to show that the KE employing the modified closure (7.27) also preserves the same three integrals with the energy correction appropriately modified (Gramstad and Stiassnie, 2012). Note that Janssen's equation (7.25) does not conserve energy.

7.4.3. Concluding remarks

Here, we briefly summarize our understanding of the phenomenon of fast evolution of wave turbulence and discuss the main implications and open questions. It has been established that evolution of weak turbulence on the scales faster than those prescribed by the standard KEs is possible and, moreover, is common in nature. This fast evolution occurs on the dynamic rather than kinetic timescale. It can be caused by relatively sharp spatial or temporal inhomogeneity of any of the multitude of exogenous factors affecting a wave field. For example, in the wind wave context, the sharp inhomogeneity might be due to wind gusts, interaction with currents and/or internal waves, obstacles, bottom topography or wave breaking. However, we reiterate that even in situations where everything is spatially and temporally uniform, endogenous fast evolution regimes, e.g. associated with spectral fronts, inevitably develop as a result of natural evolution of an arbitrary initially smooth wave field.

At present, the primary tool available for examining the phenomenon of fast evolution is the DNS. So far, the phenomenon was examined only in the water wave context, although a similar DNS technique can be applied to random wave fields of other physical nature. Nevertheless, even the most efficient DNS is an expensive tool. A more efficient way to describe wave spectra evolution on both fast and kinetic timescales can be provided by the GKE, which has been derived using a systematic asymptotic procedure. However, so far only the most basic properties of the GKE have been established. Crucially, the GKE holds the same integrals as the standard KE, with a different, more exact, expression for energy. Therefore, the basic picture of field evolution, that is the types and directions of the cascades, remains the same. To get a detailed picture of field evolution under relaxed conditions of the GKE, extensive numerical simulation of the GKE is necessary. Numerical algorithms and software for efficient integration of the GKE are now under development.

7.5. Discussion

The results surveyed in the present review lead to the qualitative change of the whole picture of wave turbulence evolution. First, the range of situations where kinetic description can be applied has been significantly widened, being extended to inhomogeneous fields. It emerged that it is common to have co-existing "fast" and "slow" manifolds with the latter being described by the standard KE. The key change in the broad-brush picture of wave evolution can be outlined as follows. In nature, the inertial intervals of wave turbulence often do not appear in the pure form, nonlinear resonant interactions co-exist and compete with forcing and/or dissipation. It is crucially important to understand *a priori* when nonlinear interactions are dominant, so that turbulent cascades are present. The corresponding order of magnitude estimates of intensity of nonlinear resonant interactions are usually obtained using the standard KE. However, the new picture implies that any departure of wave turbulence from the slow manifold governed by the standard KE leads to evolution on the dynamic time scale. That is, the effective nonlinearity is much stronger than the standard KE predicts, and there is a kind of strong restoring force, which returns the field into the slow manifold. This enables the formation of weakly turbulent cascades even when nonlinearity, as estimated from the standard KE, is weaker than forcing and/or dissipation. This fact dramatically extends the range of situations where a broadband field of random resonantly interacting weakly nonlinear waves is indeed in the weak turbulence regime.

The emerging novel picture of wave evolution leaves many questions to be addressed. First, an efficient code for the GKE remains to be developed and implemented. This would provide us with a new insight into the issue of formation of turbulent spectra. Second, although a similar generalization of wave kinetics is straightforward for wave fields with "decaying" dispersion law (where triad interactions are dominant), simulations of fast evolution of such fields have not been performed yet and there might be room for surprises. So far, we focused our attention exclusively on the evolution of wave spectra, but it should be mentioned that when wave spectra evolve on the dynamic timescale, all higher moments also follow. This opens a possibility to describe evolution of not only wave spectra, but without a great extra effort also of higher moments and probability density function. At present, a detailed discussion of the progress in this direction is probably premature.

Acknowledgments

The work was supported by UK NERC grant NE/I01229X/1 and EPSRC grant GR/T09507/01. Computations were performed on the ClusterVision computer cluster at Keele university, and on the ECMWF supercomputing facility, the access to which is gratefully acknowledged.

Bibliography

Annenkov SY, Shrira VI. 2001. Numerical modeling of water waves evolution based on the Zakharov equation. *J. Fluid Mech.* 449: 341–371.
Annenkov SY, Shrira VI. 2006a. Role of non-resonant interactions in the evolution of nonlinear random water wave fields. *J. Fluid Mech.* 561: 181–207.
Annenkov SY, Shrira VI. 2006b. Direct numerical simulation of downshift and inverse cascade for water wave turbulence. *Phys. Rev. Lett.* 96: 204501.
Annenkov SY, Shrira VI. 2009. "Fast" nonlinear evolution in wave turbulence. *Phys. Rev. Lett.* 102: 024502.
Annenkov SY, Shrira VI. 2011. Evolution of wave turbulence under "gusty" forcing. *Phys. Rev. Lett.* 107: 114502.
Autard L. 1995. Etude de la liaison entre la tension du vent à la surface et les propriétés des champs de vagues de capillarité-gravité developpés. Ph.D. Thesis, Université Aix-Marseille I and II (in French).
Babanin A. 2004. *Breaking and Dissipation of Ocean Surface Waves*, Cambridge: Cambridge University Press.
Badulin SI, Pushkarev AN, Resio D, Zakharov VE. 2005. Self-similarity of wind-driven seas. *Nonlin. Proc. Geophys.* 12: 891–945.
Badulin SI, Babanin AV, Zakharov VE, Resio D. 2007. Weakly turbulent laws of wind-wave growth. *J. Fluid Mech.* 591: 339–378.
Banner ML, Young IR. 1994. Modeling spectral dissipation in the evolution of wind waves. I: Assessment of existing model performance. *J. Phys. Oceanogr.* 24: 1550–1571.
Benoit M. 2005. Evaluation of methods to compute the nonlinear quadruplet interactions for deep-water wave spectra. *Proc. 5th ASCE Int. Symp. on Ocean Waves, Measurements and Analysis*, Madrid, Spain.

Creamer DB, Henyey F, Schult R, Wright J. 1989. Improved linear representation of ocean surface waves. *J. Fluid Mech.* 205: 135–161.

Dyachenko AI, Korotkevich AO, Zakharov VE. 2004. Weak turbulent Kolmogorov spectrum for surface gravity waves. *Phys. Rev. Lett.* 92: 134501.

Dysthe KB, Trulsen K, Krogstad HE, Socquet-Juglard H. 2003. Evolution of a narrow-band spectrum of random surface gravity waves. *J. Fluid Mech.* 478: 1–10.

Gramstad O, Stiassnie M. 2013. Phase averaged equation for water waves. *J. Fluid Mech.* 718: 280–303.

Hasselmann K. 1962. On the non-linear energy transfer in a gravity-wave spectrum. Part 1: General theory. *J. Fluid Mech.* 12: 481–500.

Hsiao SV, Shemdin OH. 1983. Measurements of wind velocity and pressure with a wave follower during MARSEN. *J. Geophys. Res.* 88: 9841–9849.

Janssen PAEM. 2003. Nonlinear four-wave interactions and freak waves. *J. Phys. Oceanogr.* 33: 863–884.

Janssen P. 2004. *The Interaction of Ocean Waves and Wind*. Cambridge: Cambridge University Press.

Kadomtsev BB. 1982. *Collective Phenomena in Plasmas*. New York: Pergamon.

Komen GJ et al. 1994. *Dynamics and Modelling of Ocean Waves*. Cambridge: Cambridge University Press.

Krasitskii VP. 1994. On reduced equations in the Hamiltonian theory of weakly nonlinear surface waves. *J. Fluid Mech.* 272: 1–20.

Korotkevich AO, Pushkarev A, Resio D, Zakharov VE. 2008. Numerical verification of the weak turbulent model for swell evolution. *Eur. J. Mech. B/Fluids* 27: 361–387.

Lavrenov IV. 2003. *Wind-Waves in Oceans: Dynamics and Numerical Simulations*. Berlin: Springer.

Litvak MM. 1960. *A Transport Equation for Magnetohydrodynamic Waves*. New York: Cornell University, p. 186.

Lvov YV, Nazarenko SV, Pokorny B. 2006. Discreteness and its effect on water-wave turbulence. *Physica D* 218: 24–35.

Nazarenko SV. 2011. *Wave Turbulence*, Lecture Notes in Physics, Vol. 825. Berlin: Springer.

Onorato M, Osborne AR, Serio M, Resio D, Pushkarev A, Zakharov VE, Brandini C. 2002. Freely decaying weak turbulence for sea surface gravity waves. *Phys. Rev. Lett.* 89: 144501.

Peierls R. 1929. Zur kinetischen Theorie der Wärmeleitung in Kristallen. *Ann. Phys.* 395: 1055–1101 (in German).

Pushkarev AN, Zakharov VE. 2000. Turbulence of capillary waves-theory and numerical simulation. *Physica D* 132: 98–116.

Stiassnie M, Shemer L. 2005. On the interaction of four water-waves. *Wave Motion* 41: 327–328.

Tanaka M. 2001. Verification of Hasselmann's energy transfer among surface gravity waves by direct numerical simulations of primitive equations. *J. Fluid Mech.* 444: 199–221.

Trulsen K, Dysthe K. 1996. A modified nonlinear Schrödinger equation for broader bandwidth gravity waves on deep water. *Wave Motion* 24: 281–289.

van Vledder GPh. 2006. The WRT method for the computation of non-linear four-wave interactions in discrete spectral wave models. *Coastal Eng.* 53: 223–242.

van Vledder GPh, Holthuijsen LH. 1993. The directional response of ocean waves to turning winds. *J. Phys. Oceanogr.* 23: 177–192.

Willemsen JF. 2001. Enhanced computational methods for nonlinear hamiltonian wave dynamics. Part II: New results. *J. Atmos. Ocean. Tech.* 18: 775–790.

Waseda T, Toba Y, Tulin MP. 2001. Adjustment of wind waves to sudden changes of wind speed. *J. Oceanogr.* 57: 519–533.

Yokoyama, N. 2004. Statistics of gravity waves obtained by direct numerical simulation. *J. Fluid Mech.* 501: 169–178.

Young IR. 1999. *Wind Generated Ocean Waves.* Amsterdam: Elsevier.

Young IR. 2006. Directional spectra of hurricane wind waves. *J. Geophys. Res.* 111: C08020.

Young IR, van Agthoven A. 1997. The response of waves to a sudden change in wind speed. In *Nonlinear Ocean Waves*, ed. W Perrie, Advances in Fluid Mechanics Series, Southampton: WIT Press, pp. 133–162.

Zakharov VE. 1968. Stability of periodic waves of finite amplitude on the surface of a deep fluid. *J. Appl. Mech. Tech. Phys.* 9: 86–94.

Zakharov VE. 1999. Statistical theory of gravity and capillary waves on the surface of a finite-depth fluid. *Eur. J. Mech. B/Fluids* 18: 327–334.

Zakharov VE, L'vov VS, Falkovich G. 1992. *Kolmogorov Spectra of Turbulence I: Wave Turbulence.* Berlin: Springer.